Managing The Small Construction Business

A Hands-On Guide

A Journal of Light Construction Book

First printing: August 1992

On the Cover: Scheduling board and job-tracking bins of Lakeview Lumber & Construction, Lake Barrington, Ill.
Photo by Jim Zorn, Stemo/Zorn Photography

Editorial Director: Steven Bliss
Managing Editor: Josie Masterson-Glen
Technical Editors: Sal Alfano, Charles Wardell
Article Editors: Clayton DeKorne, David Dobbs, Don Jackson, Marylee MacDonald, Paul Spring, Wendy Talarico
Special Projects Editor: John Wagner
Editorial Assistant: Patricia Poulin

Production/Design Director: Jim [illegible]
Designer: Annie Stewart
Illustrator: Tim Healy

Founding Editor: Michael Reitz

International Standard Book Number: 0-9632268-1-9
Library of Congress Catalog Card Number: 92-73306
Printed in the United States of America

A Journal of Light Construction Book
Journal of Light Construction Books is a tradename of Builderburg Group, Inc.

Builderburg Group, Inc.
P.O. Box 435
Richmond, VT 05477

Contents

CONTENTS

Introduction

"**H**e was a crackerjack carpenter, but not much of a businessman..." How many times have we heard that epitaph recited about a failed construction business?

The reason is simple: Most small contractors come up through the trades. By the time they take the big step of hanging out a shingle as an independent contractor, most have mastered the technical aspects of their craft. That's how they achieved enough success to launch a business in the first place.

Yet precious few have had sufficient experience on the business side to estimate accurately, negotiate a fair contract, keep adequate records, and deal with the myriad new relationships they must now master — with owners, employees, and subcontractors, not to mention architects, accountants, lawyers, and code officials.

So it comes as no surprise that so many builders find their progress impeded by one management snag after another, and why so few last long enough to learn on the job the business skills they need to survive.

This book is an attempt to guide the newcomer, or small, growing business, around many of these management pitfalls. Because no one person has all the answers, we've chosen selections from a variety of successful contractors — each writing about the aspects of the business he or she knows best. These articles represent the best business contributions to *The Journal of Light Construction* from 1986 to 1992.

Not every idea will apply to your business — each company faces its own market conditions and each person has his own business style. But the wealth of real-world experience boiled down in these 200-plus pages should provide any small contractor with a big head start in tackling the key problems he's likely to face.

Finally, our thanks to the many authors who contributed to this book. Nearly all are working contractors who have taken time out of their busy schedules to share their experience and hard-won business wisdom with the new folks coming up through the ranks. They're definitely a group worth listening to.

Steven Bliss
Editorial Director

Chapter 1 • Quality Control

Managing for Quality Control

by Michael Chapman

Many builders believe that "quality" homes mean expensive cabinets and appliances, and lavish square footage. But quality has as much to do with how you put the parts of a building together as it does with the products you choose. We've all seen top-quality products ruined by a lousy job of installation. And we've seen low-priced products installed properly and operating satisfactorily.

As a builder in Santa Fe, N.M., I've sought to establish a reputation as a quality contractor. That has forced me to think about how to control quality on my jobs. My ultimate goal is to turn over a blemish-free building to clients, and to know that they're happy about their new home and satisfied that the building process has gone smoothly. On that final walk-through, I don't want them holding back my check, and I don't want a long punchlist.

To avoid problems, I focus on two areas. First, I pay attention to building details during both design and construction. And second, I maintain good communication with clients and tend to the little things that concern them.

Keeping On Top of the Details

Nothing creates more quality-control problems than poorly planned construction details. To prevent problems, we have our construction super review and critique the plans as soon as the designer finishes them. He looks for unclear details, errors, and omissions. For instance, he might see that the plan would work better if we brought the water in at another location or reversed the swing on a door.

After any revisions, we hold a preconstruction meeting. The estimator, the designer, the construction superintendent, and the salesperson discuss the orientation on the lot, key construction details, specialty items, and specific, critical dimensions — such as the size of an alcove for an unusual piece of furniture.

After any further revisions, the super puts together a set of plans for each sub, marking notes in red pencil. The electrician's plans might be marked to highlight the exact placement of an electrical outlet or to indicate a floor outlet that needs to be roughed in before we pour the slab. Marking up each set of plans also helps the super think through the job and plan any difficult sequences.

Quality in housing relies on good construction details and good customer relations

The same subs. We build 15 to 20 custom homes a year, but unlike many companies with that volume, we do a large part of our work through subs, using the same ones on every job. We often hold open houses in

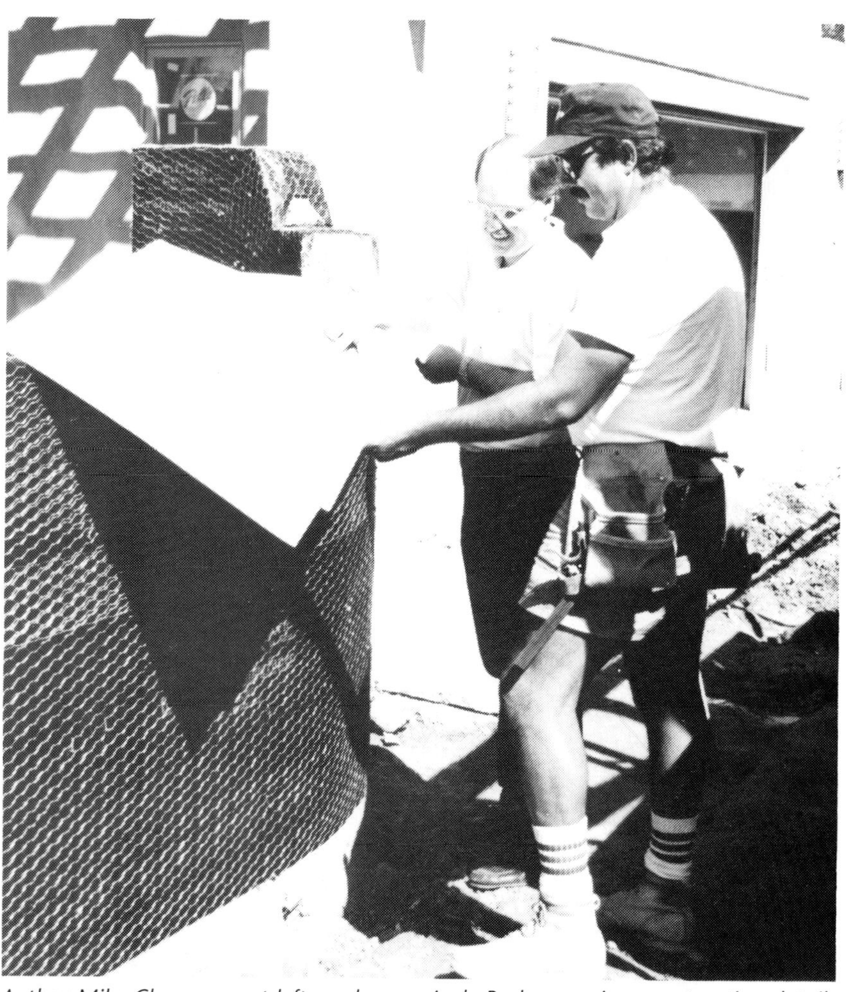

Author Mike Chapman, at left, and super Andy Barbero review construction details as a house nears completion. Barbero is on site every day to assure that the subs do high-quality work.

Punch-List Pointers

by Richard Lind

Since you the builder are held responsible for the defects typically found in new homes, it's important for you to identify any problems at a predelivery walk-through. Now is the time for you and the buyer to determine who is responsible for fixing each problem. The day of the closing, or the day before, is best for this.

With pencil and master checklist in hand, go through the house room by room, checking trim, wall surfaces, ceilings, floors, paint, glass, appliances, mechanical systems, hardware, etc., finishing up in the basement or garage. Then proceed outside to examine the exterior of the house.

If possible, have the buyer sign a simple statement acknowledging that he and you have walked through the house on a particular day and identified the items that need attention. Explain that the signature represents only an acknowledgement of the walk-through, not a release for items not on the list that might require attention in the future.

Although some buyers refuse to sign, I've already accomplished what I have set out to do: to determine between us in a businesslike manner, what, if anything at this point, requires attention; to acknowledge which items are my responsibility to correct; and to educate the buyers about their responsibilities for maintenance and care. Few new home buyers are brazen enough to ask me to repair damage they've done after the inspection, such as vinyl floor covering torn when the fridge was delivered.

Some builders fear subjecting their homes to such scrutiny, viewing it only as an opportunity for the buyer to nit-pick. But don't worry, the house indeed will be scrutinized at some point in time, and the problems will have to be corrected sooner or later. It's best to do so while you are still in control of the project.

Do Your Homework

The key to a successful builder/buyer walk-through is preparation. You should do a thorough private inspection prior to the walk-through with the client. Don't allow too much time between the two inspections or new unexpected problems might crop up; but allow enough time for you to solve the problems prior to the final walk-through. Don't leave anything out, for the sooner a defect is found, the easier it is to fix it—particularly if the subs are still handy. Follow a simple, but thorough, master checklist. Don't forget to include a shakedown of all appliances and mechanical systems.

As builders, we can almost forecast the items. We've seen the same things crop up, year after year, punch list after punch list: door striker plates that don't engage, cabinet drawers that are obstructed, garage door locks out of alignment, double-hung windows stuck with paint, and always, that tub-drain toggle.

By correcting these items well prior to the closing or completion dates, things get done your way, rather than someone else's way later on. Your way is always less expensive and just as good.

Never allow a buyer to move into a house before closing. Besides the fact that the buyer could damage the home during this time, it's just bad practice. Suppose the finances don't work out, for example. It's hard to remove someone once they are in. When the customer has title to the land, of course, this is especially difficult to prevent, so protect yourself contractually as best you can.

Finally, arrange the closing or passing of title to the completed house well prior to the expiration of the purchase and sales agreement. This leaves you in a more comfortable position to postpone the delivery and closing if something requires attention at the eleventh hour. It also puts you in a better bargaining position in the event of a dispute, such as a buyer's demand to hold back money for punch-list items.

Even So...It Won't Be Perfect

Despite your best efforts, like death and taxes there always will be some punch-list items that crop up after the closing and the buyer's occupancy. Respond to the items on the list promptly. We want to avoid any unnecessary inconvenience to our buyer. In addition, we want to take corrective measures before the problem compounds — for example, a leaky toilet seal can develop into a ruined floor.

If a delay is necessary in responding to a callback, keep the buyer posted so he doesn't feel ignored. Most buyers are reasonably patient if they know they are not being given the run-around.

Any serious difference of opinion should be followed up with careful documentation and a letter. If you should ever have to go to court, the very first question asked will be, "Has there been any correspondence?" ■

Richard Lind is a builder in the Wellesley, Mass., area.

houses under construction, and potential customers will say, "We want the same framer to build our house." With the same subs, we can guarantee uniform quality.

We don't deviate from our list unless the sub can't schedule us at the time we need him. This means the sub doesn't have to come in with the low bid on every job. When you take the low bidder and you bring your customer to the job site, you could be embarrassed by the quality of the sub's work.

The key to getting good work from subs is to communicate with them when they begin a job and every time they arrive at the site. I want a super on site a half hour every morning and every afternoon. If the sub is going to show up at 1:00 p.m., the super will be there to go through the plans. We don't allow subs or crew foremen to solve problems without the superintendent's approval.

We also have to make sure subs remember how we want the work done. For instance, with drywall I want a three-coat job, not two coats. Without a reminder from the super, details can slip a sub's mind, particularly if he's rushing off to another job. Also, unsupervised subs may take short cuts or forget change orders.

Correcting mistakes costs time and money. If the drywall contractor leaves behind sloppy work, like a buildup of joint compound in the corners, and the painters coat the walls, someone has to come back and clean up after the drywall repair and touch up the paint.

Our superintendents do walk-throughs with the subs and provide them with a punchlist of items needing attention. The subs know they're going to get our next job if they give us good service. And we expect to get taken care of.

We'll also use payments to get a sub's attention if he's behind on service calls. Subs sometimes try to collect on jobs that aren't quite finished. The boss comes over to the office to pick up the check; maybe his crew told him they were done. But our super knows when a sub has to come back to finish up. He holds the check until the work is done completely and correctly.

The Customer Is Always Right

We place just as much emphasis on building a quality relationship with the customer as we do on building a quality house. To start this relationship on the right foot, we use a contract that clearly spells out our expectations of quality and defines how disagreements will be resolved.

Contract for quality. Customers often create problems because they're nervous about quality, but not knowledgeable about construction. This can lead to clients holding back money during payouts or until the final punchlist is complete. Haggling about payouts will cause a relationship to quickly deteriorate. If you have a qualified buyer and can get construction loans, you can avoid this problem by dealing with the bank, not the client. The bank is an objective third party that knows how to evaluate job completion.

However, if the clients are financing the job themselves, we insist that the inspection be conducted by a third party who will authorize the release of funds. This can be an appraiser or an independent construction inspector. I include in the price the cost of hiring an inspector—generally $50 each time.

Our contract is written so that our draws coincide with the start of each phase of construction rather than the end. Many contractors specify a draw at the completion of framing. However, this can create a delay in getting paid if you've substantially completed the framing, but still have kitchen soffits to fur out or a few studs to straighten. If you time your draw to the start of insulation instead of the end of framing, the customer can see when the insulation installers begin their work, and you're not in a rush to get every last bit of framing complete in order to meet Friday's payroll. Working this way reassures the customer that construction is proceeding on schedule, and helps maintain the good relationship that is so important.

Scheduled walk-throughs. We require customers to attend five walk-throughs and to sign off on the specs and plans at each stage:

1. We rough stake the home prior to excavation and have the client walk the site with us to check for proper location, orientation, and tree removal.

2. Before we begin framing, we do a "color consultation," having the client take a final look at colors for tile, paint, and cabinets. This gives us time to order all necessary items.

3. We also do a framing walk-through where we review window and door place-

Hand-crafted features, such as the rounded stucco chimney (left) and exterior scrollwork (right), require good subs and close supervision.

ments and electrical, TV, and telephone outlets. We also use this opportunity to point out quality features the customer is not likely to recognize, such as caulked sill plates.

4. Once the drywall is hung and taped, we do a "trim" walk-through. Linen closet shelves and master closet shelving and rod heights are very important to our customers, so we mark them on the wall with the customer prior to installation.

5. During a final walk-through, the customer checks out the house completely. We prepare a detailed punchlist of items that need attention.

A key factor in quality control is having enough time to complete the house. To make sure you have enough time, you'll need to extend the completion date if there are legitimate delays. Walk-throughs are a good time to identify problems that might cause delays. On the foundation walk-through, for instance, we might say, "We lost a couple of days because of rain." Or when showing customers the framing, we'll say, "The windows didn't come in yet, and that will set us back a couple of days."

If the customer makes changes, our contract lets us add an appropriate number of days to the schedule. If these changes are made after the framing stage, we double the markup.

You really have to level with the client if you aren't going to make the move-in date. Don't wait until a week before closing to discover you're three weeks behind. Rushing to complete a house by an arbitrary date will detract from its quality.

Regular meetings. In addition to walk-throughs, we meet regularly with clients to resolve questions before they grow into problems. We keep the meetings short and precise. With difficult or worrisome clients, we schedule weekly meetings between them and our superintendent. These generally take only 10 to 15 minutes and can turn difficult clients into easy ones. The clients come with a list of questions that they're prepared to discuss, and the discussion gives us a chance to educate them about how we work.

Solving problems. When a problem arises, I call the clients and ask them to meet me at the job site. For example, on a recent job I ordered interior doors with operable transoms, but the doors showed up with fixed transoms. Although the door company had made the mistake, I knew I'd have to pull the doors after the clients moved in, which was only three weeks away.

I explained the problem and told the clients how I was going to solve it. But I didn't wait until the week before they were supposed to move in to spring this on them. Communication early on is the key.

Clients often worry more about minor details than about big ones. If you don't meet with them frequently, you may not notice these red flags. On one house, for example, we had an undersized beam. This was a serious structural problem that meant pulling the beam and replacing it. But when I called the client, he said, "Okay, but what are you going to do about the telephone outlet in the den?" That was really bugging him, because he thought I was going to forget it.

Punchlist performance. I schedule a final walk-through two days before closing and correct anything on the punchlist the next day. I like to close immediately after the job is complete. I won't sign a contract that has a retainage clause because I feel that shows a lack of trust between the client and builder, and trust is something I work very hard to maintain. If a guy wants to withhold $2,000 because of a late microwave delivery and a door that needs to be planed down, that's the last guy I want to get involved with.

We've already caught most of the problems by the time we do the final walk-through. The job super has already worked up a punchlist with the drywall sub, painter, electrician, hvac sub, and cabinet installers and has corrected any problems. The super has also checked appliances.

Items the client comes up with are usually minor and can be taken care of in a day. They typically involve finish details, such as a drawer or door that sticks, paint touch-up, missing caulk, or a carpet that's tucked up.

Warranty follow-up. If you've done a good job during construction and have finished the home as agreed, then the warranty items should be minor. We call the customer thirty days after closing to do one more walk-through and make any necessary corrections. We try to complete all the work in a one- or two-day period within two weeks after the walk-through. The work is generally scheduled with the customer the day of the walk-through.

We also pay attention to nonwarranty items that concern the customer. For instance, I had a gentleman explain to me that his toilet was too small. I was tempted to suggest that he sit down and I would take my tape out and see where the problem was. Instead, I told him we would see what was available. We found a larger one and replaced it, but at his expense. After all, he had signed the spec sheet. He appreciated the help with what he felt was a major problem, and I had a satisfied client.

Maintenance tips. Customers sometimes feel they haven't been given a quality house when little problems begin to show up after construction — drywall cracks due to wood shrinkage, for instance.

To address this problem, we supply new homeowners with information that educates them on what they can expect in the next year. We tell them that the house framing will shrink, and show them that they may get a crack between the plaster fireplace and the drywall above it. We also warn them that the beams, which come from the damp Northwest, will dry out during the first heating season, opening some joints. We ask them to let the building go through a winter and a summer and assure them that we'll be back then to take care of minor problems. Forewarned in this way, they don't get upset when small problems occur. They know we've predicted the problems, and that we'll fix them.

We also give them a three-ring binder of tips on how to take care of tile, grout, roofs, and floors. The binder also contains appliance warranties.

Quality in housing comes both from the structures you build and from the relationships you build with your clients. To succeed as builders, you must effectively manage both processes. ■

Michael Chapman is president of Michael Chapman Homes in Santa Fe, N.M.

Chapter 2 • Customer Relations

INTERVIEW

High-End Remodeling: Building on Customer Satisfaction

Contractor Deva Rajan, at left, succeeds at his high-end remodeling business by offering his customers an unusual degree of service. His son Abe, at right, handles estimating and project management.

Many of the remodeling dollars being spent these days come from the so-called move-up or luxury market where clients expect a high degree of service and organization from their builder. California contractor Deva Rajan has just this kind of clientele and has earned a reputation for doing, as one colleague put it, "elegant business."

He concentrates on three areas of residential remodeling: complex structural repair (foundation work, hillside stabilization), high-end kitchens ($40,000 and up) and baths, and general remodeling (additions, expansions, second stories). He runs four crews, and shares the office with a full-time office manager and his son Abe, who handles both estimating and project management. His dollar volume last year was just under $1.5 million.

Rajan has spent nearly 35 years in design and construction, starting as a laborer during summers off from college. After graduate school in design at U.C. Berkeley (and summers as a carpenter) he taught for several years at the college level. In his first ten years as a contractor, he concentrated on remodeling. In the 1970s Rajan took charge of the historical restoration of a Russian fort north of San Francisco, and then remained in that area for nearly ten years doing new construction and spec building. He has since brought his company, Canyon Construction, back full circle to full-time remodeling.

JLC: *How do you find your clients? Do you advertise?*

Rajan: We stopped advertising about three years ago; we rely on referral and repeat customers completely now. But what we've done with that advertising budget is to pour it into a very thorough *callback program* that costs the client nothing. We go back three different times after we've completed the job: at 30 days, six months, and a year. We inspect our work, and take care of any problems the client might have without their having to call.

JLC: *Is this one of the reasons clients come to you?*

Rajan: I think it's part of it. Like 98% of general contractors, we're interested in providing our clients with quality materials, good craftsmanship, and reliable subcontractors. But we also try to give them an unusual degree of service based on very thorough supervision and communication right on through to the end of the project. The teamwork approach that we use with design professionals and subs, and the care we take with estimates, scheduling, and the work itself is all part of it.

JLC: *What's your first meeting with a potential customer like?*

Rajan: In some cases we bid a project along with other contractors without even meeting the client — pretty cold stuff. But we'd like to get away from these entirely.

The second type of project comes from the clients themselves. They may have plans developed by an architect or a kitchen designer, or they may not have gotten that far yet. I take my son, who is our estimator and project manager, to the site with me to meet with the clients. We spend a lot of time just listening and asking questions about how they see the project.

JLC: *Do you suggest that they look at some of your past work?*

Rajan: We leave them a reference packet that includes ten pages of clients, at maybe 15 names per page — really stuffed. What often happens in our community is that the prospective clients will thumb through and find someone they know, and at that point you know you're way ahead.

JLC: *What's the next step?*

Rajan: We arrange for a walk-through with the subcontractors who will be key to that project. It almost always includes the plumber and electrician, but we may also ask the roofing contractor or painter, and frequently the cabinet sub. It takes about an hour, and requires the free run of the house. It gives our subs a chance to answer any questions they have: drain pipe sizes, entry service for power, things like that. At the same time, we're looking at the job for access, excavation problems, etc.

JLC: *How long will you spend preparing the estimate?*

Rajan: For a $250,000 project, the estimate will take at least a week. We'll spend five solid days doing material takeoffs and thorough labor estimates. We enter these numbers and subcontractor bids into our computer as they come in. We end up with a 12- to 15-page print-out.

JLC: *How do you present the bid?*

Rajan: We block out an hour or two with the couple — never just one of them — and go through it very thoroughly.

This is a really pivotal meeting. They already have a first impression of you, and now they are really geared up to check you out. The key thing to them is the numbers: They want to know if you're in the ball park. Assuming you are within 10% of the other bidders, the game is still wide open. We rarely get a project because we're low bidder.

JLC: *How much detail do you give the client in your estimate?*

Rajan: We literally give them everything. We use the 16 CSI codes as a format for our estimates. The first page of our bid is a summary — totals for each of the categories with our contingency, sales tax, and contractor's margin listed below that.

The second page is broken down into materials, labor, and subs. Again it's a summary page, so they can see where the costs are for each of the CSI categories. At that point we explain our markup categories: contractor's margin (10% profit, 10% overhead) and builder's contingency.

The pages that follow are detailed extensions of each one of those 16 categories. For example, category #3 is *Concrete*, so this page lists each sequential step in the work — form work, rebar, placement, stripping, etc. — and details the cost of materials, labor, and subcontractor. Backing that up are the labor and material takeoffs.

JLC: *You mentioned a figure for "builder's contingency;" what's that?*

Rajan: Some contractors refer to this as their *risk*. In remodeling, no matter how well we estimate or how much experience we have, there are always surprises waiting in the walls — pipes we can't see, more wire, less wire, whatever. But customers hate surprises; everyone hates surprises.

We take this right to the client by saying, "As seasoned builders we know that we'll run into things that we can't anticipate. We will catch any of those problems that come up and absorb them in the contingency you pay us as part of the contract."

JLC: *How do you figure the contingency?*

Rajan: We know from keeping statistics on our past jobs that our material estimates can be off by as much as 5% one way or the other. Some of this is due to price fluctuations; the rest is accounted for by waste or small mistakes in estimating. So we add a 5% contingency on materials to cover us.

On our own labor we figure 10% for discrepancies in crew performance, unpredictable field conditions, weather, and things like that. Finally, for subcontractors we add in 5%. This one puzzles some clients. But even though you get firm numbers from subs, they still manage to come up with a few extras.

JLC: *Don't clients hesitate at paying the extra cost?*

Rajan: Sure, we hear objections like "Shouldn't this be a part of your overhead or profit?" But on the whole, it's turned out to be a kind of selling point for us because clients are very afraid of surprises — afraid that they are going to be killed in extra work orders. The builder's contingency gives us a chance to say, "We won't charge you for a pipe that's discovered in the wall that we didn't see. We *will* charge you if you want to add another light or a larger window that's not on the drawing."

JLC: *Do they ever suggest that they keep whatever is left of the contingency money at the end of the job?*

Rajan: We hear that occasionally from clients or their architect. I explain that this arrangement would be an "owner's" contingency, which would mean that they are the risk takers, and that they would also be responsible for costs that *exceed* the 5%-10%-5%.

JLC: *Are there other concerns you try to deal with at this stage?*

Rajan: Clients are almost always anxious about how long the project will take or how many weeks they'll be without a kitchen. So we now give them a complete scheduling document with the estimate, but almost anything that suggests you've thought through the project in terms of its duration is a marvelous sales tool.

JLC: *If you're not relying on being low bidder, what else do you do to win the contract?*

Rajan: We're basically presenting the reasons our company is going to be better to work with. One of these is the quality of our subcontractors. We just started using actual biographies of our subs, and it's worked very well. When clients realize that you have selected the best subs there are, their confidence in your ability to build their dream grows.

This is particularly important after you leave and they begin to review the presentations of the other contractors.

JLC: *What else?*

Rajan: I think our own confidence in the numbers is very important. Being willing to talk about cost in very certain terms removes the lingering doubts in the client's mind about whether you're shooting from the hip. Because at this stage, the client *is* looking for little pockets of guessing, cheating, or whatever they suspect might be going on. The fact that we are disclosing the numbers — all of them — leaves little room for these thoughts.

We will even pull out a subcontract for some part of the job and then go back to our estimate to show that they are the same, right down to the dollar. It all reinforces that the numbers are real.

JLC: *But don't you often find out after all this work that the client just can't afford the work the way it's drawn?*

Rajan: Sure, the initial document almost always yields some surprises or draws some gasps. We look at the estimate as a working document that can be adjusted very easily, especially with computers. When the client says, "We didn't know that granite countertops were going to be that expensive," we explain how easily we can make substitutions. That's part of our job, as we see it — to fit the project to their budget. This puts us both — the client and the contractor — on the same side of the table. And that's so much better than the traditional adversarial position.

JLC: *Do you find a lot of resistance from architects when you close the distance between contractor and client?*

Rajan: There's a growing willingness on the part of architects to work with builders on a shoulder-to-shoulder basis. There's still a residual element that feels responsible for protecting the client from things getting wildly out of hand, but that's probably healthy. You don't want to work with an architect who just lies down and lets you walk all over him. He's no help at all when it gets tough.

JLC: *How many meetings are usually required with a client before you end up signing a contract?*

Rajan: It varies, but for a $250,000 project we've found it takes about five meetings for everything to settle down. Maybe by the third meeting they've decided to go with us, but it will take another two meetings before everyone is really comfortable with the price and the scope of the work.

JLC: *Is it really worth all that time?*

Rajan: Sometimes it doesn't seem like it, particularly if the project doesn't materialize after all that effort. But in sticking it out, we often find other bidders dropping away, or growing impatient and showing a side that makes the client uncomfortable.

JLC: *With all of the time and attention you give to clients before they sign the contract, don't they feel abandoned once the work begins and you're not on site?*

Rajan: We handle that by trying to get everyone involved. On a larger job, we hold a preconstruction meeting with the architect, clients, subs, and foreman. One thing that's accomplished is a chance for the subs and the architect to go over quality standards with the client listening in. Painting is a good example. We use phrases like apartment grade, standard grade, or custom finish. The owner is usually eager to know the difference and that gives the painter a chance to really talk about his work.

We also ask the clients to talk about why they're remodeling, and describe their vision of the finished project. What happens at this meeting is hard to describe, but the result is that our subs and foreman end up feeling much more committed to the job.

JLC: *Isn't this also a chance for the different trades to anticipate problems and discuss scheduling?*

Rajan: Definitely. The heating guy may say "We absolutely have to get our stuff in before the plumber, get us in there first." Or the electrician may ask the cabinetmaker to leave a raceway for his wiring in an island.

We also bring a preliminary schedule to the meeting — just a bar chart based on initial impressions and our own experience of what needs to come before what. Then we talk about whether we have the durations and sequence just right.

JLC: *How do you handle the inevitable complaints and change orders as the work progresses?*

Rajan: It depends. If the client has established a good relationship with the foreman and he has the maturity to handle it, we go with it. But a lot of stuff can pass through the cracks because the foreman isn't on the job near the end when mostly subcontractor work is happening.

This is where having a project manager in the office is so critical. Because he's there from the beginning with the clients, he has established a relationship with them and knows what they want. (This also makes it pretty difficult for clients to invent stories about what they've been promised.)

In our case the project manager, Abe, is also our estimator, so he really knows the work, the crews, and the schedule, and he is in a position to resolve problems before they escalate.

JLC: *So is your role mostly sales?*

Rajan: My major part is screening projects and talking with clients; I guess you could call it sales. I also review everything: estimates, scheduling, assignment of manpower on projects, hiring, everything.

A few years ago, I thought what I had to do was step back from the business. I tried that for awhile and found that it lets you build a larger company, but you lose that essential contact with the clients. In our case, we have to deliver that; our clients come to us for that. So I've stepped back in.

JLC: *Do you think remodeling contractors are taking that direction generally?*

Rajan: I think what we're going to be seeing in the next decade is general contractors having less and less crew — carpenters, laborers, and so on — and more and more subs. This will result in larger companies, or at least loose partnerships of generals and subs which may even include things like profit sharing.

JLC: *Is that where Canyon Construction is headed?*

Rajan: We will be putting an effort into working even closer with our subs. And not just plumbers and electricians. I see us using general contractors as subs — foremen who have come up through our ranks, gotten their licenses, and come back to work with us.

It's difficult at the age of 30 to start a business from scratch. You may have complete skills as a craftsman, but to put all the elements of the business together and get good jobs is very difficult.

What we can offer a real hot general contractor, age 30, who wants to do good projects and be his own boss, is to team up with him. We'll take care of the organizational and business support (accounting, billing, all that stuff), and provide good jobs that will bring him that paycheck every week. And he gets the gravy by being able to concentrate on the work itself. ■

The Art of Client Management

by Martin King

Clients' expectations and demands will run wild — if you don't establish clear ground rules

We all know of jobs where an earnest contractor strives to produce a workmanlike job, and the client, instead of gratefully accepting his product, tries to skin him alive. Although such stories raise a builder's blood pressure, it's important that he calm down and rationally assess how things got that way. Only then can he develop strategies to avoid such trouble in the future.

Keep in mind that at the signing of the contract, the interest of the owner and contractor closely coincided. From that moment, however, those interests started to diverge — until they reached a point where mutual satisfaction was out of reach. From that point on, only one side can win — the other loses.

Even that might be acceptable if the sides were evenly matched. Unfortunately, they're not. By the time major differences emerge, the client usually possesses the completed work, control of the site, and the money. No matter how generous the payment schedule, a major portion of the contractor's profit is locked in those final draws and retainage.

What power does the contractor have? Reason? Fairness? A disputed contract and verbal commitments that no one can prove? It's no contest. The owner extracts, some would say "extorts," concessions — and the contractor often feels that he can do nothing.

It doesn't have to end that way, however. This article is based on the notion that divergent interests can be managed. Every contractor knows about project management; this is *client management*. We can define client management as "procedures that increase the chance of completing a project to the satisfaction of both parties."

A shorter definition might be: how to satisfy the client without losing your shirt. Because of the unequal strengths of the parties involved, the contractor must accomplish this by disarming potential conflicts before they arise. It has been our experience that many contractor/client problems originate in (1) unrealistic expectations, (2) the absence of ground rules, and (3) the wrong signals.

Client Expectations

For openers, take an unhurried look at the clients' point of view: Why do clients start off smiling and end up snarling? Sift through your own experiences, both good and bad. You may conclude, as we have, that a major source of some clients' distress is their unsatisfied expectations. It boils down to disappointment: What the clients see falls short of the image they were carrying in their heads.

How, you might ask, can you be responsible for the clients' imagination? The fact is, the clients' expectations are usually unrealistic. Remember that it's not easy to see the reality behind a designer's rendering. Real life is always smaller and less elegant than a pretty drawing. The real question is: What are you going to do about it?

Rather than chasing down complaints generated by disappointment, it is a lot easier to bring the expectations down to earth. How? Start by not fueling those expectations with your own words. *Don't* tell them the design is one of the nicest you've seen in years. *Don't* say, "You're going to love the way this expands your living (recreation-office-assembly-storage) area." These and dozens of other small comments tend to raise expectations. Such statements may seem harmless. Down the line, however, you will pay dearly for the expectations you excite.

Instead of building high expectations, find opportunities to prepare the clients for reality. State clearly that drawings excite the imagination and always look better – and larger – than the finished project. Meet the clients at the site after the footers are poured and show where the partitions will be. Let them visualize how the furnishings will fit. Tell them, "This is probably going to seem smaller when it is built than you visualized." Openly mention the danger of unrealistic expectations, especially if this is their first construction experience. Without disparaging their commitment or your performance, do whatever you can to wean them away from their imagination. Your credibility as a seasoned pro will be enhanced, not diminished, by down-to-earth realism.

Above all, do not promise anything! Promises increase expectations and diminish satisfaction. Even if you choose to upgrade an item out of your own pocket, do it quietly — and don't be surprised if the clients respond with even higher expectations.

In brief, to satisfy clients you must exceed their expectations. That means you must keep them reasonable. When do you start? As soon as the contract is signed.

Ground Rules

Contracts tend to tilt against the contractor. They either bind him in a straitjacket (the Standard AIA Agreement) or they leave things so loose that any question becomes a battle of wills. Both types fail to address important questions that arise time after time.

In order to avoid tedious legalistic documents, a contractor might find it useful to prepare a list of "ground rules" under which the parties agree to operate. These rules should be written to defuse the typical problems before they arise. As the contractor, you should know what those problems are going to be — the clients haven't been there yet.

The time to introduce your ground rules is after the price and specifications have been agreed to, or immediately after the contract is signed. Both the clients and contractor should initial or sign them. The rules should be written in ordinary language, and their "legality" is not too important. Most people will abide by a reasonable agreement simply because they have signed it; enforcement through the legal system is seldom feasible in any event.

If a particular ground rule is questioned, you may choose to explain the reasoning behind it, or you may simply state that "this is our company policy." One does not have to understand the basis for company policy — it just exists. If an objection arises, the rule can be negotiated and modified on the spot. (The contractor should pay attention to why the objection is raised, of course.) Of paramount importance is the clients' perception that the contractor is governed by well-established standards, not by the pressures of the moment.

Here are some examples of contract clauses which have proven useful for our company in establishing ground rules:

> **"Changes in the work specifications, including material upgrades, are subject to the contractor's written acceptance, and additional charges are payable in full at the time of the request."**

The purpose of this rule is to (1) discourage change orders, which break the rhythm of the job and severely erode profits, (2) eliminate haggling over the charges, which are always much higher than the direct labor and material costs involved, and (3) remove the clients' incentive to find credits to offset the additional charges at the end of the project.

Another major source of interruption is when the owners fail to make timely decisions, then blame the contractor for the delays. The following provision places the responsibility where it belongs:

> **"The owner will submit all material and color selections by ____ (date), including an alternate choice for each item. Except by prior agreement, all selections will be standard materials available from local sources."**

If the selection date is not met, the contractor should notify the owners by registered letter that they are in violation of the agreement and may incur additional charges due to the stoppage of work or scheduling delays.

Every contractor should be ready to warrant his work for some period of time. With the following clause, the warranty provides powerful leverage in the event that payment is arbitrarily withheld:

> **"Limited Warranty. All labor and material for work covered by this contract is guaranteed against defects for a period of ___ from substantial completion or use by owner unless otherwise stated. This warranty shall not apply if final payment is not received in accordance with the terms of the agreement."**

If payment is withheld, a registered letter should state, "Payment has not been received in accordance with the terms of the agreement, and you are notified that all warranties and guarantees are hereby cancelled."

The mysterious disappearance of materials and tools from the work site is always a problem, but tends to diminish with the following clause:

> **"Owner is responsible for the protection of work completed and all equipment and materials delivered to the site."**

Anyone can make a claim. Contractors who have been subjected to excessive retainage and payment delays for real or spurious claims will recognize the utility of this provision:

> **"Any claim for damage by the owner will first be addressed through the owner's insurance, and under no circumstances will the existence of such a claim delay payment of monies otherwise due."**

It's always better to have one individual to deal with, and frequent inspections with written acceptances are a good way to limit future problems. The following clause addresses this issue:

> **"The owner appoints _____ to inspect and approve work on the owner's behalf; inspections will be made in a timely manner upon the contractor's request."**

Another issue that often goes unaddressed is access to the site. Including an access clause can foreclose this heavily traveled path to claims and payment delays:

> **"The Agreement is based on using existing access to the work area. Unless otherwise specified, no costs have been included for repairs to access areas such as driveways, walks, shrubs, trees, lawns or entries. The contractor will use normal care and procedures in moving and storing materials."**

The contractor's responsibility for cleanup of the work site should be limited to a reasonable level. A surgically clean site may suit the owner, but it can be expensive to attain. Use this provision to clarify the topic of cleanup:

> **"Upon substantial completion the contractor will remove all construction debris and leave the premises in a broom-clean condition."**

Perhaps the single most important provision of all guarantees the timely resolution of intractable problems through a swift and economical binding arbitration procedure. I use the following clause to steer conflicts toward arbitration rather than the courtroom:

> **"All claims or disputes arising out of this Agreement shall be decided in accordance with the rules of the American Arbitration Association, unless the parties mutually agree otherwise."**

Wrong Signals

The cost of construction is intimidating to most buyers, who generally lack a clear idea of how the costs are allocated. Where so much money is involved, they suspect a lot of fat must also be present — so they probe to see if they can tap it. Initially they try to upgrade materials and request minor changes in the scope of the work. If the contractor lets them succeed, he sends the wrong signal.

A game begins — like a child testing overly permissive parents: The clients push as far as they can before meeting resistance. The further they get, the more difficult they are to stop. When the contractor finally tries to apply the brakes, he appears to be imposing arbitrary and unfair constraints.

The "deep-pockets" illusion can be corrected at the outset by the simple act of charging for every upgrade, no matter how small the amount. In fact, the smaller it is, the better. If the clients are asked to pay for a wallpaper upgrade of $1.50 per roll on a $54,000 project, they immediately perceive that money is important. Far from alienating them, this action provides assurance that the $54,000 is well spent.

Because they are paying for the project, clients sometimes get the idea that the contractor will bear all the responsibility and deliver a completed project into their waiting hands. This is another misconception that leads to a poor working relationship. Words and actions that say, "We'll take care of everything" send the wrong signal. It is unhealthy for the clients to be passive spectators on their own project. It gives them time to scrutinize every detail and bird-dog the work. Most importantly, it robs them of the chance to participate.

The best message for the contractor to send is not, "Look what a wonderful project we are going to produce for you," but rather, "Let's see how fine a project we can produce together." The clients who feel they have participated in a joint effort with the contractor are far more likely to be satisfied with the result.

It's up to the contractor to bring the clients into the project in a productive way. A good strategy is to start off with something like this: "Our construction experience has shown us that it's important for our clients to understand that we both have responsibilities in making this a successful project. One of your responsibilities is to help us get off to a good start by making the material and color selections within two weeks. We are also going to need your help in performing inspections when we need them, and in preparing for the progress payments so we can move forward without breaking stride."

Follow the clients' performance and don't be shy about reminding them when they're not following through. But also make a point of bringing them into the process by providing an occasional update on the schedule and the performance of the subs.

When materials are delivered, it's not a bad idea to look them over at the site with the clients. When questions arise, try to share the decision in a way that engenders a feeling of participation. Even if an architect is monitoring the work, it doesn't hurt to maintain a direct relationship with the owners.

On commercial work, try to establish contact with the supervisors and personnel who will occupy the new space. Show them how it's going to look. Solicit their comments. Early participation means fewer problems down the line. You will bleed off resentments and lower punch-out costs.

Tactics

The contractor/client balance of power shifts at different stages of the project. At the beginning, the contractor has considerable leverage, greater than he will enjoy later. The clients want the project badly enough to pay all that money to get it. They have finally selected the company to perform the job, have negotiated a contract, paid a deposit, and initially, have nothing to show for it. In contrast, the contractor has received a substantial payment on a promise to perform. He's in control. As the work progresses, their positions reverse: The clients possess the completed work, and control the remaining funds. The contractor is vulnerable. If there is a conflict at this point, the contractor loses.

Client management offers a way of improving the odds for a successful project. However, if the contractor intends to provide a positive working relationship, he must act early, while he has leverage. For that particular project the opportunity will not come again.

Summary

While concentrating on production and company management, contractors may overlook personal factors that are critical to a project's success. Client management can increase the chances of producing a mutually satisfactory result by using the contractor's initially strong position to (1) generate realistic expectations, (2) establish favorable ground rules, and (3) send the right signals to the clients. A variety of statements and ground rules have been included to illustrate specific points. The reader should modify the content and wording of these to fit his own situation. ■

Martin King is president of Martin Churchill Associates, Inc. in Arlington, Va. After 28 years of construction and estimating experience, he now specializes in investigating and consulting on structural damage and appraising reconstruction costs.

Preconstruction Conferences

by Tom Swartz

Use a preconstruction meeting with the clients to follow up the sale, review the contract and plans, and avoid any misconceptions

After too many bad experiences, we've learned that the best way to get a project straight is to hold a meeting with clients before work begins. These preconstruction conferences have saved us an incredible amount of work and confusion. They've also saved us plenty of money.

At our company, a typical job passes through two different people before work begins. The salesman works with clients to determine what they expect from a project, what it will look like, and how it will function. But once the bid is accepted, the production manager and the lead carpenter take over. Without good communication between these two and the sales rep, clients will find themselves explaining their ideas again to the production crew.

As the job progresses from an idea to working drawings and specifications, the various people involved form different perceptions. Clients may wonder how much of the information they discussed with the salesman was filtered through to production. In turn, the production manager might wonder exactly what the salesman promised.

Unless a remodeling job is small enough for our Handyman Service, where the actual craftsman is also the salesman, we hold a preconstruction conference seven to ten days after the contract is signed. Timing it this way works well for two reasons: It's a good follow-up to the sale, and it gives us a chance to measure for and order any special materials that may take several weeks for delivery.

Usually held at the job site (typically the client's home), the conference includes our sales representative, production manager, and lead craftsman, and, in the case of an insurance loss, the adjuster. If the customer is a couple, we encourage both of them to attend since this is also a good time to make sure they are in agreement about the job.

The conferences last anywhere from 45 minutes to one-and-a-half hours. This may seem like a lot of expensive time, but redoing work and, worse yet, leaving customers dissatisfied, costs much more.

Passing The Baton

The preconstruction conference provides a good opportunity for us to transfer control from the salesman to the production people. Early in the relationship, the salesman explains our company policies, and tells the client that the production manager and lead craftsman will assume control of the project once the contract has been signed. Then, at the start of the preconstruction conference, the sales rep introduces these people and lets them take over.

Because the customer expects this transition, there's no breach of confidence. At the same time, the production team can ask informed questions because they've reviewed the plans with the salesman before the meeting and have a good, basic knowledge of the project.

The conference is also a good time to review the contract, specifications, and plans. It's amazing how many misconceptions can surface during this part of the meeting. For instance, the sales rep may have specified a 5-foot window on one wall, but the clients had visualized a floor-to-ceiling window. At this point, it's still easy to make this kind of change since materials haven't been ordered and work has not yet begun.

Job-Site Details

To make a project run smoothly, we've developed a list of particulars that we discuss during the preconstruction conference. These are the kinds of details that can make the job pleasant for our crew and our clients. We include the following:

- What time the work is to start and end each day. Our normal hours are 7:30 a.m. to 4:30 p.m. This is okay for the majority of our clients, but some ask that we arrive an hour later.
- How change orders are handled. We make it clear that change orders are costly and will delay the project.
- Any particulars about pets. For instance, if the cat is declawed and should not get out of the house, we need to know it.
- The dangers involved in the project. These should be explained to children who are home when their parents are not. We also ask parents to discourage their youngsters from "helping" the painter paint or the carpenter cut plywood.
- Which toilet facilities, if any, may be used by

Getting Off on the Right Foot

by Linda Case and Walter Stoepplewerth

Among the many topics we discuss at the preconstruction conference are the following:

Storage of materials. An area about the size of a garage bay is needed for storage of finishing materials.

Dust. On an interior remodel, temporary moving or storage of the homeowner's furniture, rugs, curtains, piano, etc., must be arranged. Realistically discuss dust protection of adjoining rooms and whether a household member has an allergy to dust. Also, dust is a computer's enemy: Customers should be warned to move a computer out of the range of the inevitable dust.

Bathrooms. If a bathroom is being remodeled, there will be three or four days during which it will be unusable. Assure the customers that you'll give them adequate notice and create minimal downtime.

Kitchens. If the kitchen is being gutted and remodeled, there will be no working sink, appliances, or space for a period of time. Where should the refrigerator be temporarily relocated? Will the owner keep the removed cabinets or dispose of them?

Electrical. Discuss the routine interruptions in electric power that characterize construction jobs. This can be a real nuisance, particularly if the customer uses a computer during the day. Also, new wiring often requires opening and patching of walls. Explain that it is virtually impossible to get a perfect match on any touch-up painting and that the entire room may need to be repainted for a consistent color.

Plumbing. If you're adding or remodeling a bathroom in a home with galvanized water pipes, the contract should suggest changing all accessible piping to copper during construction. Galvanized pipes rust from the inside out and ultimately clog and rust out. When a new bathroom makes an additional demand for water, there will be a noticeable drop in pressure, and clients should know this up front.

Also explain that plumbing work may require the opening and patching of walls, and as with electrical work, this may result in mismatching paint where repairs are made.

HVAC. If the contract calls for copper-baseboard heating in a new addition but there are cast-iron radiators in the existing house, the contractor should explain that radiators work best at 140° to 150°F while baseboard heating works best at 170° to 180°F. A compromise is needed when setting the aqua-stat for the temperature in the boiler.

When a heat pump is to be installed as part of the contract, the contractor should explain that the temperature of the air at the registers will feel cool but that this is normal with a heat pump.

Missing walls. Opening up a house is just a construction problem to the remodeler, but it is a heating, air conditioning, privacy, housekeeping, and security problem to the customer. If a wall is to be removed between the existing house and a new addition, the timing is of utmost importance.

Dormers. If a dormer is being erected, requiring the doubling of 2x6 floor joists on the attic floor, nail pops or plaster damage is likely below. It's best to warn the customer to expect this problem and the likely need for repainting at the completion of the project.

Match existing. If any matching is being done, for mortaring, roof shingles, floor tiles, special molding, windows, doors, etc., inform your customers of any differences and show them samples if possible. Or send them to visit the supplier. The contract language should be pointed out. It might read: "match as closely as possible from stock materials at existing local sources of supply."

Staining. If staining a pine door is specified, the contractor should point out the different degrees of hardness of wood used for doors and explain that staining a pine door may produce a splotched effect.

Landscaping. Consider rough grading, finish grading, sodding, etc. If the customer is a gardener or landscape enthusiast, the backhoe digging a basement or footings in the existing or future yard will affect more than the actual dimensions of an addition.

One person might have an absolute fixation about no hammer marks on the trim, but be completely oblivious to whether or not the insulation is installed properly. On the other hand, another person might be extremely concerned about air leakage and energy conservation. This does not mean that the customer should set the standard of quality for a remodeler, but it is advantageous to know what areas should receive extra care and attention if the customer is to be satisfied. ■

Walter W. Stoeppelwerth is a founder of HomeTech, Inc., a remodeling and home-inspection consulting firm in Bethesda, Md. Linda W. Case, CRA, is the owner of Remodeling Consulting Services, in Silver Spring, Md.

our employees and subs.
- Which telephones can be used.
- Whether smoking is permitted in the house or yard.
- The reason for lien waivers, which we supply after each progress payment and at the final payment.
- Payment schedules and procedures. We've found that customers will often send the check before we ever invoice them because we went over the payment schedule at the conference.

In addition to all of this, we make it a point to review our employee policies. For example, our employees and subs should not accept food or even coffee from our customers, and we let everybody know it. We also have a strict policy against drinking on the job, even the occasional beer offered by the client.

We also discuss our effect on the neighborhood. We explain that lawn signs are important because they help suppliers and subcontractors find the job site. Of course, this is also good advertising for us. Some clients prefer that we don't post any signs, and in some neighborhoods they may be forbidden.

We make sure to ask our clients for their permission to send a letter to their neighbors informing them of our presence in the area. In the letter, we warn the neighbors that there will be strange trucks on the street. We also ask that they report any problems, like debris flying into their prize rose gardens, to us and not to their neighbor. Once again, this is good advertising for us.

Since many of our customers are away during the day, we always designate a communication box. It's a place where notes, partial invoices, and payments can be left. This saves a lot of trouble if you need to let the customer know you won't be on the job because you're waiting on an inspection or a sub.

Finally, we discuss the phenomenon we call "remodeling fever." This is the sudden realization by the clients, which can strike at any time, that they are tired of all the inconveniences that come with a remodeling project. During the conference, we try to prepare them for what's to come, and encourage them to look beyond the disruptions to the finished product. This is especially important for those who've suffered a fire loss. They didn't ask for the remodeling to happen and they are traumatized before the project begins.

All of this information is also included in the presentation book we give to potential clients. That way they can see the kind of attention to detail we provide. In this sense, our preconstruction conferences not only help us once we've got the job, they also serve as a sales tool. ■

Tom Swartz is president of J.J. Swartz Company, a 70-year-old remodeling firm with offices in Decatur and Bloomington, Ill.

Don't Make Yourself at Home

by Sal Alfano

Use common sense and courtesy to build a good working relationship with your customers

It's important to make a good first impression with the owners when you sell a job. But that impression is a fragile thing and must be nurtured throughout the project. This means that you and your employees must project an image of yourselves that will inspire confidence and set the owners at ease. The more comfortable they are with your crew, the happier they'll be with the outcome of the job. They may even find it a little easier to overlook small problems or imperfections if they generally approve of the way the work was conducted. If you can't be charming every minute of every day, at least try to follow a few simple do's and don't's.

Thank You for Not Smoking

Second-hand smoke is a big issue among office workers these days, but it's just as important in construction, especially in renovation, where work is often done in the owner's home. I always look around for signs of die-hard nonsmokers; an American Lung Association sticker in a window, a "smoke-busters" decal, no ashtrays. I once had a client who worked for the American Cancer Society; I didn't bother to ask.

If you do ask and receive permission to smoke on the site, don't be a slob; use an ashtray. Even an owner who smokes may be put off by coming home every night to cigarette butts lying around in soda cans and half-empty styrofoam cups. Above all, don't leave butts floating in the toilet bowl. Among other things, you may get blamed for future sewer line problems. And when the day is over, empty the ashtrays yourself; the owner shouldn't have to clean up after you.

Oddly, many builders think none of this applies to new construction sites, especially in rural settings. But consider the fact that the owners are very excited and possessive about their new house, and that butt you just extinguished on their subfloor may tarnish their image of their new home. To them, a floor is a floor, not an ashtray. Also, they may be worried about fire, and with good reason. It's easier and safer to confine your smoking to the great outdoors and keep your ashes out of the owners' sight.

What's Mine Is Yours

When it comes to the owners' personal possessions, it's not unreasonable to ask permission to use the phone, the bathroom, and the domestic water, but that's as far as it should go. If they offer the use of a radio or stereo, fine, but don't ask. It puts them on the spot, and even though they may say yes, it could be that they're embarrassed to say no, and the situation breeds discomfort.

Even if you get the okay, you're better off leaving their stuff alone. It's amazing what you can get blamed for. If you absolutely have to have some tunes, and the owner has offered his radio, be careful. Don't spill paint or joint compound on it. Don't drop it. And leave it where you found it. This goes for furniture and whatever else you have to touch or move in the course of your work.

Your general policy should be "hands off." Don't read their newspapers or books, don't watch the soaps on their tube, and stay out of the refrigerator. It's hard for some people to have strangers in their house all day and not suspect them of rifling the drawers.

Above all, be discreet. If you have to do some work in their bathroom, and it's not up to your standards of cleanliness, keep your opinions to yourself. Most people will fall all over themselves apologizing for "the mess" even when the place is spotless. You can make them feel more comfortable by keeping your criticisms under your hat. This also applies to comments about the decor, their taste in colors or modern art, or that funny smell coming from behind the stove.

And don't use their tools unless it's an emergency. Their vacuum cleaner, their mop, their shovels and hoes are all off limits.

Colorful Language

One of my subcontractors occasionally gets letters from clients thanking him for doing such a wonderful job, and quite often they mention the surprising absence of foul language among his workers. Construction workers are renowned for their liberal use of four-letter words, but it's obvious that clients are very aware of, and often offended by, "colorful" language. Despite the state of modern movie scripts and literature, most people are not accustomed to hearing an endless stream of oaths. Some words are more offensive than others. Find substitutes. I haven't heard "gee whiz" or "golly bum" in a while.

Roach Coach

Food on the site can create a sticky situation. I make it a point to adhere to a rigid schedule of coffee and lunch breaks. For one thing, the owner gets nervous if these become extended, wondering when the work will get done. And on a cost-plus job, I always explain that the crew gets paid for coffee break, but not for lunch. It saves arguing over labor charges.

Clean up after yourself. We all do work in winter, but we sometimes forget that the snow melts in the spring. I once had a client tell me that he found over 75 styrofoam cups blown all over his property. One of my subs had dumped their coffee cups into the snow the whole time they were working at the site, and in the spring they came up like mushrooms. Always provide trash containers, and empty them yourself. The owner has enough of his own garbage to contend with.

Good Times

Punctuality pays dividends. If the crew is supposed to be there at 8 a.m. and they don't roll in until 9, the owner may begin to wonder if there are other details they're not attending to. I always ask about the owner's routine as well, and if I can adjust my starting time to fit his or her schedule, I do. Not everyone is accustomed to the sound of a tablesaw at 7:05 in the morning.

You Are What You Wear

I've never had a need for a strict dress code, but the potential is there. You don't have to buy printed T-shirts or company uniforms, but avoid the raggedy look. Torn clothing will make you look more like a bum than "Flash Carpenter." Certain kinds of clothing may identify your crew in the minds of the owners with unsavory characters stereotyped in movies and on TV. Sleeveless black T-shirts cut off above the navel, for example, may remind them more of the Hell's Angels than a skilled craftsman. T-shirts with witty sayings are okay, but anything bordering on the obscene ought to be censored.

Miller Time

Alcohol has no business in the workplace. It's dangerous to everyone on the site to have someone working with staging or power equipment while they're impaired. While it's easy to convince your crews that they can't drink on the site, it's a bit harder to tell them they shouldn't drink before they get there, or even the night before. But a heavy hangover can be just as much of a problem.

A six-pack on Friday afternoon after all the week's work is completed is a popular ritual with some carpenters. If you allow this, make sure the mess is cleaned up afterward. If the owners come out to the site Saturday morning and find a lot of empties lying around, their imaginations will run wild.

Common Sense

Some things are obvious: don't flirt with their teenage daughter; don't tolerate theft; don't use the end table for a sawhorse. A little common sense and good manners will go a long way toward building good will. ■

Sal Alfano, formerly a general contractor in East Calais, Vt., is an associate editor with The Journal of Light Construction.

Chapter 3 • Personnel

Working Hard for Your Employees

by Michael Davis

Trying to keep framers motivated to produce, to keep conscious of quality, and to stay committed to you for more than a few paychecks is what makes life interesting for framing contractors. Or it can be a nightmare. I carry between 40 and 50 full-time framers on my payroll these days, and over time I've discovered a few things that work, and a lot of things that don't.

Perhaps most important is realizing what service you're selling. About 70% of my company's work is custom residential framing; the balance is light commercial and multifamily. But overall, more than two-thirds of our total volume of $1.5 million is in labor-only contracts.

That means our primary product is labor, and labor is nothing more than people. How my guys feel about working for me, therefore, has a lot to do with my bottom line. So I invest a lot of time and money in making my company the best place they've ever worked.

The Old School

This attitude wasn't one I saw much of when I was a framer, nor is it one I understood when I was first running my own crews. My introduction to framing was working for a piece-crew that built Motel 6 units all over the Southwest. We worked six to seven days a week and used up every hour of sunlight. We lived out of our trucks, hotels, and the occasional tool trailer. It was as rough as it sounds, but we made good money and we had a lot of fun.

Those years taught me some valuable lessons. I learned to hustle, I realized that time really is money, and I saw that no supervisor could ever hope to push an employee as hard as an employee will push himself if he's properly motivated.

Some years later, I found myself running framing crews here in Albuquerque. At first, I hired piece workers (that was the way I'd always seen it done), but I always seemed to end up on the short end of the stick. I didn't feel as though I had enough control over the work, and the quality I was getting showed that. So I switched over to hourly crews.

Although I worked hard, I wasn't sure how to motivate others to do the same without the leverage of piece work. A couple of close friends were the lead men on the crews. We were all young, and none of us had a clue about management. We basically tried to rule with peer pressure and intimidation.

Speed and quality in production depend on the people who work for you. Give them all you've got.

Author Mike Davis divides his time between bidding in the office (above) and visiting job sites (below). He produces his weekly newsletter on the Macintosh seen here.

I remember hiring a group of framers once, and telling them "We have purposely overhired. We don't need all of you, and so, at the end of the day the slowest man is down the road." Thinking back, it's a wonder that they didn't just kill me on the spot. But it was 1982, and construction in New Mexico was at a standstill. Needless to say, these techniques were less than effective. We were running about 20 to 25 men then, and we'd go through four to five hundred employees a year due to turnover.

Although construction is still pretty slow here, good framers are hard to find. If I lose a framer, I know he'll be very difficult, if not impossible, to replace. The obvious answer is "Just pay 'em better than the next guy, and you'll keep 'em," but money isn't the only thing. Last year, some of my guys were offered more money by a competitor and they turned it down.

Why? A combination of little things. There isn't any magic formula you can follow to make your employees happy. A lot of it is breaking out of your old attitudes, and treating the guys who work for you like real people. Here are some of the things I do along that line.

It Starts With the Boss

An important part of managing people is communication. I once worked for a guy who told me that the key to successful management was never to socialize with your employees — if you get too close they'll start thinking they can take advantage of you.

It's true that when you get out there and start mixing with the guys you can find yourself in some awkward situations. For instance, the new guys will always try to hit you up for a raise, but that's all part of the business. Employees need to know what's going on in a business to feel as if they're part of it. And they want to hear about it from the guy who's making the big decisions. This means getting out there, getting to know the guys on your crews, and giving them a chance to get to know you. Those of us who are tradesmen turned

Managing The Managers

Since most of the people management in our company is done by crew leaders, it's important to train them to manage effectively. And since our crew leaders are promoted from the ranks, we need to help them make the transition.

A supervisor's first duty is to supervise. This is a very difficult message to get across to a tradesperson who is used to performing physical work. Supervising means making sure that every member of the crew understands exactly what he has to do, and that he has the tools and materials needed for the job. Then, *if* there's any time left over, the crew leader can beat in a few nails.

To start our crew leaders thinking about how to work with people instead of with wood, we give them all copies of the book *The One Minute Manager*, by Kenneth Blanchard and Spencer Johnson (Berkley Publishing Group, New York, 1981), and we demand that they read it. Some of them aren't real keen on having to do "homework," but with lots of peer pressure, we get them to read it. Most now say that the book has helped them manage the people on their crews.

According to the authors, a manager has three responsibilities. First, a manager must communicate effectively so that workers know exactly what he wants done. Second, he must set clear production goals. Third, he must keep workers informed about where they stand.

Communicating instructions. A crew leader has to make sure he communicates his instructions very clearly. It's critical that a crew leader take the time to do this. Typically an employee doesn't want to appear stupid. When confronted, most will say they understand a job, whether they do or not. Therefore, when giving instructions, we teach our crew leaders to have a carpenter repeat the instructions in his own words. This gives the crew leaders a feel for exactly what the carpenter understands, so they can fine-tune any misunderstood points.

In one of our weekly meetings recently, a crew leader described how he instructed a carpenter to head out an opening in a TJI roof system. He showed the guy the two pieces of stock he wanted him to use, showed him how to lay it out, and left him alone to do it. Nonetheless, the carpenter botched the job and the crew leader lost his temper with the guy.

When this was discussed, everyone at the meeting agreed that it wasn't the carpenter's fault. If the crew leader had taken more time to be sure that his instructions had been understood, the problem could have been avoided.

Setting production goals. Goal setting requires that you estimate a reasonable time frame for job com-

businessmen have a real advantage here.

Now I make a special effort to know the name of every employee, and to say hello when I'm on the job site — not as the boss to his subordinate, but as one framer to another. I don't have to fake this respect. I know they work hard, I know they're out there in the heat and the cold, and I appreciate the fact that they are choosing to work for me.

Communication

Since we have a lot of people on a lot of different job sites, the way we keep in touch with everyone is with a weekly newsletter produced on my Macintosh computer. It gets handed out with the paychecks.

The first thing we cover in the newsletter is scheduling. We list the jobs that are under way, and jobs that are coming up. If this does nothing else, it provides the guys with a sense of job security and lets them know where they'll be going next. The newsletter also talks freely about jobs we're hoping to get, and how that fits into our plans as a company.

Another topic for every newsletter is how current projects are going. We're pretty frank — we applaud the winners and analyze the losers. We don't point fingers at individuals or crews, but if we're falling short when it comes to safety or quality, the newsletter points it out as a company problem that we all need to solve. We talk about both our strengths and our weaknesses as a company, and we ask the guys for suggestions on how we can make things more efficient.

We get some of our best ideas from our crew members. They're the ones out there doing the work, so no one should know better how to make it go more smoothly. And they appreciate that they have a say in how things are done.

While we want all of our workers to be team players, we also want to reward our top performers. Everyone has that dual need: on one hand, to be part of a winning team or a great company, but on the other, to be noticed and to stand out as someone special.

pletion. This starts in the office. We provide each crew leader with an estimate of the worker-hours required for each job. In turn, the crew leader shares this information with his crew, so everyone has an idea of how the crew is doing. At the end of each day, they can look at how many worker-hours they've used up and see if they're ahead of schedule or behind, and whether they have to hustle to make up for lost time, or just to beat the estimate.

On an individual basis, it's up to established crew members to give new employees examples of good performance. Most of this responsibility lies with the crew leader because he's the most experienced, but we also encourage advanced carpenters to help bring less skilled guys along.

It's important to get crew members involved in policing themselves. Getting good performance when the boss is around is easy. The true measure of performance is what happens when the boss isn't around, which is most of the time.

One minute praise. The principle behind the *The One Minute Manager* is summarized by the phrase "Feedback is the breakfast of champions." You let someone know how they are doing with "one minute praisings" and "one minute reprimands." It's important to give feedback to every employee, and it's especially important when working with new employees.

A new employee is one of a crew leader's biggest challenges. The crew leader has to take special care explaining the task at hand, and then watch closely to see if the employee meets the expected goals. The idea isn't to intimidate the guy. The goal is to watch the new employee until you can catch him doing something *right*. Then let the employee know exactly what he did right and encourage him to continue.

If you're not getting the performance you want, you first have to decide whether this is an "I can't" or an "I won't" problem. If it's "I can't," then you have to redirect the person. This is just a training problem. If it's an "I won't" then you have an attitude problem and a reprimand is in order.

When reprimanding an employee, tell him what he's done wrong, but don't verbally assault or humiliate him. Call him aside and discuss the problem in a professional manner. It's important to remember that you are criticizing the person's behavior, not the person himself. Also, let him know he has value, despite his problem behavior. Finally make sure that he understands that the reprimand is over. This way you can both get back to doing your jobs and not waste time fuming. — *M.D.*

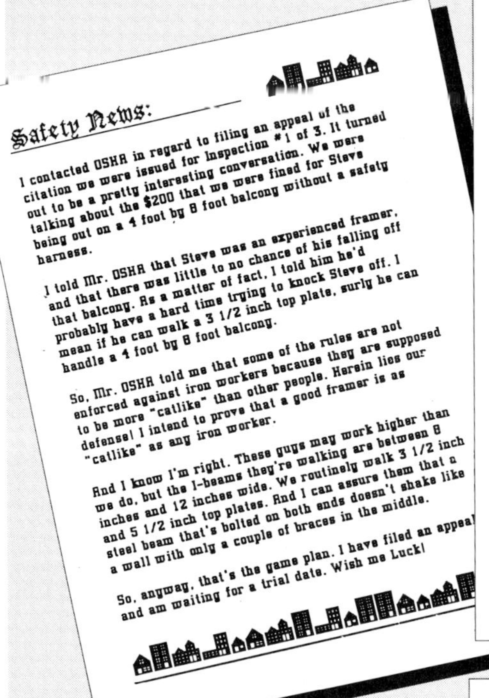

Safety News:

I contacted OSHA in regard to filing an appeal of the citation we were issued for Inspection #1 of 3. It turned out to be a pretty interesting conversation. We were talking about the $200 that we were fined for Steve being out on a 4 foot by 8 foot balcony without a safety harness.

I told Mr. OSHA that Steve was an experienced framer, and that there was little to no chance of his falling off that balcony. As a matter of fact, I told him he'd probably have a hard time trying to knock Steve off. I mean if he can walk a 3 1/2 inch top plate, surly he can handle a 4 foot by 8 foot balcony.

So, Mr. OSHA told me that some of the rules are not enforced against iron workers because they are supposed to be more "catlike" than other people. Herein lies our defense! I intend to prove that a good framer is as "catlike" as any iron worker.

And I know I'm right. These guys may work higher than we do, but the I-beams they're walking are between 8 inches and 12 inches wide. We routinely walk 3 1/2 inch and 5 1/2 inch top plates. And I can assure them that a steel beam that's bolted on both ends doesn't shake like a wall with only a couple of braces in the middle.

So, anyway, that's the game plan. I have filed an appeal and am waiting for a trial date. Wish me Luck!

Soft Ball Report

WE WON!! THE BOYS DID IT!

They beat the Interns!

This Wednesday Evening at 9:15 PM
Their last game of the year:

THE FRAMING SQUARE SOFTBALL ASSOCIATION LTD
VS.
Alamo Rental Car

You know, the teams we played last season had more imaginative names. You could have a little fun writing about teams like "Uranus" and "The Noids," But Alamo Rental Car? Of course, I can't talk. Framing Square isn't such a brain storm of a name. But hey, I figure if the Company is paying for it, and soft ball schedules are hanging by little magnetic bananas on refrigerator doors in half the homes in Albuquerque, then we might as well get some advertisement out of it. I guess Alamo Rental Car figured it that way too!

BIDDING WARS RAGE ON!

Yes, winter is upon us, and the Contractors in Albuquerque are trying hard to tie up enough work to get them through till summer; and their doing it by bidding low.

In the last couple weeks we have bid several jobs, and have been underbid like you wouldn't believe:

- We priced out a little school addition. Our bid was right at $25,000, with about $1,600 in there for profit. WE GOT BEAT BY $2,000.
- We bid a 40-unit retirement community for $121,000. We figured to make about $5,000 in profit if everything went well. WE GOT BEAT BY $4,000.00.

And the list goes on ... The moral of the story is this: Contractors are taking jobs at or below our cost. We are going to have to lower our prices if we hope to get enough work to keep us busy through the winter. That means that our margins will be narrower, and I'll be putting a lot more pressure on the crew leaders to get the jobs done. That of course means they'll be pushing you harder to produce.

If your Crew Leader starts riding your butt a little harder than usual, don't get pissed off. He's just passing on what you're getting from me.

Framing Construction

SAFETY · QUALITY · PRODUCTION

Company News: 

It's Bonus Time Again!

We had a pretty good month, all in all. We finished the Residence Inn, we kicked butt on the Days Inn, we cranked out St. Michael's ...

Newsletter Keeps Crews in Touch

Framing contractor Mike Davis uses a weekly newsletter to keep his 40 to 50 framers in touch with what's going on. In addition to a lead page that gives a detailed list of projects, their start date, and crew leaders, he focuses on crew members and the life of the company. Above are some excerpts from several editions.

Framing Square's bonus incentive plan doesn't reward just crew leaders—it applies to all members of the crew.

Along these lines, we talk in the newsletter about the successes of individual crew members. If someone learns something new, like how to cut stairs or figure rafter lengths, we write it up. Basically, we brag about him a little to let him know that we appreciate the effort he has made.

In addition, we have a program called "Framer of the Month." Every week the carpenters on each crew nominate the crew member they think has been working the hardest. The nominees are listed in the newsletter. At the end of each month, the framer with the most nominations wins a cash bonus of fifty dollars and recognition in the newsletter.

We also recognize employees on their anniversaries with the company, and use birthdays and children's births as good excuses to joke with them in print. This helps promote a team spirit.

And speaking of teams, we play league softball as a company. We pay all the fees and buy jerseys and balls so it doesn't cost the crew members anything to play. They're not the top team in the league, but they have a great time. It's a chance to get families out there, and for all of us to socialize a little bit.

Insurance and Retirement

We have a major medical plan ($250 deductible), and anyone who has been with us for a year or more can get coverage. We pay half. For a single man, the cost of coverage is around $8 a week.

When I set the plan up, I thought it would be a big plus, but as it turned out, few of our people participate in it. Our guys range in age from about 25 to 30 — still young enough that they don't feel threatened by illness.

We also have a profit-sharing trust that contributes money towards employees' retirement in years when the company makes a profit. We started the program back in 1987, but construction has been pretty slow since then, and we've only accumulated $45,000 to date. That's not going to give any of our guys thoughts of retiring once it's divided up, but it's enough to show that we're interested in their future as well as the company's.

The Bonus Program

The latest program we've initiated is the Bonus Incentive Plan. Unlike retirement, which is a pretty distant concept for some of our guys, the bonus is based on dividing up 10% of the company's profits each month among the employees. The money goes to everyone in the company from top to bottom, not to just the crew leaders.

This system rewards the performance of the company instead of the performance of one crew or one person. This helps to build a company identity, because everyone feels they are on the same team, instead of competing against each other.

Here's how we make the split. First, we figure out how much we made for the month and come up with a pool of dollars based on 10% of the profits. Then we divide each crew member's earnings by the total payroll for the month to come up with their individual percentage. We multiply this percentage by the amount we have allocated for bonuses and come out with how much to pay each employee. It works like this:

The bonus money available
for August is $2,350

Kramer Framer made $1,000 in August

Total payroll in August = $29,382.68

$1,000 divided by $29,382.68 = 3.4%

So Kramer gets 3.4% of the
$2,350 bonus pool

$2,350 x .034 = $79.90

By basing the bonus on an individual's wages, it reflects each worker's merit. By using total wages earned rather than hourly rate as a gauge, this system also credits the person who showed up every day. This may sound as though we're complicating the simple idea of giving a bonus for hard work, but bonuses can be tricky. Although the guys are all pals, there is a pecking order within the company and on each crew, and there are no secrets when it comes to how much everyone makes. So when it comes time to give a bonus, you don't want to cross those established lines or you can end up creating serious resentments.

Structured Meetings

While these motivational tools are important, they would be meaningless if the company wasn't well organized.

We have nine crews, each functioning as its own little company under the umbrella of the parent company. Once we have a project under contract, we schedule it and assign it to a crew leader, who takes over from there. Crew leaders order their own materials, schedule their own equipment, work directly with clients, and negotiate all change orders. They also do their own hiring and firing. The office staff is there to give any kind of support the crew leaders may need, but unless there's a problem, we stay out of their way on day-to-day affairs.

To keep all nine crews in sync with each other, we have weekly meetings every Wednesday evening. All crew leaders get paid to attend. In addition, we invite a representative who is nominated from the ranks of each crew to sit in.

The meeting is carefully structured. We start by going around the room, crew leader by crew leader, discussing individual jobs. The crew leaders turn in their time sheets. We review each crew's particular needs, such as equipment, materials, and manpower. As we go, we compile a list of what's needed. Once we've heard from everyone, we make adjustments. If one crew is shorthanded and another has too many people, we switch guys around until we get the balance we want.

After we have all the men and equipment lined up, we look at the jobs that we're bidding and negotiating, along with any payment problems, safety concerns, changes in insurance requirements, legal issues, and other general information. We also discuss financial matters. Everyone in the company participates in the bonus plan, so we all keep a watchful eye on the bottom line.

We then give the crew reps a chance to speak up. The crew rep is there to relay information back to each crew. In turn, we ask for input from them. Whether it's a production issue, a safety concern, or a policy matter, we want to know how we can do better.

We share a great deal of information with our people, far more than most companies. But we want our people involved in the decision making process, and they can't make informed decisions without the information.

Must Be Honest

These programs are expensive, but we have the happiest crews we've ever had. Our turnover rate has dropped dramatically in the past few years, and the quality of our work is up. I'm also enjoying my job more now, and that's worth something. I've even received several calls from former employees who wanted to let me know how much they enjoyed working for us and how good the newsletter made them feel. One of those framers was a guy I had fired—quite a switch from the death threats I used to get.

There is a risk in this approach, though. It doesn't work if it's not honest. If you don't believe what you're telling your crews, they'll know, and no program of perks is going to help you. If your heart's not in it, leave well enough alone — or you'll undoubtedly do yourself more harm than good. ■

Michael Davis employs nearly 50 framers as the owner and president of Framing Square Construction in Albuquerque, N.M.

Keeping Employees Motivated

by Gail Hermann

A good employee incentive program stimulates workers' productivity as well as pride in their jobs

Keeping good employees is key to the success of any business. Incompetent and unmotivated employees are a company's "weak link" in both production and image.

Our company started in 1979 when my husband quit his job to design and reconstruct our home after a serious house fire. Over the next few years, we became very busy remodeling summer homes and building second-story additions.

By 1987, we had grown so much that we split our company into two separate corporations. At the same time we initiated an employee incentive program. Our goal was to motivate our employees to be productive and to take pride in their work. We tried a number of practices and found several that worked well for us:

- Hourly wages are kept relatively low. Workers earn extra income through bonuses and incentives. This way when business is slow, we can still afford to keep our people on the payroll.
- All employees are given company business cards. If an employee speaks with a potential client, the party is given a card and told to call us if interested. There is no hard sell. If we get the project, the worker receives 2% of the gross. Furthermore, employees receive 2% of all jobs referred from their original lead — so staff members can benefit from a "chain of recommendations" throughout their employment. This income is discontinued if they leave the company.
- If a job is completed sooner than anticipated on the preproject estimate, the workers receive extra pay equivalent to half of the time saved. For example, if an estimated six-week job is finished in five weeks, each person receives two-and-a-half days worth of additional wages.
- We give employees bonuses on each job that progresses according to plan.
- We supply our workers with full uniforms. This keeps them looking professional and is another way to spread the company's name. (An alternative would be to establish a dress code, which might prohibit shorts or muscle shirts, or require specific colors for shirts and pants.)

Rewards

There are other methods of rewarding employees that cost the company little or no money. We suggest the following:

- Select an "employee of the month." One can even place this person's picture in the lobby or office where it can be seen by customers, sales reps, and staff members. This practice promotes a real sense of pride in employees.
- Reserve the best parking space in the area for the employee of the month. He or she will continue to work hard to keep that space.
- Use a monthly newsletter to announce the employee of the month, introduce new workers, and boost morale on lagging projects.
- Send staff members to seminars to learn from professionals about good business practices and their own strengths and weaknesses. Workers usually accept new information more easily from a third party in a classroom than from their supervisor. And they're more likely to gain insight into their work habits and make the necessary changes.
- Provide employees with personal subscriptions to selected trade magazines.
- Stop giving your employees Christmas gifts. Most workers are exhausted by this season and so their productivity decreases. Why reward them now? To us this seems like common sense, but many companies feel obligated, whatever the reason. Never give a bonus to someone who has not earned it.
- Remember how it felt to be an employee with a fixed income and no recognition for a job well done. A pat on the back, an encouraging word, or a thank-you can go a long way toward making people feel appreciated.

CARPENTER WANTED: Talented person looking for a rewarding position in a growing company. Qualifications: Common sense, ability to read a ruler, experience with power tools, desire to do the job only once (right the first time). Must Have: Own transportation to company headquarters (not job sites), the desire to work 40 hours every week. No butchers allowed. Interested? Call 123-4567.

In addition to rewarding workers, we have found that these methods successfully weed out the slackers. Since each person's performance affects everyone else's income, good employees don't want to carry those who aren't towing the line.

Rewards and acknowledgements for jobs well done bring the company increased productivity, higher morale, lower worker turnover, happier clients, and more referrals. Before introducing such programs to your company, however, do a financial projection to see what you can realistically afford.

Most of all, be honest with yourself, your employees, and your clients. Do not become greedy or find fault where bonuses are concerned. Incentives should not reflect your overhead cost increases, your mispricing a project, or other job failures. Do not punish staff members for your own mistakes. ■

Gail Hermann is a principal of Ultimate Living, Inc., a remodeling firm located in Foster, R.I.

The One-Man Crew

One person working alone is usually the most efficient way to build, but special management policies are needed

by Timothy Faller

We have all joked about the inefficiency of the stereotypical road crew where two people are leaning on shovels while a third does all the work. But when we stop to think about it, our own practices aren't much different: There are times when too many workers on one job creates labor cost overruns, and other times when not enough workers creates a slowdown. In the remodeling business, labor mismanagement can break any budget and ultimately cause business failure. We're all looking for ways to manage our crews for maximum production, and regulating crew size is one way to do this.

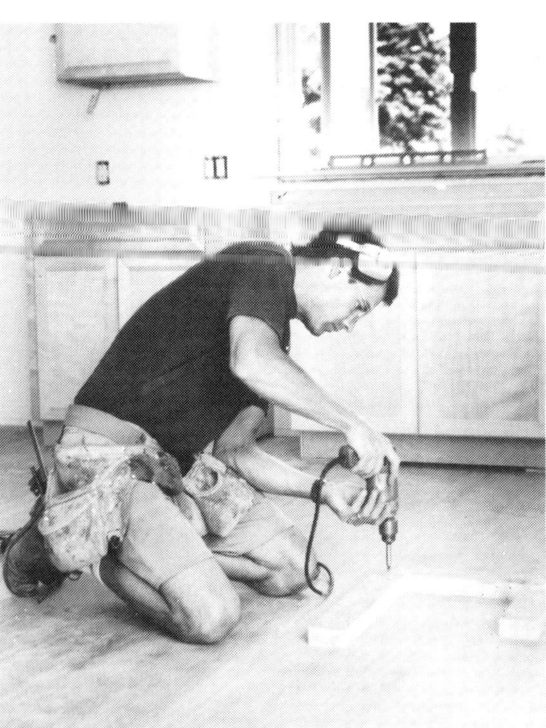

Two's A Crowd

Holding the line on the labor budget often determines the success of a remodeling company. The smaller the crew size, the better your chances are of producing a quality product and making a profit. This holds true right down to a one-man crew. If a job can be physically handled by one person, productivity will be greater than when two people are doing the same job.

Ownership. The reasons for this lie more in human nature than in job skills. As a carpenter working alone on a job assumes more responsibility, the sense of "ownership" grows. This sense that the success of a job directly depends on his or her performance increases productivity. Carpenters find the challenge of ownership hard to resist because they have traditionally been considered a dispensable part of the team.

Who's in charge? When two people are assigned to a single task, one of them inevitably becomes the supervisor and the other the supervisee. When one takes time away from work to look over another person's shoulder, productivity suffers. It's reasonable for the lead carpenter to inspect the work at various intervals to ensure quality control, but that is vastly different from two people running baseboard and both of them checking to see if the miter fits. Stopping and starting to answer questions, check work, or simply talk about what you did last night wastes valuable production time. And as crew size increases, coffee breaks seem to consume more time, and each day's startup and cleanup take longer as well.

People skills. Carpenters, by nature, are not usually good managers. They spend years learning their craft, but almost no time at all learning how to handle people. Good carpenters can often bring out the best in themselves, but not in others. Matching tasks to

Crew Size vs. Productivity					
Original crew	Cost of original crew per hour	Cost of added worker per hour	Cost of new crew per hour	Value of production per $100 spent on new crew	
				Worst case	Best case
1 Carpenter	$20	$13 (plus 1 helper)	$33	$61	$100
1 Carpenter 1 Helper	$33	$13 (plus 1 helper)	$46	$72	$100
1 Carpenter 1 Helper	$33	$20 (plus 1 carpenter)	$53	$62	$100

Figure 1. *In general, the larger the crew, the less efficient the work—particularly when you add to a crew just to keep another worker busy. The "Worst-Case" scenario assumes the added worker stands idle while the core crew works efficiently.*

people of varying abilities and motivating them to stay within the budget while still producing a quality product are skills that must be learned. Carpenters should receive some training before they are asked to manage other workers. Personality conflicts, prejudices, and bad moods complicate matters, and inexperienced managers will tend to throw up their hands and quit trying. The minute you hear your carpenters say things like "this is not in my job description," you know you're about to throw away large chunks of money from your labor budget.

Over the years, we have had people who were excellent craftsmen, but poor managers. In one case, on a large job that required three or four people, the lead carpenter didn't produce anything himself because he couldn't let the rest of the crew do their work without constantly looking over their shoulders. In another case, a lead carpenter got it exactly backwards, assigning a helper to hang a door and giving himself the job of tearing out a tile wall. But the classic example is the carpenter who left a helper to set cabinets while he took lunch orders and went to the store. This is mismanagement at its worst: $20 an hour goes shopping while $13 an hour works.

Numbers Don't Lie

Maybe some figures will help to illustrate more clearly how this affects the budget. Start with a crew of one carpenter at a cost of $20 per hour. When you add a $13-per-hour helper to the crew, you increase your labor cost by 65% to $33 per hour. Assuming that you get maximum production out of the carpenter working alone, the value of adding the helper can vary considerably (see Figure 1).

In the best of circumstances, both will produce at full capacity and you will get $33 worth of production for your $33. It is more likely, however, that at least some of the time, the helper will simply "hold the shovel" while the carpenter works. In this case, you will spend $33 for just $20 worth of production, which works out to about 61¢ on the dollar.

This means that for every production dollar you spend on the two-man crew, you will get between 61¢ and $1 worth of work — often closer to the 61¢. (In the absolute worst-case scenario, the helper works and the carpenter holds the shovel, in which case you get a mere 40¢ on your production dollar.)

The fact is that each additional person on a crew tends to lower the crew's productivity. So unless the additional person is absolutely necessary on that job, the risk to the budget is too great to keep him on. On an average job, losing $10 to $20 out of every $100 spent on labor could add up to a large sum of money.

Managing More Than One

One person can handle an enormous

WEEKLY REQUEST FORM

JOB BROWN LEAD CARPENTER JIM DATE 3/27/92

SUBCONTRACTOR SCHEDULE

	SUBCONTRACTOR	WORK REQUIRED	DATE WANTED	DATE SCH
1.	RILEY ELECTRICAL	ROUGH ELECT.	4/6/92	4/6/92
2.	BRODER PLUMBING	ROUGH HVAC	4/2/92	4/6/92
3.				
4.				
5.				

SPECIAL ORDER ITEMS

	ITEM SPECS	DATE WANTED	DATE SCH
1.	DINING ROOM LIGHT FIXTURE	5/25/92	6/1/92
2.	VANITY TILE (SECOND FLOOR BATH)	4/27/92	5/11/92
3.			
4.			
5.			
6.			

INFORMATION NEEDED

1. KITCHEN APPLIANCE SPECS:
2. OVEN
3. RANGE HOOD
4. UNDERCABINET MICROWAVE
5.
6.
7.

PERSONNEL NEEDED HELPER WHEN 4/2/92 WHAT TASK UNLOAD FIXTURES

PAPER WORK NEEDED

Time Cards_____ Order Forms_____ Job Logs_____

Weekly Request _____ Hours Report_____ Budget Report X

Figure 2 *The author's lead carpenters submit this form every week to keep him up-to-date on their needs for material, manpower, and information.*

amount of work, including tasks that have traditionally been done by a crew of two or more. Almost all of the work on small additions, for example, can be done by one person. Even kitchen cabinets, which are almost always installed by two people, can be leveled, fitted, positioned, and screwed in place by a lone carpenter skilled in the use of jigs and temporary supports. If the countertop requires two people, send a helper, but only after the cabinets are in place.

But some portions of a job, such as installing the rafters and sheathing, require a second or third person. Whenever it is necessary for more than one person to be on the job, follow a few simple rules.

Matchmaker. Until the lead carpenter has demonstrated an ability to manage people, take responsibility yourself for matching people to specific tasks and defining each person's role. This will help the lead carpenter learn how to use individual workers effectively; it will also eliminate power struggles or the jockeying for position that often occurs when two carpenters are assigned to the same job.

Get in and get out. The second or third person should be sent to the job as close as possible to the time he is needed and taken off the job as soon as he is no longer useful. This complicates scheduling, but it can save the bottom line. Too often we leave a person on a job long after the need for him is satisfied simply because we have nowhere else for him to work. The harsh reality is that if you don't have a place for someone to work productively, you should give him a little time off.

Request forms. Finally, use a weekly request form to handle changing manpower requirements (see Figure 2, at left). These forms should be turned in one week in advance to give office personnel time to evaluate the situation. Using forms will also force your lead carpenters to plan their manpower needs more carefully.

Time and Space

The time frame in which a job must be completed often affects manpower scheduling. Even though one carpenter working alone will be better for your budget, time constraints may require a larger crew. Two people can complete a job faster than one, but the trick is to maintain efficiency. The best way to do this is to divide the job into smaller tasks and create several one-man crews. This gives each person ownership of a piece of the job and encourages him to work independently. For instance, while a mason lays block for the foundation, a carpenter can take care of the interior demolition. A large project can be organized by room or by individual tasks, like all door trim or all baseboard. Make lists of tasks so that each person can move quickly from one task to another without having to stop and think about what comes next. Write this list down on paper (not a scrap of 2x4) and post it on a clipboard where everyone can see it and check off tasks as they are completed.

The size of the work space also dictates crew size on many remodeling jobs. Whoever is responsible for scheduling manpower must take into account that only a certain number of people can work in one space at a time. Kitchens and baths are usually too small for more than one worker, and you waste money when you have two people working in a one-person space.

Lead Carpenter

A one-man crew works best, but only if you don't juggle your personnel too much. We solve this problem by using lead carpenters on every job. These people can have a variety of titles, but whatever you call them, they carry the job from start to finish as on-site working supervisors.

Because a lead carpenter is on the site all day every day, he acts as a go-between for office personnel and helps the company respond to the client quickly and effectively. In the traditional management approach, in which a series of different crews moves through the job, the client doesn't feel comfortable enough with any one person to discuss problems or ask questions. When this happens, clients tend to bury their concerns until later. In the meantime, small problems may grow into major points of contention.

In addition to performing a substantial amount of the actual construction, the lead carpenter is responsible for ordering material, setting the schedule, calling subs, and finishing up all punchlist details. This frees up the production manager or company owner to focus on other aspects of the business — company growth, job costs, special-order items, change orders, and customer service — instead of having to be on the job to ensure that the work is being done.

The biggest problem with this system is that as a job nears completion, the lead carpenter will become anxious to move on and will "forget" the details. The temptation is to leave punchlist items to another carpenter while the lead starts a new project. But if this is allowed to happen, the job will drag on for several weeks after it should have been completed. It's better for both morale and the budget when the lead carpenter stays on the job until all the details are finished.

Motivation

Every carpenter is different, and you must find out what motivates and challenges each person. Some carpenters are spurred by a compliment, others by fear. Approaching each person on a level that they will understand and respond to will result in greater productivity.

Responsibility. In our company, we introduced financial incentives to try to boost productivity, but it had only a marginal effect because the carpenters felt they were already working as well as they could. Since money was tight and we couldn't raise wages, we abandoned the bonus system in favor of giving the carpenters more responsibility for the jobs they were doing. The jury is still out on this plan, but it has created lots of enthusiasm and we expect to see greater profitability.

Freedom. When you give people responsibility, you must also give them the freedom to get the job done in their own way. If you micromanage, it will drive your carpenters crazy. As long as they build safely and according to plan, we allow them the freedom to use their own techniques and abilities, including problem solving. Not only does this improve morale, it also makes good use of an individual's experience to help get through the tough situations that come up regularly.

Information. For this system to work properly, however, you have to provide your lead carpenters with as much information as possible. They should have all the paperwork associated with their projects, including budget, contracts, and price lists. This also means telling them how much profit you expect the job to generate, and how you plan to earn it. This is an issue of trust, but if a person believes they've been entrusted with inside information, their production will improve. And the more they know, the easier it will be for them to make decisions that will be good for the company.

Quality control. At the same time, you have to maintain quality. Make regular checks and learn how to correct without criticizing. Use a checklist at each major step in the production process to inspect the work for conformance to the plans and specs. This will eliminate many problems down the road.

Attitude. Your enthusiasm can be catching. The way you or your field supervisor approach troublesome jobs will affect the attitude of the lead carpenters. If they are beaten down by the problems on their jobs, it will show in their work. The first rule should be "no bad-mouthing the client," no matter how hard the job becomes. I've had carpenters describe their projects as "the job from hell," and I have heard clients called worse; it has always had a bad effect on morale. An optimistic attitude will also leave its mark.

The lead carpenter concept creates a new class of carpenters who no longer feel expendable. Because they are a critical link in the success of the company, they have a more positive, optimistic outlook on the job. This leads to better productivity and improved efficiency. ■

Timothy Faller is the production manager of Hopkins & Porter Construction in Potomac, Md., where he oversees a field crew of between 8 and 15 carpenters and helpers.

Writing an Employee Manual

Clear policies make for good employee relations

by Ward Smyth

At the heart of any good employer/employee relationship is a clear definition of expectations—of what's acceptable, and what's not. As a company grows, this clarity becomes even more important. When you have more than one crew, with different foremen—each making disciplinary and policy decisions—the potential for misunderstanding, favoritism, and just plain poor management begins to multiply. In addition, without clear policies, a single manager may respond differently to a given problem depending on whether a job is going well or going poorly.

When our building and remodeling company began to grow, we realized that we needed consistency. With 18 people in five crews, we decided it was time to write an employee's manual. Now if someone walks in the door wanting to know what it's like to work at Salmon Creek, they can get a clear and uniform response.

EMPLOYEE HANDBOOK

Salmon Creek Builders

Putting It Together

A good manual provides clear guidelines for behavior yet leaves people room to be themselves. Some foremen, for instance, are more discipline-oriented, others are more flexible. And each crew tends to have its own "personality"—some listen to rock-and-roll, some to Vivaldi. There may be no difference in production rates, but there is in style. Running a business with multiple crews is not that different from running a family. We all need to have the rules defined, yet nobody wants to work in an oppressive environment. Finding that balance is important; we didn't want to write a book that everyone would have to consult daily.

Every business is different, so before writing our manual, we did quite a bit of research. We ordered a mail-order book on the topic and used it as a reference. It provided us with key definitions and warned us about common pitfalls. We also begged and borrowed employee manuals from all sorts of large and small companies.

When finished, our manual was just 24 photocopied and stapled pages. Some employees have said it's a bit wordy, but there's not a lot we would cut. One of the problems we had in putting it together was that there were hundreds of other things we could have put in, but we wanted to make it something that our people wouldn't throw in the trash; we wanted to offer a format where employees could easily find answers to specific questions. We have made amendments to it, mostly in response to suggestions made by foremen, and requests from workers. Our manual is a two-way street. We have as much obligation to follow it as do our employees.

We insert a statement at the end of the manual in which the employee acknowledges that he or she has read and understood it. We included the statement so that people would take the manual seriously. An employee manual becomes a contract in most states, so we wanted to have it signed. A duplicate copy of the statement goes into the employee's job file.

The current version of our employee manual is divided into three basic sections: benefits, employment policy, and rules and regulations.

Benefits

For compensation we had two options: we could either pay a high wage with no benefits, or pay good wages with a good benefit program. The person who comes to you because you offer the highest wages will leave immediately upon being offered more. To attract stable, career-oriented people, we offer the latter: competitive wages and a healthy benefit package.

The major benefit in our package is health and life insurance. After much pondering, we decided to pay 100% of the employee's insurance bill. If a person wants to put his spouse on the policy, he can do it as a pay deduction. Other than a brief description, we've kept the details of the plans out of the manual. The insurance company supplies that, and we don't want to document something that's subject to change from time to time.

Another of our better benefits is our tool-purchase program. I view it as a real two-way street. The carpenter whose circular saw takes a bounder off the roof probably doesn't have $185 in his pocket to replace it. But most likely he needs it the next day at work. Under this program, he can go to any store where we have an account and charge a tool to the company (workers can charge up to $200 without prior authorization). We take that amount out of his next four paychecks. It helps take some of the bite out of unforeseen expenses. The other advantage of this benefit is that we probably have guys using much better tools than you'd see generally.

We also offer membership in a credit union, and I don't believe this is typical for a company our size. Again, we were hoping this would be an enticement to the long-term employee with a stable background. We offer this benefit through ABC (Associated Builders and Contractors) but we have no requirement that anyone use it. The credit union helps some employees to establish credit, while providing others with a vehicle for savings.

In terms of sick days and vacation days, we don't carry any over at the end of the year. Instead, we multiply an employee's December pay rate by his unused time and make a deposit into his credit union account. We provide seven paid holidays, but we do have a requirement that employees work the day before and the day after a holiday (if they are scheduled workdays). Our vacation policy is to grant one week after one year of employment, two weeks after two years, and three weeks after seven years.

The major advantage of all of these worker benefits is that Salmon Creek attracts higher caliber people. The investment is substantial. The basic social security, workers comp, etc., cost about 22% to 25% of base salary, and the other benefits probably add another 10% to that, but we feel that it's worth it.

EMPLOYEE WARNING NOTICE ☐ 1st NOTICE ☐ 2nd NOTICE

NAME	CLOCK NO.	DEPARTMENT	DATE

NATURE OF VIIOLATION	ADDITIONAL EXPLANATION
☐ Substandard Work	
☐ Safety	
☐ Conduct	
☐ Tardiness	
☐ Absence	
☐ Absence	
☐ Attitude	
☐ Disobedience	
☐ Carelessness	
☐ Other—See Explanation Area	

I Have read this notice and understand it. ▶	EMPLOYEE'S SIGNATURE	DATE

WARNED BY	APPROVED BY

Employment Policy

Most of our policies are standard: we promote from within; require neat, business-like dress; and provide equal opportunity.

We do have a list of required tools — something I haven't seen much in other small home-building companies. There's nothing terribly exotic on the list, but it solves the problem of the company having to serve as arbitrator when Sam doesn't have a tool, then borrows someone else's and breaks it. And this policy works nicely with our tool-purchase program.

All new hires are put on a 90-day probationary period during which we can review new employees and they can review us. You can't really gauge a tradesman's skill until you see him at work. We chose 90 days over 60 days because we wanted to see if people were going to stay before giving them the full benefit package. This period also provides a standard time limit for the new employee to obtain the required tools on the tool list.

We do a written performance review of everyone in the company (including office personnel) twice a year. We use a standard form that's four pages long and evaluates everything from work habits to how employees get along with others—all graded on a one-to-five scale. Each foreman reviews everyone in the company and when the reviews are done, I summarize them using symbols instead of names so the employees don't worry about who said what about whom. One copy of the review goes to the employee and one copy goes into his job file. These reviews let our workers know how we feel about them. Also, they know that the boss is aware of their performance and that they can anticipate potential raises accordingly.

Rules And Regulations

Our rules are pretty straightforward. No gambling, no drinking, no abusive language on the site, etc. As I say during the job interview, this part of the manual tells you how to get fired. We basically work on a three-offense system (although some behaviors, such as walking off the job, will result in immediate dismissal). The first offense is a verbal warning, the second is a written warning and a maximum of three days suspension, and the third means dismissal. We use a little three-part slip that we buy from a business form company. To avoid inconsistencies, we actually write up verbal warnings and keep a record of them in the worker's job file.

If you set up a good program, and hire good people, the rules section won't be used very much, which is as it should be. Some of the rules we've never had to use at all. But simply having some regulations in place provides a clear frame of reference for everyone.

Finally, there are three issues we address in our rules that bear further discussion: substance abuse, safety, and moonlighting.

Drugs and alcohol. We don't allow drugs or alcohol on the job. And although we don't follow people to and from work, we make it clear that it's not acceptable to drink before work or after work on or near the job site. If someone actually has a problem with alcohol or drugs, we handle it in a variety of ways, depending on the worker's attitude and how long he's been with us. We had one fellow who was saving enormous amounts of money in his credit union account so that he could buy cocaine. He was a younger guy so we talked to both him and his parents. He had no interest in helping himself, so we ended up dismissing him. But we had another worker who was able to commit himself to a rehabilitation program. His family had tried unsuccessfully for some time to get him to do just that, but it was only through his foreman's encouragement (literally to the point of driving him to the rehab center) that he finally took steps toward his eventual recovery.

Safety. If you don't make a commitment to safety it gets lost, so we're pretty strong on the subject. For example, not using safety devices on the job is an offense. And I make it clear that I don't want things like loose boards left around the site. If anyone gets hurt I expect them to report it — even if it's only something in their eye.

We had the American Red Cross come in and give their standard first-aid course to all workers and we plan to do it again. (It's actually an OSHA requirement that all foremen have a current first-aid card.) The course was made up of six two-hour sessions. The first hour of each session was on payroll, but the second hour was on the employee's time. The course wasn't "mandatory," but workers who didn't attend lost an hour's pay for each session. (We had only one person who didn't do it.) Everyone was a little cynical at first, and there was a lot of joking about the thing, especially about CPR. But after the course, the more self-confident workers actually admitted that they felt better about it.

Moonlighting. In the manual, we state that a worker "may not work full- or part-time for another company engaged in the general contracting business without written approval from your construction supervisor." We also forbid our employees to work full- or part-time for any client of Salmon Creek. Either activity will be "considered cause for termination." The point is not to forbid moonlighting altogether, but rather to control it.

This rule came about after an experience we had a while ago. A client asked one of our carpenters to cut down some cabinets on his own time. Well, they weren't just cabinets — they were $8,000 dollars worth of Japanese lacquered cabinets. While this fellow is a good framing carpenter, he isn't a finish carpenter. So the next thing we knew, he had ruined the cabinets and the customer was deducting $8,000 from *our* bill.

So, as long as it is within the boundaries described in the manual, I actually encourage moonlighting. If I think someone will be responsible about it, I'll help him get work, because we get requests for miscellaneous projects. Moonlighting is beneficial from an educational standpoint, because there's nothing like being your own boss, and having to figure things out by yourself, to help you really understand the construction process.

Valuing The Employee

As you can see, our employee's manual contains the core of our personnel policy. In addition to the manual, we try to keep communication open — but not by showing up at the end of the day and having a beer with the crew, like many people do. Instead, we meet monthly, on the payroll, to discuss whatever comes up. At the most, we structure 15 to 20 minutes of the hour-long meeting. The rest is unstructured time for comparing notes about equipment and jobs. You'd be surprised what we learn from these meetings. Two different guys from different crews will mention trouble they've had with the company truck, and we'll realize that the problem is serious. The meetings help to get things done, and to get information out.

My basic philosophy is to try to pay as well as we can and give the best benefits possible without killing the goose. I don't want our company to be too formalized, but I think that it's valuable for all of us to understand our standards and procedures. Finally, don't forget that the company consists first and foremost of the people that work for it. And they'll perform best if they feel they're an important part of the company. ■

Ward Smyth is the former owner of Salmon Creek Builders, in Salisbury, Conn.

Getting Practical About Job-Site Safety

by Michael Davis

To prevent serious accidents, the boss must set the right tone, spell out the policies, and enforce them

As most of us know from personal experience, construction is dangerous work. That means some of the people who work for you are going to get hurt sooner or later. But how often and how badly they're injured isn't just a matter of luck; you can strongly influence the outcome by taking a realistic and consistent approach toward on-site safety.

A Few Sobering Facts

If someone gets hurt on one of your job sites because of a hazard you could have corrected, it's defined as negligence, and it can cost you everything you have in a legal judgment. If someone is killed on your job and it's defined as criminal negligence, OSHA will do everything within its considerable power to see that you go to jail for up to 18 months and pay fines up to $50,000. And it doesn't matter that you weren't there. Someone has to be held responsible, and most states protect site foremen from prosecution.

Think it can't happen to you? Try these numbers on for size. An employee sustains an on-the-job injury every 18 seconds in this country, and a workman is killed every 47 minutes. The most dangerous profession according to OSHA statistics? Construction.

But don't think you can just drive out to the job tomorrow morning and tell your foreman that you want a safe site and expect it to happen. No one cares as much about your business as you do. You've got to set the tone, spell out the standards you want enforced, and keep hammering at it.

The trick is to find that fine line between what you can do to make your job safer, and what you have to do to get the job done. You've got to tailor your safety program to fit your jobs and your people, or it won't be taken seriously at the level where it counts.

I put my greatest safety efforts into four major areas: hiring, training, drug and alcohol policies, and equipment.

Safe Hiring

Statistics tell us that 5% of all accidents are caused by unsafe conditions; the other 95% are caused by unsafe actions. That means that you can go out to your job and hang up nets, install big red warning signs, even pad the walls, and unsafe workers will still find a way to hurt themselves.

The first step toward changing this is extensive applicant screening, something most big companies do. You start by checking the employment references of all candidates. Then you ask about preexisting

Davis warns his framers to release the nail-gun trigger between shots. In the past, one framer shattered a finger by "bounce nailing." Another had to have a 16d nail surgically removed from the ball of his ankle.

injuries and handicaps, and run a physical screen and a drug screen. This process costs a few hundred dollars per applicant and takes about a week.

Good advice? You bet, but it's not realistic for most small contractors. We've tried to find a happy medium. We used to stand there and shoot the breeze with anybody who wandered on site with nailbags looking for the foreman, which often resulted in our hiring the guys whose best skills were as talkers, not workers. Now we use a three-page, preemployment screening packet we've developed.

The first page tells prospective employees a little about our company and what we have to offer. It starts out, "We're about to ask a bunch of questions about you. So we figure that it's only fair that we answer some of the questions that you might have about us." It gives an overview of our policies, tells when payday is, and that sort of thing.

The next sheet is my pride and joy. It's a simple, 15-question quiz that we've worked up on framing. The first few questions are on materials: "How big is a 2x4?" etc. Getting these right qualifies you as a board hauler. Then the questions get into rough openings and stud heights. If you get that far, you might make a decent nail beater. From there

Straight Talk About Safe Tool Use

After writing $4,000 checks to your insurance carrier for workers compensation month after month, you get to the point where you see a job site not as a workplace, but as a series of accidents waiting to happen. The three biggest hazards on my sites without question are saws, nail guns, and scaffolding.

Here are the main tool safety tips my crew bosses and I give a guy on his first day and thereafter if he doesn't pay attention.

Saws. I figure the 6-inch scar on my right thigh gives me the right to talk about not getting lax with a power saw. Most carpenters know how to cut safely, but they get in a hurry. We make it clear that nobody gets fired for taking the time to set up a cut.

There are two ways most people get hurt, and both of them involve binding the saw. First, someone will try to rip a 2x4 while holding on to it with his other hand. The saw will bind and kick back, and he'll lose a finger or two. Second, someone will block up a piece of lumber on a saw horse improperly and when he tries to cut it, the saw will bind and kick back and bury itself in his leg.

To prevent saw injuries, we make the following recommendations:

- If you're cutting a large board, put a block under it and use your foot to hold it down. If you're cutting a small piece, tack it to a saw horse.
- Unplug the saw before changing blades or fishing for that 1/4-inch slice of fir that wedged itself between the blade and the guard.
- Avoid cutting nails, especially with a carbide-tipped saw blade. The nail head or carbide chunk coming your way will not only be very hot, but will be traveling at a high rate of speed.
- Always adjust the depth of the blade so it's cutting no more than 1/4-inch deeper than the material. This is easier on the saw and leaves less blade exposed.
- Make sure all saws and power cords are grounded. You don't want to get a jolt in the middle of a cut, and I don't want to have to pay the fine OSHA levies if they happen to drop by and the cords don't pass muster.
- Never operate a power saw

it goes into layout, stairs, and rafter cutting. This form has saved us a ton of money by weeding out unqualified applicants.

The third and final sheet of our "hello" packet is a medical history questionnaire. It asks simple questions like, "Have you ever been injured on a job site?" and "Is there any medical reason you can't perform the job you're applying for?" If the applicant has a bad back, he's required to tell you. If he's been hurt on every job he's worked, then you know there's a pretty good chance he'd get hurt on your job, too.

Training

Every new employee should go through an orientation that includes a tour of the site, an introduction to the crew, and most importantly, a review of what safety practices and equipment are expected. You may lose a few guys by insisting on this stuff — the really cool ones that would rather die than conform — but you will cut down on the most common accidents like foot and eye injuries, and saw cuts.

Never just assume that a new employee knows how to operate your equipment. You may feel a little embarrassed telling an experienced framer how to use a nail gun or operate a radial-arm saw, but there are lots of different ways of working, and many of them aren't safe.

Small crews should think through risky operations, like raising this scarf-joined beam, with safety in mind. The crew leader needs to set the tone.

And to really cover your butt, you should provide every employee with a written copy of your company's safety policies. Then you have them sign a document stating that they have read it, understand it, and agree to follow the procedures it outlines. We're still working on this phase; our last attempt

unless you're wearing safety glasses. No *ifs*, *ands*, or *buts*.

Nailers and staplers. These tools are great, but they're called "guns" for a reason: Depending on the size of the compressor and how much hose is being used, these tools push between 90 and 120 pounds of pressure. I've seen a sheathing nail that missed its stud tear right through the plywood and bury itself in someone's thigh on the other side. I've also had a guy shatter the bone in his finger when he accidentally shot his hand. Here are some of the rules we have laid down to prevent these kinds of accidents:

- No one operates air equipment until everyone in the immediate area is wearing protective eye glasses.
- Don't even *think* about holding the safety back and shooting nails or staples into the air.
- Aim all air nailers away from you when plugging them into an air hose; they'll sometimes fire a nail when the pressure comes up inside the cylinder.
- Watch out for safeties that stick in the open position, and worn trigger mechanisms that will double fire. I've even had one gun go fully "automatic" on me.
- When nailing studs to a plate, it's important to shoot the bottom nail first, then move your hand back and shoot the upper nail. This way, if the top nail splits the plate or curls out of a knot, you won't take it in your hand.
- Don't keep the trigger depressed. This may sound a little extreme, but some of the worst injuries we've seen have come from ignoring this rule when working in tight quarters or up on a scaffolding. In one case a 16d nail penetrated the skull of a Minnesota carpenter who had the misfortune of coming up a ladder just as his partner — with his finger on the trigger of a gun — was turning to climb down.

This last principle regarding the trigger also applies to *bounce-nailing* subfloors and roof decks. I know it sounds like a machine gun and it looks really cool. But

continued on next page

On large, multistory commercial sites, Davis's framers wear hard hats. He requires boots and safety glasses on all his sites to minimize small injuries like nail punctures and sawdust in the eyes.

ended up looking like the yellow pages for some major city.

Drugs And Alcohol

Everyone who gets a paycheck from us understands that we will not tolerate drugs or alcohol on our job sites. Although we can't afford to use testing as part of the hiring process, our application requires anyone who works for us to submit to a drug test if he is involved in an accident.

We use our weekly newsletter to underscore what happens to an employee if he's hurt on the job and tests positive for drugs. Here in New Mexico it voids his workers comp coverage. That means the insurance company won't pay his medical bills and he can't collect disability. We find that this is a great deterrent. Guys who are drug users won't sign the application, and guys who work for us don't use drugs because they don't want to lose their benefits if something happens.

Safely Equipped

Framers love to work in tennis shoes, shorts, and little else. That was my standard uniform when I worked in the field, and I know a few guys who'd rather be unemployed than wear anything more. But that's just too bad. They aren't paying the insurance premiums and they don't have to pay the bills if they get hurt.

We used to have a real problem with people stepping on nails. Someone would get one in the foot every week or so and it cost us a hundred bucks each time. Finally, one guy took a nail in a way that damaged some nerves. It required surgery and cost us a bundle, so now we require boots. But we try to be flexible; there are some boots that look and fit like sneakers, but have hard soles and offer the necessary protection.

continued from previous page

unless the framer is an exceptional hand with a nail gun, his nail spacing will be irregular, he's going to produce a lot of "shiners" (misses), and sooner or later he'll end up injuring himself. I know this is true because I once nailed the third toe of my right foot to the floor with an 8d nail. We also had a guy brush his leg with a nail gun, not realizing he was holding the trigger down at the time. The 16d nail had to be surgically removed from the ball of his ankle.

One of the more bizarre stories about getting careless with a pneumatic tool involves our vice president, Tom Wood. He was stapling off wall sheathing a few years ago. He didn't want to have to drag a sawhorse all the way around the house, so he was jumping up to staple along the top plate, while pushing the sheathing tight to the wall with his other hand. Well, he misjudged his jump once, and he stapled his finger to the top plate. It left him hanging there with his toes barely touching the ground, yelling for someone to come help him down.

Scaffolding. Scaffolding didn't used to worry me, because I kept to two basic rules. Never put an inexperienced worker out on a scaffold, and always have the person working on the scaffold build it himself.

A bad accident changed my thinking last year. One of my crew leaders was doing some beam work from a wall scaffold he had built. During the process of "plumbing and lining," the wood scaffold jacks worked loose from the wall. In a hurry, he jumped up on the plank without checking the scaffold. As it gave way, he grabbed at a beam to catch himself, but fell backwards about four feet and landed on a concrete stem wall. The beam struck him full in the face, followed by the scaffold plank. He suffered a severe concussion.

The scaffold issue is a tough one for residential contractors. Spiking some blocks to the wall, nailing triangular jacks to them, and adding a few planks is fast, cheap, and effective. The problem is they will not stand up to the scrutiny of an insurance or OSHA inspection (OSHA requirements are so strict I don't think I've ever seen a site-built scaffold that would pass). If someone gets hurt on one of these homemade scaffolds, the owner of the business is in a very bad spot.

Our current policy is to use tubular metal scaffolding, but you still have to make sure it's dug in level and securely anchored. And if you're going up more than one section, you must use railings. If you're running siding or sheeting off the scaffold, attach them on the three open sides. If you're working overhead, then close in all four.

—M. D.

Typical Job-Site Injuries and Their Associated Medical Costs*

Injury/Event	Associated Medical Cost
Slipped on the ice, a minor sprain resulting	$195
Hit the thumb or fingers with a hammer	$350
Ladder not secured and tipped over, minor sprain resulting	$30
Slipped on job-place material, resulting in major back injury	$38,800 plus permit settlement
Slipped, concussion resulting	$9,800
Cut finger while drilling	$300
Foreign body in eye	$75
Splashed concrete into the eye while shoveling	$120
While installing door hardware, pinched finger	$265
Strained back while lifting	$300

The associated medical costs are actual costs based on the experience of one construction company in New England (1989); they may vary depending on your location and other circumstances.

Another frequent problem was sawdust in the eyes. Again, this meant hundred dollar trips to the doctor's office just to get the guy's eyes washed out. But in this case we didn't wait for a disaster to occur, such as someone losing an eye to a nail. We now require safety glasses for everyone. We've made a deal with a local safety supply house to buy large quantities of good-looking safety glasses that we provide to the guys at a good price.

Then we come to every renegade framer's pet peeve: blade guards. I've heard every argument in the world against them. One of the more convincing ones is that if you use a blade guard on your saw, you get sloppy. You think it's always going to be there and then one day it sticks and you get cut. And there's the argument that you can't make angle cuts or take a quarter inch off a 2x4 with the guard on. These may be legitimate points, but the bottom line is that they are required — by OSHA, by my insurance carrier, and most importantly, by me.

Making It Work

A safety program isn't something you can just write up and distribute; it's an ongoing conversation between a contractor, his crew leaders, and the guys on site. We have two primary ways of keeping this conversation going: meetings and a company newsletter.

We use "tailgate" safety meetings when we're working a commercial site, especially if we've hired some new people. We try to keep them to 15 minutes; usually right after lunch. The idea is to get everyone thinking about safety. Although you may not have discussed saws in particular, if your guys are thinking safety, maybe they'll hesitate before making that overhead pocket cut from a shaky ladder.

We haven't found "tailgate" lectures as effective for a small residential crew where the same six guys work shoulder-to-shoulder everyday. In this situation, a dangerous move by someone on the crew is going to be met by a crew leader screaming "Don't do that!"

Getting your lead men to take responsibility for safety is a key component. We give out bonuses to the crew leaders if they bring a job in under budget, but the costs of an injury are treated as a direct job cost. If one of our guys brings in a job $1,000 under budget, but he racks up $1,000 in doctor bills because of an injury, then it's a wash and he gets coal in his stocking.

When we do have accidents, the crew leaders must fill out and submit a report at the next weekly payroll meeting. All crew leaders are present and we go over the accident in detail. We try to isolate the causes and take steps to prevent another incident. This often means having the crew leaders discuss it with their crews, and writing it up in our newsletter so the guys can also hear it from the "horse's mouth."

Safety is a serious issue. The way I see it, you either develop a workable safety policy that keeps your people healthy and limits your exposure, or you're just wasting your time waiting for bad luck to catch up with you and put you out of business. ■

Michael Davis employs nearly 50 framers as the owner and president of Framing Square Construction in Albuquerque, N.M. After spending more than a decade as a production framer, he can still count to ten on his fingers.

Chapter 4 • Subcontractors

INTERVIEW

Hiring and Handling Subs

Clear rules, fairness, and give-and-take keep subs on track and on the team

Steve Farrell is a fourth generation builder who grew up in the family remodeling business. All of the Farrell Company's $1.5 million of work last year was residential remodeling in the heart of California's upscale Silicon Valley.

Steve "runs the field" for the company, visiting each of their average four job sites daily. He keeps in constant contact with his subs and his crew — two carpenters and a helper — via the phone in his truck, and even puts on the nail belt occasionally. His father handles initial client contact and bidding, while Steve's wife takes care of the books.

Like most contractors, Farrell sees his subs as vital to his success. He works hard at maintaining clear, consistent relationships with them. Here are his thoughts on this aspect of the business.

JLC: *How important are subs to your company?*

Farrell: Their work represents about 70% of our volume, so a lot of our profit comes from the 20% we mark them up.

JLC: *Where do you use subs?*

Farrell: We do all our own foundations except for flatwork, all framing, siding, and interior trim. We'll do drywall on a small job, but everything else is subbed out. That includes the usual trades plus specialties like shower doors, security systems, special paint finishes, etc.

JLC: *Do you have just one subcontractor bid on each trade?*

Farrell: In most cases, we have two. First, this allows us to compare cost and specs during the bidding process to make sure *we're* not forgetting anything. Second, if one of our subs gets injured or sick, we've got someone who is familiar with the job to step in. Third, since we're not the only contractor these guys work for, it means we won't have to stall a job because of scheduling problems. And fourth, it allows us to choose different skills, approaches, and temperaments for different kinds of jobs.

JLC: *Do you ever have more than two subs bidding on a single trade?*

Farrell: No, it's not fair or good business to round up fifteen guys, sign the lowest bidder — the guy who made the mistake — and then babysit him through the entire job. In fact, I'll even tell a guy if I don't have anybody bidding against him. I have an idea of what the job will run, and they know that if they nail me on a cost, I'll take it as a personal shot.

JLC: *How do you find your subs?*

Project manager Farrell, at right, reviews the coming week of framing with his carpenters, and coordinates the start dates for his subs.

Farrell: The best source is our other subs; my electrician found both my tile man and one of my plumbers.

JLC: *What do you look for?*

Farrell: We want subs that are established, but fairly small. That way our relationship is with the owner. And smaller subs are usually more responsive since we're providing them with a good part of their total business.

A few months ago we hired a big drywall firm for a monster house we were doing. Because our job represented less than 1% of their gross, they did the project on *their* time schedule at *their* level of quality.

JLC: *How do you judge things like reliability, trade skills, and attitude toward clients when interviewing subs?*

Farrell: I look for enthusiasm, for how a sub dresses, how well he carries himself, if he listens carefully, the condition of his truck.

I also want to know how he came up in the business, and even how much education he's had. Our clients are successful, and they're used to working with people who are sharp — they demand it.

JLC: *Who are your most valuable subs?*

Farrell: The guy who does my demolition and hauling, my tile man, and my cabinetmaker are all pretty important. But the sub that makes or breaks any remodeling contractor is the painter. It's the last layer, and the one the clients always pick on. We've used the same guy for 12 years, and he's the only subcontractor with whom I have a personal as well as a professional relationship.

JLC: *Do you have your subs inspect the job prior to bidding?*

Farrell: Usually, though I don't like to drag them through a job unless I think we have a good chance of getting it.

But a walk-through can really nip problems in the bud. For instance, I can ask my plumber how he's going to get his pipes through a bearing wall, while letting him know he's going to be looking at a glulam the size of Ohio (and I will have him shot if he gets a drill anywhere near it).

Or my electrician may find he has to upgrade a box, even though the plans don't show it. This way there are no surprises, and we can go back to the client and say, "If you choose us for your project, we aren't going to hit you with a lot of extras later on. Our price is higher, but it's because we've done our homework."

JLC: *Do you include both subs who are bidding in the walk-through?*

Farrell: This is where it gets tricky. Usually only one guy goes out, but if we discover a problem that's not on the plans, I'll tell the other sub who's bidding. If I don't, I'm essentially sticking the guy who didn't come out with the financial consequences.

The sub that was on the inspection can say, "Hey, I killed my time coming out and you gave all the information to the other guy." That's a hard point to argue, but I'll give him the job if he's within 5%.

JLC: *Do you get written bids?*

Farrell: I ask them to phone in a number, but then back it up with a "scope of work" description so I'm covered later on when we get into the job.

JLC: *Do you award the contract based solely on price?*

Farrell: I start with price. If there's a big difference between the two bids, I call up the guy with the lower number and ask him if he feels comfortable with it. If he says yes, I tell him I'm going to hold him to that number even if he comes crying to me midway through the job. But I'll also give him a chance to refigure it if he feels at all hesitant.

If the difference between the two bids is minor, 5% or less, and the guy with the lower figure is sounding wishy-washy, I'll just go with the higher guy. But I also try to match the job with the sub.

JLC: *Once the client has signed, do you hold preconstruction meetings?*

Farrell: Yes, but they're not long and drawn out. I just introduce three or four of my primary subs — electrician, plumber, painter, and sometimes my tile man — to the homeowners. This increases the clients' confidence level once the job starts because it's no longer a bunch of strangers walking into their house.

JLC: *How do you let your subs know when they're going to start?*

Farrell: After my bid has been accepted I sit down and draw up a schedule. I use a relatively simple bar chart and go over it with the homeowner. Then I call up each subcontractor and let him know exactly when he's starting.

JLC: *What about changes?*

Farrell: They're a fact of life. When I see a change coming up, I'll call my subs and inform them. I'm in contact with my major subs every day, and they can always reach me in my truck.

JLC: *Do you intentionally overlap your subs?*

Farrell: Yes. Everybody has an effect on the next guy and I overlap them slightly so they're communicating. For instance, I run my sheet-metal sub through while my carpenters are still on the job, to straighten out any problems.

JLC: *Is the sheet-metal sub the first to rough in?*

Farrell: No, my plumber is always first. Getting his pipes around everything is tough. The sheet-metal guy is next; he also needs a break. The electrician is last, because his material is a lot more flexible. Worst case, he's going to use an extra 100 feet of Romex and complain a little more because he's last in.

JLC: *What about the order of finish?*

Farrell: That's just as critical. For instance, you don't want your hardwood flooring sub in there while the painter is driving the humidity through the roof; and you want the cabinets sealed before the tilesetter turns the wood black with mortar.

I let my painter finish inside, then while he works outdoors, my electrician and plumber trim out. After that, the flooring sub can come in, followed by a final touch-up tour by the painter.

JLC: *Sounds great in theory, but aren't there conflicts?*

Farrell: Sure. If I have subs growling at each other, I'll just say to one of them, "Hey, this isn't working, come back next Tuesday when the other guy won't be in your face."

And sometimes you just need to turn the site over to one guy to get him out of there. Drywall is a good example. It's messy and dusty and nobody likes being there while it's happening. I tell the homeowner to visit relatives, and give my drywall sub whatever he needs to get him the hell out of there. If the other guys don't have anything to do during that time, I'll get them started on one of my other jobs.

JLC: *But subs have scheduling problems too, and can't always start when they promised.*

Farrell: True. They often get loaded with extras on the job before yours. When this comes up, I can usually give them slack on the front end, but I set an absolute finish date. If they're going to have trouble meeting that, we figure out a schedule of weekend work, or I help out by picking up material, etc.

Sometimes they're just tight financially, and are trying to finish up their current job so they can buy materials for my job. In that case, I'll give them a check ahead of time. But when I do, I expect them there the next day.

JLC: *Does your lead carpenter handle problems with subs when you're not on site?*

Farrell: No, I try to take that load off my guys. They'll make other decisions on their own work — frame, finish, whatever — but typically not with the subs.

I'm on every job at least once a day. If it's critical, I'll get all my other jobs out of the way early, and spend the rest of the day on one site. The subs know that, and they know the decisions all have to be made by me, because I'm the only guy who is carrying all the relevant information in his head.

JLC: *Any problems with homeowners asking your subs for changes?*

Farrell: It's the same deal. All of this stuff has to funnel through me or it just doesn't work. If the homeowner requests something and the sub goes ahead, he just did that work for free. The sub's contract is with me, not the homeowner. I'm flexible on a lot of stuff, but not this.

JLC: *How do you handle change orders?*

Farrell: When the client asks for a change, I go to my sub immediately and get a rough guesstimate, right on the spot. Then I'll double it and go back to the homeowner to get a "go" or "no go."

This does two things. First, it's fast, so that change orders don't end up shutting down the job. Second, if we do the work, I bill the client for actual costs plus mark-up, and we end up as "good guys" when we come in below the estimate. With all the stories homeowners hear about remodelers hosing their clients on extras, this is more important than you might think.

JLC: *What about the gray area between a change order produced by hidden conditions, and an item the sub missed when he bid the job?*

Farrell: If my electrician comes to me and says he's going to have to put in a new service box and didn't bid it, I might split it with him if I'm doing okay on the job to that point. Technically, it's not my problem, but by giving a little I end up with a happier sub and a little leverage with the guy that I didn't have before.

However, I don't do this often; I expect these guys to be professionals and know their trade.

JLC: *How about conflicts between trades about who does certain tasks?*

Farrell: The trick is to pick these up right away — before bidding — and circle them on the plans. Where I see it most often is between the plumber and the hvac or sheet-metal sub over venting. The way I handle it, bath fans are provided and hooked up by the electrician, but ducted by the hvac sub. My plumber usually vents his own water heater, and provides roof jacks for his vent pipes.

And I don't let anyone but my plumber run gas lines, even to the furnace. He also hooks up all appliances — another area where I don't want mistakes.

JLC: *How do you proceed when you find a problem with a sub's work?*

Farrell: I tell the tradesman on the job so he can correct it without having to look bad in front of his boss. But I expect him to turn to it right away, not finish what he's doing and come back. That way he's not making the correction a separate project, and the problem is less likely to mushroom.

JLC: *What if he shrugs it off or argues the point?*

Farrell: I tell him to pack it up and I call his boss. I'm real demanding — these guys are always one bad job away from not working for us again and they know it. It's important to have a little leverage with these guys, but I also know I can *never* abuse it. I've got to be clear in communicating what I want, and absolutely consistent and fair in how I deal with them.

JLC: *How about callbacks?*

Farrell: Ideally, I do a walk-through before the subs leave, but if not, I'm still holding that final payment. Since I do all the scheduling and okay every invoice before it's paid, it won't slip past me.

JLC: *Have you been "liened" by subs very often?*

Farrell: Never. We pay our subs as soon as we're square on the work and they've invoiced us. In fact, most of our guys don't even send the pro forma, pre-lien notices to the homeowner because they trust us. However, I never object if they do; we all need to protect ourselves by getting things down on paper. But I do remind the guys who sent notices to issue a lien release the day they get my payment; you can't do just the paperwork that's advantageous to you.

JLC: *Do you do anything special to reward your subs? Christmas parties, bonuses?*

Farrell: Not really. We provide them with a steady source of good work and we keep our word. That's a lot. If we throw a big Christmas party and invite our subs, it'll cost us $1,000 and they'll remember it for about four months. But if we offer them a solid business relationship where the rules are consistent and there's some give and take, they'll remember that for a lot longer. ■

Success With Subs

Find the best, communicate clearly—and pay them what they're worth

by Richard Lind

Few business owners are presented with as many operational and management dilemmas as are home builders. The builder often has little control over his toughest problems. On-again, off-again market conditions, diminishing land supply, and strangling environmental regulations frequently require that the builder think more in terms of survival than profit. The old saying "When you're up to your tail in alligators, you almost forget you set out to drain the swamp," couldn't be more appropriate.

Builders do, however, have control over the way they operate their businesses. Chief among their management decisions is the selection of subcontractors. Maintaining healthy, businesslike relationships with subs certainly increases (if it doesn't guarantee) the likelihood of business success.

It is common industry practice to subcontract nearly all the trades that would normally work at a builder's site, regardless of the size or volume of the operation. In fact, many builders operate successfully with a totally subcontracted operation, or with only, perhaps, some part-time general labor. By coordinating the talents of various trades and individuals, we can accomplish what we set out do to *and* achieve a pleasant, productive, and continuing relationship between builder and subcontractor.

Finding Good Subcontractors

Successful working relationships rarely "just happen." They almost always result from a well-thought-out approach.

For starters, find the best subcontractors you can. They can be hard to find at times, since many subs do quite well by word-of-mouth, and therefore don't bother to advertise. But you can find pretty good leads from other builders (especially if you are involved in a local home builders association), from other subcontractors (in another field), from suppliers to the sub's specific trade (plumbing supply firms, for example), and from the bulletin boards at your general building supply house. And don't forget the local coffee shop at 6:30 in the morning. Sometimes I think that coffee shops survive exclusively on the construction business. In fact, if you want a good indicator of how the market is in your area, check out how the local coffee shop is doing.

Get an Early Start

Allow sufficient lead time to line up the subs. The time needed will depend on the specific trade, the market and economic conditions, and the weather. Getting an early start will help you to avoid unnecessary delays, higher "rush" charges, and promises made under pressure that can't be kept.

Check Them Out

There is absolutely no excuse for not checking out a subcontractor that you haven't worked with before. Secrets are a rare commodity in the building community and bad news travels fast. But in this instance, it's a blessing. A few well-placed phone calls can tell you a great deal about a subcontractor: his performance, credit rating, and call-back reliability, for example. Is the sub accessible? Are phone calls returned? Few things can be more damaging to the relationship between a builder and home buyer than the failure of a subcontractor to respond to complaints promptly. The question is rarely one of competence in the trade, but more of how a sub runs his business. You can generally get a pretty good idea from other builders a sub has worked for, as most people are familiar with the standards of their peers. Since consistency seems to be part of human nature, it's reasonable to expect your relationship with a particular sub would be similar to what others have experienced with him. So a little effort to check around up-front can save time, expense, and a lot of heartache later on.

Keep It Clear and Simple

When soliciting bids for a job, be accurate, open, and above board with the job description, so that the sub will know exactly what he is bidding on and so there is no question as to what your specifications are. This puts all bidders on a level playing field and enables you to make apples-to-apples comparisons when reviewing bids. Don't hesitate to ask for suggestions on where to cut costs if the specifications are flexible, as they might very well be on a spec house. Let the sub perform his job the way he usually does, unless it compromises the specs: It will generally cost less. This situation typically occurs in the bidding process for mechanical installations, where savings can result from using the equipment the subcontractor usually installs, as long as there is no reduction in quality.

Most people in business are honest and well intentioned and disputes generally result from communication breakdowns and inadequate direction. Communicate clearly so that the sub can figure the costs for the job. Be careful to avoid coming across in such a complicated manner that the sub is frightened away or pads his price to protect himself. Items that are clearly understood before a job begins rarely present problems later on.

When you agree on the contract price, make sure your contract includes a payment schedule. The timing and amount of progress payments should be predetermined, based on work completed. Avoid a per-diem payout, except perhaps with specialty work where the time required to do the work is unknown but there is a definite beginning and end to the task.

In addition to clear job descriptions and specifications, subs need to know when they are expected to start, and you need to know how long it will take them to perform the job. Make sure, as much as you can, that the sub has sufficient crew available to do the job, so that your schedule is maintained.

Schedule the various trades realistically so that the job is ready for them when you said it would be, and do whatever you have to in order to meet your commitment. For instance, if through no fault of yours, temporary electric service has not been connected, provide a generator for the interim. This not only keeps the job moving, but sends an unmistakable message of your determination to meet the schedule, and establishes your dependability in the build-

What I Look For in K&B Subs

by Lynn Comeskey

My philosophy is to locate good subs and work with them exclusively; my tile setter has been with me for over ten years. I couldn't and wouldn't ask the sub whom I used only when he was the lowest bidder to do the things I ask my regular subs to do. It may cost more to do business this way, but the level of service is much higher.

The Right Stuff

Although most of us are looking for the perfect sub, we'll settle for quite a bit less. Here's a list of the qualities I look for in order of their importance:

Technically competent. Most of my jobs are high-end, and my clients demand top-notch work so I need to trust that my subs' work will be consistently good.

In some of the trades such as plumbing and electrical, it's fairly easy to judge. But in a trade like tile setting, you get into subjective areas: Is this the best layout for tile of this size on this countertop? Will the grout still look good five years from now?

I have become a better judge of who is "good" with experience. For example, my hardwood flooring contractor is expensive. However, he has been in the business for years. He has seen most problems before, and often has an esoteric tool or technique that can fix a problem without ripping out the first attempt. Having him available to solve problems is well worth the cost.

Caring. We can usually find competent people, but if they don't go the extra step, it doesn't work for me. I want my sub to take care of things that need fixing. For instance, if my plumbing sub sees copper attached directly to galvanized, I want him to replace it with a dielectric union without even asking.

Reliable. I've got to be able to count on these guys to turn in a bid, show up on time, and finish when they say they will. You can't maintain a schedule if your subs don't.

I know a cabinetmaker whom I really enjoy working with because of the quality of his product and his sense of humor, but who routinely gives me bids late or not at all. I now ask him to bid only when his work is specified by the customer.

Clean and neat. If you're going to succeed in the kitchen and bath remodeling business, you'd better be clean, neat, and careful. You can't have a glazier who routinely scratches the tub or the ceiling when installing full-height doors. Nor should there be metal filings on the floor when he leaves.

These guys should also look professional and consider the impact they are having on the customer. For instance, most of my customers don't smoke, and I don't smoke. So I look for employees who don't smoke, and I apply the same yardsticks to subs.

Available. Accessibility and availability are the next most important criteria. I stopped dealing with one painter because he never returned my calls — even to get a bid. Given a choice, I prefer to deal with a sub big enough to have an office with a secretary who can relay a message during the day to someone who can make a decision.

Another advantage of dealing with a larger company is that they are better able to get someone to the job when we need them. We are willing to give up the choice of which tradesman actually does the job for this flexibility. This sometimes means that the quality varies. But if you have a sub who stands behind his work — and you're one of his larger accounts — he'll come in and make any necessary corrections.

Good client skills. Next, we come to the "gorilla factor." We want our employees and our sub's employees to be pleasant and courteous. Although they don't work directly for the client, about 50% of our work—one-half the impression the client is left with—is subbed out.

Interested in the long term. I want to work with a sub who wants to form a relationship. This means there will be give and take. If my sub clearly makes a mistake, I expect him to cover it. Usually, the fault is not clear, and we agree to split the cost.

Priced right. I purposely placed price last. I do have a price pain threshold, but it is relatively high. If someone is above the threshold, I rely on the traits I mentioned above. ■

Lynn Comeskey is a remodeling contractor who specializes in kitchens and baths. His firm, Mac & Lou Construction, is based in Palo Alto, Calif.

ing community. Assuming no work interference, schedule several subs with their respective trades on the site simultaneously; it not only keeps the job moving, but livens the pace, as one person works alongside another, rather than alone.

Require subs to visit the job well in advance of the start of their work, to make their needs known (when relevant) to the prior trades. This usually involves them with the framer. We have all witnessed at one time or another a "scene" that took place when a justifiably irate framer was asked to open up (and later replace) a door frame in order to place a bulky fiberglass tub unit, or to widen the bay between two floor joists, after the fact, to accommodate a warm-air duct. Had the framer been alerted early on, the disruption could have been avoided. Any requirements for cleanup imposed by the builder should be agreed upon, as well as conditions the sub requires of the builder, such as equipment to be provided, job access, or heat in season.

It is good practice to meet with all subs at the site when their work begins to review the game plan and layout of the work. As the work progresses, make frequent inspections to keep on top of things. Be available throughout the job, so that if unforeseen problems or emergencies arise, you will be there to make a decision. By keeping an eye on the work and making frequent inspections, errors often can be nipped in the bud.

After the foundation is in, the first major sub on the job will be the framer. Since trade schedules tend to overlap on the job, the first subcontractor's performance will set the stage and establish the pace, appearance, and discipline for all that follow. Anything less than good performance will necessarily carry on to the next trade. So make it your business to start out on the right foot.

Contracts and Other Paperwork

Your contractual arrangements with your sub, whether written or oral (so common, unfortunately, in residential building) should be the result of an informed business management decision. It should be established beyond the shadow of a doubt that you are, indeed, dealing on a builder/subcontractor basis and not on an employer/employee basis that could make you liable for F.I.C.A. taxes, workers comp coverage, and the like. Have legal counsel hold your hand while preparing the form that the relationship will take. The subcontractor agreement should be simple, but specific. Each trade has different requirements. (See sample contract, next page.)

Don't expect your lawyer to know the intricacies of a specific trade: You will have to let the lawyer know what you want included in three major areas. First, you will want to define the scope of the work. If you are contracting with a framer, you will need to state that he will be responsible for the framing from the top of the foundation up, and to the point of installing the mechanicals. Second, you will need to make clear who is responsible for supplying the materials. For instance, plumbers normally provide their own materials, whereas the builder usually supplies the concrete for the foundation, and the wood for the frame. Finally, you will need to determine who is responsible for coordinating municipal permitting and inspections. The builder usually organizes these items for the frame, but the hvac sub will usually coordinate the fire chief's inspection and permits himself, for example.

It's good practice to sit down from time to time with your lawyer and insurance advisor to review overall business direction and strategy, as well as any liability exposure. Make sure you include a discussion of subcontractor involvement with your company.

Every state in the union has a workers compensation statute, and although the laws vary a bit from state to state, the statutes typically require that all employees be covered by workers compensation insurance. Accordingly, every employer — including subcontractors — must provide coverage for his employees. If a sub fails to do so, the builder becomes liable in the event of a claim for injuries received on the job by a sub or his employees. In addition, the builder's insurance company will raise his premium to cover the sub's work. No subcontractor works on my job unless he has first provided me with a certificate issued by his insurance carrier stating the workers comp coverage. A prudent builder (even one without employees) will carry his own workers comp policy as back-up protection.
(*continued on page 56*)

SUBCONTRACTOR AGREEMENT (CONTRACTOR/SUBCONTRACTOR)

Agreement made between the Contractor: ______________________________

____

____

and the Subcontractor: ______________________________

____

____

1. General project description: ______________________________

____

____

2. The specific work of this Agreement shall consist of: ______________________________

____

____

and shall be in accordance with the specifications, drawings, or as reasonably inferable from either or both, and other contract documents.

3. The work shall begin no later than ______________________________
and be completed no later than ______________________________
subject to adjustments as provided in the General Conditions of the Construction Agreement.

4. The Contractor shall pay the Subcontractor the sum of ______________________________
______________________________ Dollars ($ ____________________),
for the work under this Agreement, subject to additions and deductions as stipulated in the General Conditions.

5. The Contractor shall make progress payments to the Subcontractor according to the following schedule:

Completed Work	Amount to be Paid*
______________________________	____________
______________________________	____________
______________________________	____________
______________________________	____________
______________________________	____________
______________________________	____________
Total	____________

* An amount equal to ten (10) percent of the payment shall be retained by the Contractor in accordance with Section 9 of the General Conditions.

Courtesy of J. Hardy LeGwin & Associates / *Project Notebook*

SUBCONTRACTOR AGREEMENT (CONT'D)

Final payment shall be made when the work has been completed and the Agreement fully performed to the Contractor's satisfaction.

6. The Subcontractor agrees to be bound by the construction documents as follows:

   a. The Construction Agreement is the executed agreement between the Owner and the Contractor dated ______________________________ .

   b. The General Conditions of the Construction Agreement executed and dated

   c. The Specifications contained in the Project Manual and dated ______________________________ .

   d. The Drawings as follows: ____ .

   e. Other documents as follows:

7. Other provisions:

This agreement entered into this ____________ day of ________________ , 19 ____ .

______________________________	______________________________
Signature of Contractor	Signature of Subcontractor
______________________________	______________________________
Printed Name	Printed Name

Builders should also familiarize themselves with the federal "right to know" law, as it applies to hazardous substances in common usage on building sites, as well as any similar state laws. And anyone running a business today should be well aware of the obligations under Occupational Safety and Health Administration (OSHA) regulations, not only to provide safe working conditions, but also to limit liability. In view of ever-increasing regulations (and unfortunately, a society ever-ready to sue), it's important to document your compliance with these rules by keeping good records — copies of correspondence, a daily log, lots and lots of snapshots, and anything else that might tend to support your position if called upon. All too often, when a problem or a serious accident gets a lot of publicity in an area, the agencies are prone to make sweeps of the territory looking for scalps.

Finally, but perhaps most significant: When you negotiate your contract, allow your subs to make a profit on the job. This is not just being a "good guy" — it's simply good business judgement. It is totally unrealistic to expect good performance, including callback dependability, unless the job has been profitable for the subcontractor. There are exceptions to every rule, but it is unreasonable to think that you wouldn't have to eventually pay some price for low bids. A contract is good only if all parties to the agreement are happy.

When Things Go Wrong

In spite of your best planning efforts and investigation, there will be occasions when your relationship with your subcontractor sours and has to come to an end. End the marriage with a handshake and no hard feelings. This is not inconsistent with your displeasure, for whatever reason, but a decent and businesslike way to react. And you just never know when your paths will cross again. Let them go with a kiss.

There will be times, too, when you feel you are the only one making a contribution to the marriage, such as cooperation, prompt payments, having the job ready — and you don't seem to be getting the same in return. Don't be a glutton for punishment; do what has to be done, part company, and find another sub that meets your criteria. However, if it seems to be a recurring problem, it's a clear indication that you're doing something wrong. Take stock of your methods and make the corrections you need to get back on track.

Subs Are a Goldmine

Besides their obvious function, subs can be a tremendous source of information and ideas. Most subs, like builders, really enjoy their work and are eager to talk about it, given the chance. Seize the opportunity and pick their brains; find out as much as you can about their trade, product, or system. Also, subs who travel from builder to builder can be an enormous source of fresh ideas carried over from job to job: little tips learned on another job that can make your project go a bit easier, or some new and cost-saving approach that you might benefit from. And, of course, rely on the sub's experience in his own trade to help you get an optimum installation or layout. In contract work, it's not unusual for customers to provide blueprints of overdesigned (and pricey) mechanical installations. Your hvac sub can help you avoid doing a disservice to that customer, who is ultimately paying the bill.

In their travels from development to development, subs often glean other useful tidbits, such as what is "hot" and what is not. They can tell you what price ranges are currently selling, and what buyers seem to be looking for or are willing to forego in making choices based on price. They often know where lots may be available, or where there is a brick-box apartment building for sale that is ripe for conversion or rehab. They can tell you what town is talking moritoria, so that you might hustle with your permit applications.

These extras make good working relationships with your subs even more appealing. But of course the bottom line is that you want your job to go smoothly, and your customers to be happy. A healthy relationship with your subs can go a long way towards doing just that. ■

Richard Lind has been a builder in the Wellesley, Mass. area and a member of the Builders Association of Greater Boston (BAGB) for over 30 years.

Sub Or Employee?

by Clayton DeKorne

The law is gray, but the risks are clear if you hire "subs" who are really employees

Most people think of small builders and remodelers as general contractors who maintain crews of carpenters and only sub out special trades. Many firms, however, now subcontract major portions of their work. Often this includes tasks traditionally performed by employees, such as framing and finish carpentry. At one extreme are firms that even sub out the job management to a construction-management firm, which in turn manages the other subs. With the persistent and growing shortage of skilled labor, this reliance on "non-traditional" subs has spread from sophisticated companies to more conventional firms that also maintain a core crew. But in many cases, the legal and tax status of such subcontractor relationships lies in a gray area.

Are Your Subs Legitimate?

Consider a journeyman-level carpenter who becomes a sub on his way towards going out on his own. He stays with his old crew. Few things change. He takes home a better paycheck (one without withholdings) and he feels more independent. And you, the hiring contractor, go along willingly. Why shouldn't you? You hang on to your skilled labor without the burden of payroll taxes and workers compensation insurance.

But if you maintain the subcontractor relationship in name only, you face considerable risks. The courts, in the event of a lawsuit, or the IRS, in an audit, will always investigate this working relationship. If they reclassify a worker, you could be liable for third-party claims — claims made against you by someone injured by the subcontractor's actions — or claims made by a sub who injures himself. Moreover, you can be assessed for back payroll taxes including *both halves* of social security, unemployment taxes, federal income tax withholding, and penalties.

There are ways to legally change the status of a worker from employee to sub. But you have to actually change the way you do business with the person.

Insurance Matters

Since insurance companies routinely audit their clients, the insurance risks of hiring subcontractors are the most apparent. Most business liability coverage designates only the owner and employees as "insured." Such a policy probably will not cover third-party claims unless you have additional coverage, such as *contingent liability*, to cover your subs. Thus, in its annual audit, your insurance company will request a certificate of insurance to show that each subcontractor you hire carries his own liability insurance. Without this piece of paper, your insurance costs will be raised to cover the risk.

A subcontract agreement is only marginally helpful in sorting out liability claims. Standard forms specify that the subcontractor is responsible for workers compensation, public liability insurance, and all income taxes and unemployment compensation. This establishes a record of your mutual intention. But a court of law often looks at a contract as just another piece of paper. It is not certain evidence that a contractor and sub have an independent business relationship. The plaintiff's attorney may try to make a case that a contractor actually functions as an employer and the "agreement" is but empty words.

In the event of an insurance claim or a lawsuit, a subcontractor's status will be examined to see how it conforms with the following common-law guidelines:

1. The extent of control, which, by the agreement, the contractor may exercise over the details of the sub's work;
2. Whether or not the sub is engaged in a distinct occupation or business;
3. The sub's occupation and whether it is one traditionally done without supervision or direction by the contractor;
4. The skills required for the work;
5. Whether the contractor or sub supplies the tools needed to perform the job;
6. The length of time the person is employed;
7. The method of payment — whether by time or by the job;
8. Whether the work performed is part of the contractor's regular business;
9. The intent of the parties in entering the relationship;
10. Whether the contractor is an established entity.

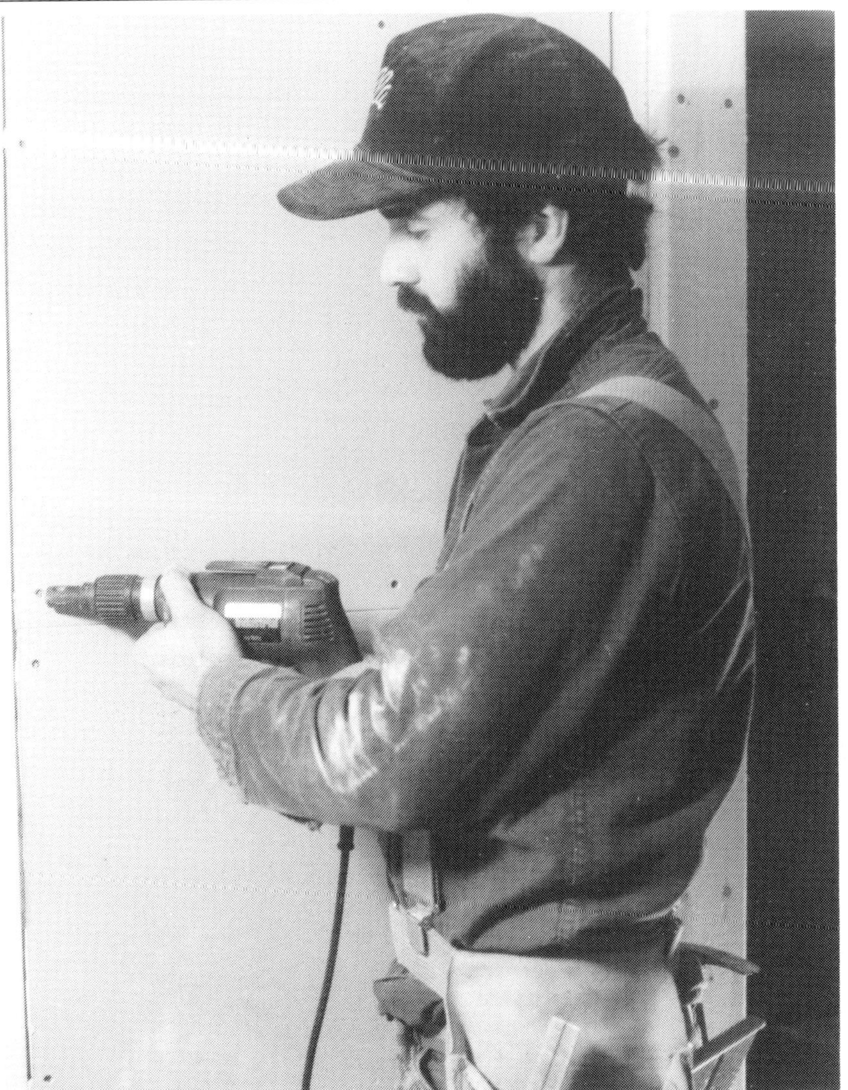

If you directly supervise this tradesman on your job, the IRS will probably consider him an employee, not a sub.

The first determinate — commonly called "the right to control" — is the most important. In a 1976 appellate court decision, the judge wrote (concerning liability for a construction site accident) "all of these factors are of varying importance in determining the type of relationship involved, and, with the exception of the element of control, not all elements need be present." This is a clear affirmation that you can legally work with nontraditional subs, even though factors #3 and #8 seem to exclude this.

This ruling has given rise to a common-law definition of employee and subcontractor. An employee is someone you direct step-by-step through the job, and you are an employer if you have control over the specific task you are hiring a worker for. An independent sub is someone you turn loose on the job after showing him what needs to done. When you subcontract out work in this way, you hand the right to control the job over to the sub. A sub must be free to negotiate when he performs the job and to choose the tools and techniques he deems best. And since the definitions are based on a worker's approach to a specific *job*, it requires paying and billing a worker out *by the job*.

Tax Concerns

In years past, the IRS had a strong incentive to classify a worker as an employee, simply because an employee used to earn more money for the Treasury than a subcontractor. Perhaps because the IRS felt it has not been able to adequately track down all the subcontractor earnings and has been losing money by this scheme, it recently increased the self-employment tax contribution. An independent subcontractor now is required to pay 15.3% of his wages to Social Security, as required by SECA (Self Employed Contributions Act). This equals the total tax contribution acquired from an employee, which includes both the employee's withholding and an employer's contribution, as required by FICA (Federal Insurance Contribution Act).

General contractors are required to file Form 1099-MISC for all nonincorporated subcontractors who are paid $600 or more each year. This enables the IRS to track the earnings of self-employed workers.

According to John Wyckoff, a business consultant from South Burlington, Vt. who specializes in counseling building professionals, a contractor should require two things of every subcontractor who walks onto the job site. First, you need verification that a subcontractor is incorporated. If the subcontractor is not incorporated, he should supply an employer identification number to facilitate filing Form 1099. In some cases this may be a social security number. And second, a certificate of insurance is needed to verify that the sub has liability insurance.

This will set the stage for establishing a true contractor-sub relationship. But if you are audited, you may be required to prove that you actually work as contractor and sub rather than as employer and employee.

The IRS Training Manuals 8463 and 3142-01 list 20 factors used to determine

IRS Subcontractor Guidelines

1. Is the individual providing services (the sub) required to comply with instruction on when, where, and how the work is to be done?
2. Is the individual provided with training to enable him or her to perform the job in a particular manner?
3. Are the services performed by the individual a part of the contractor's business operations?
4. Must the services be rendered personally?
5. Does the contractor hire, supervise, or pay assistants to help the individual performing under the contract?
6. Is the relationship between the parties a continuing one?
7. Who sets the hours of work?
8. Is the individual required to devote full time to the party for whom the services are performed?
9. Does the individual perform work on another's business premises?
10. Who directs the sequence in which the work must be done?
11. Are regular oral or written reports required?
12. What is the method of payment — hourly, weekly, commission, or by the job?
13. Are business or traveling expenses reimbursed?
14. Who furnishes the tools and materials necessary for the provision of services?
15. Does the individual have significant investment in the tools or facilities used to perform his services?
16. Can the individual providing services realize a profit or loss?
17. Can the individual providing services work for a number of firms at the same time?
18. Does the individual make his or her services available to the general public?
19. Can the individual be dismissed for reasons other than nonperformance of contract specifications?
20. Can the individual providing services terminate his or her relationship at any time without incurring a liability for failure to complete a job?

whether an individual is an independent contractor or an employee (see list above). These questions have been compiled to cover all industries. Some of the questions, such as #4, #11, and #13, often are not relevant to the building trades. Others, like #7, #9, and part of #14, are invariably gray areas. Setting the hours of work is often a mutual agreement based on a project's schedule. The business premises—a job site—are common to the principal and his subs. And often a contractor will supply all the materials from the concrete to the shingles, even though jobbers will install them.

Factor #6 looks gray, too, because general contractors often have continuing relationships with their subcontractors. But, according to Wyckoff, the rule is specific. A "continuing relationship" is interpreted as a steady, rather than a repeated, relationship. For example, a contractor might always hire the same plumber. But the plumber also works for other contractors. Similarly, a finish carpenter can work repeatedly for the same contractor. If he works for only one, however, he is, without dispute, that contractor's employee.

Wyckoff asserts that a few other factors are key to establishing a true contractor-sub relationship with nontraditional subs in the building trades:

- *Skills and independence.* Above all else, a subcontractor should have the necessary skills to perform the work on his own. If a worker receives training or supervision from the contractor, he is considered an employee. Similarly, a subcontractor must hire his own helpers. He cannot use a contractor's helper on a routine basis.
- *Risk of profit or loss.* Often this factor can validate a subcontract agreement. It confirms that a sub operates a distinct business.
- *Furnishing tools.* The larger a worker's investment in tools, the more likely he is considered to operate an independent business.
- *The method of payment.* An independent subcontractor should be paid by the job.

There are no rules on how many factors a worker must conform with to be classified as an independent contractor. A contractor should set things up to conform with as many conditions as possible. He then has a fighting chance in the world of lawsuits and tax audits. ■

Clayton DeKorne is an associate editor with The Journal of Light Construction.

Chapter 5 • Estimating

Estimating for Profit In Remodeling

by Martin King

I prefer to open estimating seminars with two questions:

Question 1: "What is the most *profit* you can possibly make on a $15,000 remodeling job?"

Answers usually start at $3,000 and peak at $8,000 with a lot of haggling. Of course, even if your labor and materials were free, and you were the world's most efficient company, you could not exceed a profit of $15,000.

Question 2 is easier: "What is the most money you can *lose* on a $15,000 job?"

Every contractor knows that answer: There is no limit on how much you can lose. A single project can consume six months worth of profits...and more.

That paradox is the reason why good estimates are so important. Profit is limited but losses are not.

The goal of estimating is to produce an appropriate profit, so the consideration of profit is an appropriate place to start a discussion of construction estimating.

Stephen Cridland Courtesy: American Plywood Association

By identifying areas of uncertainty in your remodeling estimates, you can anticipate the unexpected — and charge for it.

Profits Are Produced, Losses Occur

We produce profits consciously, by taking specific actions. The production of profit requires experience, foresight, and planning — things that minimize uncertainty and risk.

Losses flow from everything else. Losses require no conscious effort, and are simply the monetary expression of Murphy's law.

Unless an estimate contains a conspicuous omission, we tend to blame losses on production factors. That's where the costs appear that can overwhelm profits. When we do lose money, we scrutinize the production area for ways to prevent similar losses in the future. Profitable jobs seldom receive the same scrutiny, even though they usually include unexpected costs as well. Whether the job was profitable or not, however, unexpected costs always eat into profits.

This article will examine some factors that contribute to unexpected costs and will show how a good estimator manages them. It is the estimator's job to identify areas of uncertainty, and to compensate for them.

Closed and Open Costs

One set of problems arises from the way that estimates are prepared. A typical estimate lists the material and labor costs, along with subcontractor quotes. The unit material and labor costs are multiplied by the quantities required. To the sum of these costs are added factors for supervision, contingencies, overhead, and profit. This orderly array is really a mixed bag of both predictable costs and highly variable costs.

The material costs are predictable and easy to find. You look at recent invoices, call suppliers, or send them a material takeoff and let them quote you a price. If you track job costs you may transfer material prices from other projects. Except for outright estimator error, there is no reason for variation between estimated and actual material costs. Similarly, if the specs are right, a firm quote by a responsible sub should match his billing.

These precisely defined costs can be termed *closed costs*. They stay put, and don't start to grow as soon as you look the other way. Closed costs are easy to calculate, and tend to get more emphasis than they deserve. If estimating were just a matter of listing closed costs, contractors would rack up profits with unwavering regularity. The fact that they don't means that there must be another kind of cost.

We can call these other costs *open costs*. Unrecognized open costs are a major reason why year after year, some contractors have so little to show for their efforts.

For contractors, the principle open costs are labor, supervision, overhead, and sales costs. Two other open costs that can be significant are financing and insurance. Open costs are *time-sensitive*. The longer it takes to perform a procedure, the higher its overhead, supervision and other costs become.

A successful estimator looks at all the variables that may affect open costs and provides for them in his estimate. His provisions may take the form of either higher prices for specific categories or of limiting clauses in the contract document. One way or another, the estimator uses informed judgement to provide for open costs.

Labor

Even though pricing manuals and data bases show standard unit costs for labor, it is a mistake to plug labor into an estimate as a closed cost. Published unit-labor costs are usually based on large quantities and ideal site conditions. They also assume specific levels of skill, equipment availability, and supervision that may not apply to the company and project at hand. There is no procedure so simple that production costs cannot increase from two to ten times because of difficult conditions, incompetence, or other unforeseen problems. As a major component of cost, labor requires the estimator's close scrutiny. Some of the variables affecting labor cost are:

Project duration. Mobilization and setup, as well as closedown and cleanup, are part of most operations, and their time is distributed over the total quantity of work performed. Small quantities bear a greater portion of this labor "overhead," and will exhibit a higher unit cost. In addition, small or divided work units deprive tradesmen of time to develop efficient work rhythms. The result will be an unexpectedly high production cost.

Unfamiliar materials. New materials come on line all the time, and architects love to spec them. The anticipated cost advantage of a new material will not materialize until installers learn how to handle it and understand the manufacturer's storage, preparation, and installation standards. First time around, someone has to pay for this training — the estimator's job is to be sure it's the buyer.

Complex or unorthodox design. The possible impact of complex or unorthodox design on performance time is devastating. Standard production rates go out the window when drawings must be scrutinized at every step and when work is delayed while questions are answered. And unorthodox designs almost invite mistakes. This in turn means tearing out completed work, ordering more material, paying additional labor, and rearranging the work schedule.

When confronted by an unorthodox project, the estimator should review the plans with experienced superintendents and tradesmen, as well as with manufacturers' reps. Such consultations help educate the estimator about new construction techniques that can lead to more competitive quotes. They also may help him to see unforeseen complications. Either way, the project benefits.

Incomplete plans or specs. Whenever the estimator encounters an ambiguity in the specs, he has two choices. He can sidestep the issue or he can meet it head-on by asking for a time-consuming clarification. Facing problems head-on is difficult, but sidestepping them means that they will have to be resolved later, probably at the worst possible moment, by personnel who may be unskilled in negotiations.

An example of this is the customer who can't decide on cabinetry, appliances, and other materials. It's tempting to sidestep such choices by including "allowances" for them in the estimate. But such allowances guarantee that decisions will be postponed until the last possible moment. At that point, however, selections may complicate or delay other work underway, or conflict with work already completed. Leaving these questions unresolved may make it easier to get the contract signed, but they are time-bombs waiting to blast holes in your profit. Smart estimators avoid them.

Supervision

On-site supervision frequently becomes an unexpected open cost. Some contractors don't show it at all. Others include it as a percentage of the total job price. But we have not found a relationship between the job price and the amount of supervision required. Jobs that use many trades, unorthodox design elements, or an extended production period are supervisor-intensive. Other projects may be large and simple, requiring little supervision. In other words, the price of a job may or may not reflect the amount of supervision required.

A more accurate way to estimate supervision cost is to project the actual quantity of superintendent days required, based on the length of the project, and charge them out at a superintendent's rate. Site mobilization and punch-out also involve supervision costs, which the estimator should evaluate and include as specific items.

The important fact to notice is that *anything* prolonging the project also increases supervisory cost. This hidden multiplier effect can make delays extremely costly.

Overhead

Overhead is office rent, telephone, utilities, the secretary, the boss's salary, the estimator's salary, the company's cars, trucks, and insurance: everything that's not a direct project cost. Estimators should not treat overhead lightly. The goal is to have all costs covered by year's end, leaving profits intact. For this to play out, *all* costs not covered as estimate line-items must be included in overhead.

Overhead is especially high in smaller businesses. For many contractors, insufficient overhead allocation is a chronic drain on profits. Some contractors add an arbitrary overhead factor gleaned from a pricing manual. If you enjoy groping in the dark, that's one way to do it. Others look at last year's sales and divide their total overhead expense by that number. This provides a factor that expresses overhead as a function of sales. Multiply that factor by the price of the job and you will find an overhead allowance for that job.

Even this is haphazard, however, because overhead is really a time-sensitive open cost. Each of us has 260 days a year to cover our overhead. How much overhead a project uses depends on the length of time it runs, not its dollar amount.

The result of calculating overhead as a function of job cost is to earn an unexpected profit on one project (the excess overhead allowance), then spend that profit to cover the inadequate overhead allowances of other jobs. Successful business owners believe in retaining the profit they make, not using it to subsidize other work.

Estimator Accuracy

The profit margin of a given company is based on decisions made by the management and sales departments. Whether management decides on a 23% profit factor or a 2.3% profit factor, it has no effect on the accuracy of the estimate.

An estimate that yielded an additional 20% profit would probably be considered a superb estimate by most standards, and one that lost 20% would be a disaster. However, from the professional estimator's point of view, both estimates were equally inaccurate, and a conscientious estimator would analyze both to see where he went wrong. Success for the estimator is a zero difference between estimated and actual costs.

Tying It Together

The purpose of a business is to earn and retain a profit for the owner. Activity not directed to that end is not business — it's a charitable endeavor, a hobby, or something else.

Again, the construction estimator's job is to anticipate the costs of a project as accurately as possible so that the business can earn an appropriate profit. To simplify this process he splits the project into its component parts and assigns costs to each part. Some of these costs are precisely known, closed costs. Others are variable open costs.

We have found that a major cause of lost profit lies in the estimator's failure to provide for unexpected open costs. In order to avoid this failure, the estimator must first recognize all the variables in a project. Then he must evaluate their probable effect, and express that effect in estimated dollars.

Because many costs — overhead, labor, and supervision in particular — vary with the length of the project, they should be estimated as open costs in order to produce a consistent and predictable profit. ■

Martin King is president of Martin Churchill Associates, Inc. in Arlington, Va. After 28 years of construction and estimating experience, he now specializes in investigating and consulting on structural damage and appraising reconstruction costs.

Pricing for Small Jobs

Use over-the-phone estimates to streamline your estimating and sales of small jobs

by Tom Swartz

For years our small jobs endured the same sales and administrative routine as our larger jobs. Whether we were installing a $100 screen door or doing a $20,000 kitchen remodel, we met with the customer, wrote up the estimate, issued a contract, and went through the scheduling process. As a result, the time and money we spent on estimating small jobs was eating up our profit. In many cases we were even losing money.

In 1988, we launched our Handyman Service, a new division of our company designed to handle small jobs and make them profitable. We defined small jobs as anything that could be completed by one person in a day or two. We didn't set a strict dollar volume since costs are greatly affected by materials, but the majority of these jobs fall in the $100 to $300 range. Few are higher than $2,000.

Creating this service entailed finding new ways to streamline our sales and estimating process. We realized that many of our calls were for the same sorts of small jobs, such as patching drywall, repairing or cleaning gutters, trimming interior doors, or recaulking windows. Developing a manual with standardized prices for these typical jobs would keep us from reestimating them every time they came in. It would also let us give customers an estimate during our first contact, usually over the phone. Finally, having a price book would mean that anyone, with a little bit of training, could give an accurate price.

The ability to quote a price quickly and then stick to it tells our customers we are experienced and gives them confidence in us. By comparison, pricing on time and materials, something we still have to do occasionally on some small jobs, means warning the customer that the price is approximate. It also practically guarantees an argument when costs run significantly higher than the initial quote.

Giving a price up front is also an instant qualifier. We no longer find ourselves running out on sales calls only to find that the customers are just "shopping" and can have their nephew do the work for $20.

Some remodelers might argue that paying a sales call in person, even on a $200 job, is a great opportunity to get to know a client. This may be true if your company is young or if business is slow. But last year we did about 1,500 small jobs that earned our company $550,000, or 25% of our total sales. With this kind of volume, we can't afford to send a salesman out to each site.

Creating A Database

The first step in this process was to choose software. We decided to go with a program called First Choice (Spinnaker Software Corp., 201 Broadway, Cambridge, MA 02139; 617/494-1200), but similar software, such as Dbase (Borland

International, 1800 Green Hills Rd., Scotts Valley, CA 95066; 408/438-5300) or Lotus (Lotus Development Corp., 61 Medford St., Somerville, MA 02143; 800/872-3387), would work just as well. In some cases, an estimating program that lets you combine the tasks involved in a small job might do the trick. Timberline's Precision Estimating Plus (Timberline Software Corp., 9600 S.W. Nimbus, Beaverton, OR 97005; 503/644-8155) is a good example.

After choosing our software, we reviewed the most recent two years of files and pulled all the invoices on jobs that totaled less than $2,000 and didn't involve subs. This gave us about 1,000 invoices to work from.

We were then ready to start inputting data. We established three fields on which to sort our data. We called the first field "Type." This gives the general nature of the work, such as "Carpentry," "Painting," or "Plumbing." We called the second field "Classification," which further focuses the job. If the type is "Carpentry," for example, the classification might be "Exterior Doors." The next field is "Phase." In this example, the phase would be "Install storm door."

We also created a series of fields to help us analyze job costs. Our database originally included labor hours, labor costs, material costs, and what we charged for the job. This information gave us a job history and helped us set our handyman prices. To avoid confusion, we erased all this information once we came up with our price.

In our last field, called "Description," we input as much detailed information as possible about each type of job. For example, under "Install storm door" we might write, "Remove existing aluminum storm door, install new 3'-0"x6'-8" Larson storm door (number 273-55) supplied by customer with three standard hinges, latchset, and door closer (all of which come with the door)."

As we've refined our database, we've

HANDYMAN PRICE LIST

CLASSIFICATION	PHASE	DESCRIPTION	PRICE	CLARIFY
Deadbolt	Install	Install single cylinder deadbolts - deadbolt to be Schlage #G120	$98.50	
Ext. Door	Install	Install exterior door, 1 3/4 solid 210 core door	$210.00	
	Weather Strip	Furnish and install weatherstripping around the door	$65.00	
	Screen	Replace screen door and jamb	$295.00	Ea. $150.00
	Lockset	Install lockset using Schlage #3 finish	$79.00	
	Closure	Repair rear door closer	$55.00	
	Break-in	Repair sprung door due to break-in	$75.00	
	Threshold	Furnish and install a new aluminum threshold	$85.00	
	Cut Off	Trim bottom of front door, put in high rug threshold to fit the door opening	$105.00	
	Adjust	Put longer woodscrews in hinges so service door will close properly	$55.00	
	Install	Install a new exterior door unit – labor only – pre-hung or slab only	$185.00	
Garage Door	Repair	Repair overhead garage door cable	$75.00	
	Opener	Install garage-door-opener hardware	$100.00	
Int. Door	Cut Off	Cut off doors to clear carpet	$50.00	1st door
	Cut Off	Cut off doors to clear carpet	$35.00	Ea.additional door
Show/Doors	Install	Inst. shower doors	$75.00	

The author's Handyman Price List makes it possible for his small-job salesmen to give accurate telephone quotes. Specific descriptions of completed work make it possible for the salesman to compare the job being quoted with similar jobs done previously. The book is updated regularly and is accessible in the company's computer database.

shortened our descriptions enough that they can be read quickly. But we've left enough information that the salesman can make sure the job that's being quoted is close to the one described in the book.

Finally we've printed out copies of the database for our small jobs salesman, who gives about 95% of the phone quotes, as well as for our office staff. Copies also go to our handymen, who may use the book to give a quote to a customer while they're in the field. The current version is about 20 pages long and includes 164 items. We use a three-ring notebook with tab indexes to make the book easy to use. Our staff can also access the database on the computer by entering key words, prices, or any other information.

Setting Prices

While inputting all this information is fairly straightforward, things get complicated when it's time to sort through it all on the computer and decide which jobs to keep in the database and how to price them. My first step was to eliminate all the jobs that were unique or unusual, such as getting squirrels out of the attic or hanging a tie rack on someone's closet door. Next I consolidated similar jobs. For instance, we had dozens of different descriptions of gutter repairs. I consolidated these into four jobs: maintenance, repair, and cleaning (one-story house or two-story house).

As a rule we mark up labor and materials 100% to achieve a 50% gross margin on small jobs. But this varies from job to job. It's not unusual for our book to have two different prices for each line item: one that we charge for installing the materials and one that we charge for both installing and supplying the materials. If we're not supplying the materials, we may raise the price to cover us in case we damage the materials and have to replace them. This extra also helps compensate for the fact that we're not making anything on materials.

For example, we allow three hours of labor to install a storm door even though the job is likely to take only two-and-a-half hours or less. Our cost per hour is $18, including taxes and insurance. With a 100% markup, that's $36 an hour, times three hours, for a total cost of $108. If we're supplying the door, we can get only about a 40% markup on it. We've found that it's hard to go higher than this on big ticket items since people generally know what these items cost and prefer to buy them directly. Our charge to install the door we supply is $348. That includes $90 for two-and-a-half hours labor, $10 for disposing of the old door, and $218 for the new storm door.

We also charge more for jobs that are done under adverse conditions. Anyone who's worked in an attic in the Midwest on a steamy day in July knows that it takes more time to get the job done. The same goes for a moldy, muddy crawlspace in the middle of a rainy April.

In some cases I've substituted a unit of measurement to cost by. For example, we charge $50 for every 8 feet of soffit we install on a one-story house. This includes all of our labor and materials. (In this case we mark up our materials by 100%.) We also use a combination of unit pricing and flat rate charges for some jobs. To fill hairline cracks in brick veneer, we charge $185 for 200 square feet. We then charge another 60¢ for each additional square foot.

The problem with developing a price book is that once it's done, you have to start updating it. We try to adjust our prices once every quarter, based on changes in overhead, material costs, and other factors. We also add new jobs as our experience and our services grow.

Using The Book

There are a few tricks to using a price book correctly. When pricing over the phone, it's imperative that our salesman ask the customer to be very specific in describing the job so that he can visualize the work. If someone calls asking us to come out and replace some missing roof shingles, we need to find out whether the roof is leaking. If so, we'll probably need to add in the cost of patching some drywall and doing some touch-up painting. We'll also find out how many shingles are missing (people can answer this more easily than if you ask how many square feet of shingles are gone), the shingle color, whether the roof is steep, and whether any gutters need to be repaired.

Next we give them a price range for the job, between $100 and $150 to replace fewer than 20 shingles, for instance. A range gives us some room for error, since our handyman may get to the site and discover the shingles that the customer said were green are actually gray. Having some cushion in the price means the handyman can run to our supplier and pick up the correct color. It also means we can often give a bill that's lower than the price we quoted, something all customers appreciate.

If we get out there and discover the price will be significantly higher, we call the clients and explain the difference. But if we'll lose only a few dollars, we go ahead and do the job anyway. It's easier, and probably less expensive, than going to all the trouble of rescheduling.

Our price book is meant to serve as a guide, not a bible. As we've gained more expertise in pricing jobs over the phone, we've learned when to follow the book and when to send a handyman over to look at the job first. Servicing your customers well sometimes means being flexible. ■

Tom Swartz is president of J.J. Swartz Company, a 70-year-old remodeling firm with offices in Decatur and Bloomington, Ill.

Unit-Cost Estimating

by Walt Stoeppelwerth

Accurate estimates depend on thorough job costing and attention to the intangibles

No aspect of running a remodeling business is more important than accurate estimating. To be accurate and effective, your estimates must be based on your real costs of doing business, and they must be integrated into overall management of the business. Otherwise, an estimate is just a stab in the dark as to what a remodeling project will cost.

Three Ways to Estimate

There are three basic approaches to estimating. Most contractors are already familiar with the first two, which are the "guesstimate" and the stick method.

In the guesstimate (also known as the SWAG, for "sophisticated wild guess"), the contractor tries to price the job based roughly on his experience of similar jobs. While it is useful for giving ballpark figures to qualify customers for price, this is no way to give a real estimate, either for sales purposes or for your own use.

Most contractors have used the stick method: You try to identify every piece of material the job requires and estimate the cost of buying (materials) and installing it (labor). Subcontractor costs are added to the cost of this shopping list for a final expected cost; then you add your markup to get a final estimate. This method can yield accurate estimates. But it takes too long to use in a business like remodeling, where, for most remodelers, only one of every four or five estimates lands a job.

That is why I prefer the third option, the unit-cost method.

The Unit-Cost Method

The unit-cost estimate is based on units of measurement such as lineal, square, or cubic foot, or per item. Every element must be included, from lineal foot measurements for footings, to square foot measurements for walls, floor, and roof systems, to per-item counts for such elements as windows, doors, and electrical outlets.

Theoretically, you could compile your per-unit material and labor costs for each of these systems from scratch by going over your own records. In reality, it makes more sense to use one of the estimating manuals currently available, such as HomeTech's *Remodeling and Renovation Cost Estimator*, or similar pricing manuals published by R.S. Means, Craftsman, and other companies, as a basis for your costs. These manuals have unit-cost labor and material prices for typical estimating categories. Most provide yearly or quarterly updates that give current costs for materials and labor for standard unit-cost estimating categories. Some of them come with area modification factors that further refine the prices for your geographic area. Modified to reflect the way you do business, these guides can form the basis for accurate estimates.

Knowing Your Costs

The key to the success of the unit-cost method is accurately figuring out your actual costs. To accomplish this, you must keep costs by category on every single job. This is the only way to check the accuracy of your estimates later, so you know where to make adjustments next time.

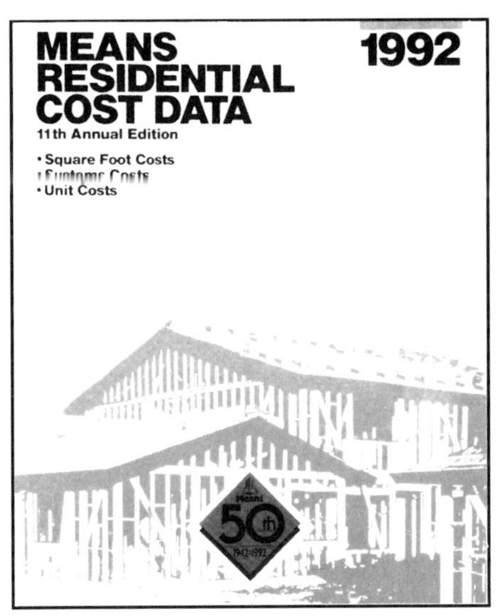

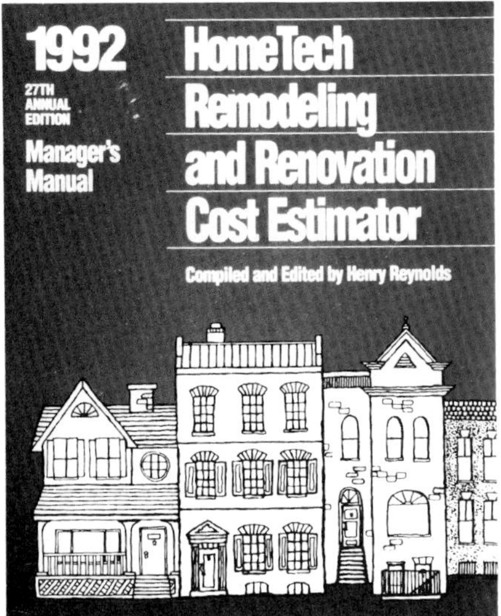

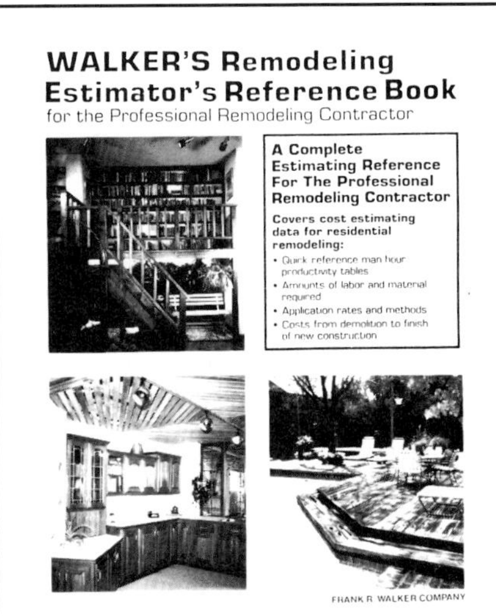

[illegible] starting point for developing your own price book. However, you need to track your company's actual costs for every job to keep your figures accurate and up-to-date.

There are two types of costs every estimate must account for:

- *Job costs* are those costs that are directly attributable to a given job.
- *Overhead costs* are costs that cannot be tied to a given job, but are part of the cost of being in business.

Direct Job Costs

In remodeling, direct job costs consist of five elements:

Labor. True labor costs include hourly wages and the "burden," including Social Security, workers compensation, unemployment insurance, and any fringes such as paid holidays, vacation, medical benefits, or truck and tool allowances. If you pay a carpenter $15 an hour, your true cost may be as much as $20 to $25 or more.

Materials. This should include not only the materials, but sales tax and delivery as well. If you have a driver who spends 30 to 40 hours a week picking up materials and delivering them to the job, this worker should keep a time card allocated by job.

Subcontractors. This is what you pay to your subs, recorded as a per-unit cost.

Plans and permits. These costs should be included in direct job costs rather than overhead, and billed and [illegible] with particular jobs.

Cleanup. Whether done by your workers or subbed out, this is a direct job cost, not overhead. Dump fees are also a direct cost.

Overall, any cost that can be applied to a particular job should be counted as a direct job cost.

How to Track Job Costs

To track unit costs, you must record, for every job, the actual costs associated with the elements of the job, such as square feet of wall demolished or built, lineal feet of cabinets installed, or individual windows framed and installed (see sample forms, facing page.) Track the costs by the same categories that you use in estimating; you needn't invent your own, but can take those that pertain to you from one of the published estimating

Tracking Job Costs

JOB COST RECORD

Date 1992		Description: Name	Item/ P.O. No.	Check No.	Total Job Cost Dr.	Distribution of Job Cost: Materials	Labor	Sub-Contractor	1 Plans Est.	2 Tear-Out	3 Excavation	4 Concrete	5 Masonry	6 Floor Framing Est.	7 Wall Framing	8 Roof Framing	9 Roof Covering	10 Exterior Trim
JUNE	3	RAY BRADBURY	PAYROLL	1234	250 —		250 —			250 —								
	5	HERSH EXCAV.	PIERS	1238	105 —			105 —			105 —							
	6	GANDIN CONC'E.	PADS/PIERS	1244	350 —			350 —				350 —						
	9	BILL NORRIS	FIREPLACE	1249	2360 —			2360 —					2360 —					
	9	ANDERSON LMBR.	MAT'LS	1265	1420 —	1420 —								550 —	270 —	600 —		
	12	ALPINE ROOFING	ROOFING	1273	880 —			880 —									880 —	
	12	ANDERSON LMBR.	MAT'LS	1275	185 —	185 —												185 —
	14	MORRIS CEDAR	MAT'LS	1281	1050 —	1050 —												
	15	RAY BRADBURY	PAYROLL	1282	798 —		798 —							220 —	160 —	318 —		100 —
	16	FRANK HERBERT	PAYROLL	1283	912 —		912 —							240 —	205 —	352 —		115 —

JOB COST AND CONTRACT RECORD

SUMMARY OF ESTIMATED AND ACTUAL COSTS

CAT. NO.	CATEGORY OF WORK	Estimated Total Cost	Extras and Allowances	Estimated Net Cost	Actual Total Cost
1	Planning, Engineering, Permits	-0-			-0-
2	Tear-Out and Demolition	220 —		220 —	250 —
3	Excavation and Grading	125 —		125 —	105 —
4	Concrete	350 —		350 —	350 —
5	Masonry	2200 —	150 —	2350 —	2350 —
6	Floor Framing	1000 —		1000 —	1010 —
7	Wall Framing	650 —		650 —	635 —
8	Roof Framing	1350 —		1350 —	1270 —
9	Roof Covering, Flashing, Gutters	800 —		800 —	880 —
10	Exterior Trim, Porches, Decks	475 —		475 —	400 —
11	Siding	1200 —		1200 —	1050 —
12	Doors and Door Trim				
13	Windows and Window Trim				
14	Plumbing				
15	Heating and Cooling				
16	Electrical				
17	Insulation				
18	Interior Wall Covering				
19	Interior Ceilings				
20	Millwork, Trim, Stairs				
21	Cabinets and Appliances				
22	Specialties				
23	Floor Covering				
24	Painting and Decorating				
25	Clean-Up and Hauling				
	TOTALS				

Courtesy of HomeTech, Inc.

The left-hand columns of the Job Cost Record (above) record all expenditures of a given job by date, payee, and check number. The middle columns identify each cost as material, labor, or subs. And the right-hand columns break down the costs into estimating categories. A summary sheet (at left) compares estimated costs with actual costs for each category.

guides mentioned earlier.

Once you have the costs for a completed job, check them against those in your estimating guide, and adjust the guide as necessary. This process of comparing actual to estimated costs, called job costing, isn't difficult; it's mainly a matter of persistence. But you will need to add some essential record-keeping steps to the way you work and track your workers' time.

Time cards. Have each worker on payroll keep a time card on a daily basis, accounting for time both by job and by category of work. For instance, Bob might spend four hours framing nonbearing interior stud walls at the Smith job, one hour blocking joists at the Wagner job, and three hours installing replacement windows at the Wagner job.

Record, in the proper categories on each job ledger, the total cost of paying Bob for those hours. Do this for each worker. At the end of each job, add up the labor spent on each category, then divide it by the number of units (such as linear or square feet of stud wall or number of windows) for a per-unit labor cost. By constantly checking your per-unit costs against your guide, you can fine-tune and update the category costs.

Using lead carpenters and one-person crews makes job labor costs easier to keep. Single workers tend to concentrate their efforts clearly on one task at a time. When crews of two and three workers are shifted regularly among jobs, tracking worker time by project and category becomes much more difficult.

Materials. In the unit-cost system, material costs for large items include the many smaller ones that accompany them to make a finished assembly. Unit prices for drywall, for instance, include the board itself, drywall nails or screws, tape, corner beads, spackling compound, and even sandpaper. The cost of a door includes both the door and the jambs, as well as butts, locksets, and casing.

Keeping track of material costs is fairly easy. The key is to require each supplier to submit a separate invoice for every job on every order. This means that when you go to the lumberyard to pick up materials for three jobs, you get three separate invoices, so the costs can be attributed directly to the proper jobs.

Each job gets its own materials folder; the charges on the invoices are then entered into the job-cost ledger by estimating category. On one particular invoice, for instance, the cost of 2x6 studs, plywood sheathing, and half the 16-penny nails might be costed as "sheathed exterior wall," while the 2x4s and the other half of the 16-penny nails get entered as "frame interior stud wall." At the job's completion, the total sum in each category can then be divided by your estimating unit, which in this case would be lineal or square feet. This way, material costs always reflect the latest prices.

Subcontractors. Have each subcontractor submit a separate invoice for every job. These costs should then be tracked according to your estimating categories. Adjust your subcontractor costs as necessary whenever your subs change their rates.

Change orders. Every change order *must* be estimated and costed as a separate job. If they are instead included in the overall estimate and job costs, they will undermine your ability to know whether your original estimate (or that of the change order) was accurate. Treat the change order as a separate job.

Running The Estimate

With your estimating manual updated where possible with your known record of job costs, you are ready to make quick and accurate estimates.

Material costs. As long as you correctly identify everything you need to build, estimating materials is fairly easy. Simply use your latest job-cost material figures. If you are faced with a category for which you don't have job cost records or that you haven't used for a long time, use the price from your estimating manual. If the unknown item is likely to be a significant part of the estimate, check the current cost directly with your supplier.

Labor costs. The hardest category to estimate is labor. Remodeling is a stop-and-start operation, and the time lost between steps can be substantial. A good estimating manual allows for this inefficiency in the labor costs it uses. The hidden costs of dead time — the quick run to the store for supplies, the dash to the truck to get the new blade — are included in the labor costs listed in these guides. Of course, if your job costing shows that your own typical costs are higher or lower, adjust accordingly.

Subcontractor costs. With subcontractors, it's most efficient to prenegotiate fixed prices for standard items of work, so that your subs don't have to visit and estimate every prospective job. On the trickier jobs, you can describe the job to them on the phone or have them come out and take a look. In some cases, you can get the sub to give a markup for extra difficulty in your prenegotiated price list.

Weighing The Intangibles

The above steps will let you develop 80% of an estimate quickly and accurately. This lets you focus on the 20% that requires more time and judgment — the intangibles that make or break most remodeling jobs. If not accounted for, these intangibles, such as difficult access or a surly client, can destroy your profit.

To account for them I recommend a four-part "judgment analysis":

Condition analysis. Do the job conditions vary significantly from the normal conditions from which your unit costs are derived? By job conditions, I mean those things surrounding a job that aren't a direct part of the construction. These include such items as access, storage of tools and materials, and the threat of theft.

For instance, consider your labor costs for a remodel on a third-floor apartment accessed by a narrow community stairway that must be kept free of materials and clutter, and in an area where the likelihood of theft forces you to remove all your tools at night. Your costs will be higher than usual, because you're paying your workers to carry their tools and materials up and down stairs every day, and to clean up often. As closely as possible, estimate the hours that will be needed to handle these extra tasks.

Project analysis. This is similar to the condition analysis, but relates more to the actual work. Will demolition be particularly tough for some reason? Is there a tricky framing tie-in between roofs of different slopes? Are there obstructions or tight framing spaces that will give the subs difficulty when installing the electric, plumbing, and heating runs? You can adjust for these factors by including additional estimated labor hours, and by getting your subs to adjust their estimates similarly.

Customer analysis. Is this client likely to be the one in five that will seem determined to deny you a profit? Is he likely to sue, or threaten to? Does he take an hour of each day asking you questions? Is he an engineer who works to fine tolerances and will scream if he sees the shims around a window frame? Or a meticulous housekeeper that insists the place always be spotless?

Analyzing customers is the most nebulous of all judgment factors. Some things, such as the cost of extra cleanup at, say, 15 extra worker-minutes a day, can be quantified for an entire job. An adjustment for spending 30 minutes every day with the customer to go over the job is also uncomplicated. But a decision to raise your costs or your markup 25% or more to account for the unpredictability — or deviousness — of a particular customer is much tougher. The ultimate judgment call is whether to walk away from the job because of such potential problems. For these variables, you must rely on your own experience and judgment. The important thing is that you recognize them and account for them somehow, either through added hours or an extra markup to protect your profit.

Company capability analysis. This has to do with your ability to efficiently handle the job. How does this job fit into your company's capacity to perform? If the job is outside your normal experience, then the learning cycle — a little head scratching and training time — must be included in the estimate. As well as possible, you must try to translate these extra tasks into additional hours.

Your judgment here will also involve issues such as your confidence that your crews can learn fast, your desire to take on new and different projects, your desire to keep crews busy in a slow time, or the likelihood of finding new or specialized workers or subs in good times.

Adjusting your estimate based on this four-part analysis can make the difference between profit and loss. But you need to make sure that the adjustments for a given job don't skew estimates for future jobs. Record your adjustments *as adjustments* — added hours or extra markup — so that at job-costing time, you can "unadjust" the figures to standardize them for costing purposes. This will also help you gauge the

Typical Remodeling Overhead

	Annual Volume			
	$100,000	$250,000	$500,000	$1,000,000
Business Expenses				
Sales	7,000	20,000	50,000	100,000
Production Supervision	5,000	15,000	30,000	60,000
Advertising	1,000	5,000	19,000	50,000
Rent	1,200	5,000	10,000	19,000
Office Staff	4,000	15,000	30,000	60,000
General Insurance	2,000	4,000	12,000	22,000
Truck (Sales, Mngmt. Only)	5,000	10,000	15,000	20,000
Telephone	1,200	2,400	3,600	5,000
Radiophone, beepers, etc.	360	720	1,080	1,440
Tools & Equipment	1,000	2,500	5,000	10,000
Office Equip., Lease/Rent	500	1,200	2,500	5,000
Office Supplies	100	250	500	1,000
Accountant	300	600	900	1,200
Legal Fees	500	1,000	1,500	3,000
Dues, Associations	250	500	500	500
Education, Seminars, Travel	250	500	750	1,000
Entertainment	100	200	500	1,000
Bad Debts	1,000	2,500	5,000	10,000
Total Overhead Costs:	$31,060	$86,250	$184,130	$370,340
Percentage of Volume:	31%	35%	37%	37%

Note: *Most remodelers don't mark up their estimates enough to cover the many items that make up their overhead. That's the main reason most remodeling companies fail.*

accuracy of your adjustments.

Accounting For Overhead

Once you've estimated the direct job costs and intangibles of a job, you must mark up that total by some percentage to obtain a final estimate and sales price. Observing the remodeling industry for many years has convinced me — and this is proven in the marketplace every day — that the minimum markup a residential remodeling contractor can charge to stay in business is 50%, which will yield a 33% gross profit. (Marking up a $100 job 50% will give a total price of $150, 33% of which is profit.)

The 50% markup applies to contractors who do less than $500,000 annual volume. For businesses that gross over $500,000, the minimum markup is 67%, for a 40% gross profit.

These high markups are needed because overhead for a remodeling contractor runs at least 25% and may be 35% or more for medium-sized and large companies, which have more nonproduction workers such as administrators and salespeople (see table, facing page).

Unfortunately, most remodeling contractors do not understand how many things overhead must cover, and thus use too small a markup. Overhead is more than your rent, utilities, and phone; it is everything that can't be directly job-costed. That includes sales efforts, production supervision, advertising, truck expenses (payments, insurance, gas, and repairs), office equipment and supplies, tool purchases and repairs, office work (whether done by administrative personnel, you, or your spouse), and your accountant, bookkeeper, and attorney fees. You must also pay yourself for whatever functions you perform that aren't direct costs, such as sales and general administration.

These costs add up. Over 90% of remodeling contractors in this country charge too little — that is, less than a 50% markup. This is why there is a 50% fatality rate in remodeling every year and a 96% fatality rate in five years. They simply don't charge enough to survive.

Other Principles of Estimating

Some other estimating fundamentals will help you keep estimates on target.

Estimate one job at a time. You should go out on the job, meet with the customer, then return directly to the office and complete that estimate before doing anything else. If you wait a day, you won't remember all the aspects of a job, particularly the details that feed your judgment analysis.

Include everything. You must include a price for every item and task on the job. Something as small as a 4-foot section of base molding must be priced, or moving a hose bib 10 feet. Leaving several small things out hurts you worse than misjudging the cost of bigger items. If an item that is 5% of the job is underestimated by 20%, it only affects the total estimate 1%, and may well be compensated for by another item that is overestimated. But if a similar item (or items) is forgotten completely, the oversight could take away all the net profit on the job.

Round up. Don't estimate for a 20x7$^1/_2$-foot wall, estimate for one that's 20x8 feet. Do not subtract windows and doors from framing estimates on walls, but figure them solid. If you are figuring a gable end which is a triangle, figure it a solid rectangle for studding and sheathing and simply reduce the siding figure by 10%. By always rounding up and figuring areas solid, you help account for material waste; you'll almost certainly have to buy that much material anyway.

If there is a condition to quantify, do it on the spot. If you must plaster patch a wall, for instance, don't just write "patch plaster," or you may later misremember how big the patch is. Everything should be measured and quantified on the spot.

Have someone else check the completed estimate. This will turn up simple oversights and mistakes. One advantage of the unit-cost system is that the second person need not be experienced in construction, but can check the estimate against a master checklist or your estimating guide. The same can't be said for the stick method, where someone who doesn't know precisely how you build things will have a difficult time checking your figures. ■

Walt Stoeppelwerth is a Founder of HomeTech, Inc., a remodeling and home-inspection consulting firm in Bethesda, Md.

The Hidden Costs of Renovation

To make a profit on renovation jobs, try to anticipate the cost of intangibles— from bad weather to extra job-site security

by Sal Alfano

Renovation work presents a number of special problems for the estimator. Not only are materials and labor harder to price than on new construction, but renovation involves a number of intangibles that can throw an estimate way out of line. These come into play in all types of renovation, but I will limit this discussion to substantial structural and cosmetic changes to the interior of a single-family residence. Whatever the scope of the work, the goal is to anticipate expenditures and to price the job accurately to avoid cost overruns.

Demolition

Whenever you are installing new materials, the existing materials first must be demolished and removed from the site, or they must be salvaged and saved for reuse. A few simple but important considerations can prevent catastrophe.

Protection. Everything that is not going to be demolished must be protected, especially if the house is occupied. If floors will not be refinished, a layer of polyethylene or resin paper taped at the edges and seams should provide enough protection. You can lay inexpensive sheet goods — paneling, waferboard, even 1/4-inch drywall — over the paper to prevent scratches and scuffing.

Consider temporarily boarding over doors and windows close to heavy demolition work. If the owners cannot put their furniture into storage, keep it out of the way and covered well enough to keep dust out.

Cleanup. In estimating, remember that protective measures may have to be reapplied several times over the course of the job, and ongoing cleanup is essential. Nothing lasts forever, and the heavy demolition early on may wear out the coverings. Every job should be swept — vacuumed, if possible — every day to prevent the accumulation of debris and extend the life of the protection you have provided. It makes things a lot more pleasant for the occupants too.

Speaking of occupants, make sure your clients realize that, in spite of your most conscientious efforts, into their lives a little dust will fall. Most surfaces will not be damaged by a light dusting, but if they are particularly worried about specific pieces of furniture or areas of the house, take whatever steps you deem necessary. Or, better yet, ask them to handle it. In the case of upholstery, drapery, or carpeting, even a little dust or a stray dollop of joint compound can be hard to clean. Try to determine what expectations your clients have for the cleanup, and then adjust your prices accordingly.

Security. Finally, when an exterior wall is demolished, there may be a period when the old wall or roof is down but the new one is not yet up. Depending on the location and season, the building and its occupants will have to be protected from rain, snow, mud, cold, heat, insects, animals, vandals, and thieves. Determine which of these applies and provide in your estimate for securing the site. Protection from intruders may be as simple as affixing a piece of plywood every night and removing it every morning, but it can be as complicated and expensive as hanging a temporary door complete with lock set and multiple sets of keys.

Cold weather is especially difficult. Even if you use a temporary tent, you may need portable heaters or increased use of the existing heating system. You also may need to supply auxiliary heating to speed the drying of joint compound or paint.

Whatever you do, have a plan at the time you do the estimate; don't wait until you're on the job facing an emergency. If the heating system is going to be replaced or temporarily shut down, for example, make sure water supply lines and traps can be conveniently drained. If you end up fooling with the pump or the city water connection, it can get expensive.

Salvage. Carefully identify and mark salvaged materials to prevent them from being discarded accidentally. And take special measures to protect materials to be reused. Reinstalling existing woodwork will save money on materials, especially for intricate moldings, but labor costs will rise — sometimes dramatically.

In addition to being tedious, it's very difficult to remove wood trim without causing an occasional scratch or split. Later, these will require touching up or special millings to match damaged pieces.

If much woodwork is to be reused, mark individual pieces according to location to simplify reinstallation. Allow time for these steps in your estimate.

Plumbing and light fixtures are commonly salvaged for reuse. Unfortunately, they are especially susceptible to damage. Mark them, as with woodwork, so that no one — including subcontractors — destroys them. Inspect the fixtures three times: once before removal to determine existing damage; again after removal in case there's hidden damage or damage from removal; and once again during reinstallation to ensure that no change in their condition has taken place.

In any case, it pays to price the job as if some damage or difficult reassembly will occur. We commonly price salvaged fixtures at 80 to 90 percent of their cost new. This covers the higher labor costs for installing salvaged material.

Rubbish. Spend some time planning how debris will be handled. If you're working on the second story or higher, plan how you will remove debris from the building. A rubbish chute often works well and can be built quickly and cheaply, but it usually requires renting a large trash container.

If carrying trash by hand is the best alternative, consider using empty feed bags instead of plastic trash bags. They're a bit smaller but hold just as much because they don't tear as easily.

Finally, make sure the appropriate vehicle can reach the site; otherwise, allow for the added expense of moving the rubbish pile closer to the pick-up point.

Working Conditions

Many New England contractors schedule renovation to fill the slack winter months — when bad weather affects job costs the most. In many cases, particularly when the building is unoccupied, you will have to provide for snow removal.

The local pickup-with-plow competes well with hand shoveling, but reliability is important: If you start your workday at 7 a.m. and the plow doesn't come until 9:30, you're in trouble. Remember also that exterior work will be more difficult in deep snow, especially when staging is used, and that concrete work will become increasingly risky and expensive.

Extremes in temperature also will take their toll on tools, materials, and personnel, reducing productivity and eroding morale. Add time for these inevitable delays if you're working in bad weather.

Storage. Storage presents special problems, too. In bad weather, exterior storage involves more time and expense. Even in good weather, the location of your storage point will affect your rate of work.

If materials are stored outside but your work is inside, workers will spend a lot of time squeezing through doorways carrying cumbersome sticks of lumber. If there is room to store materials inside, keep it close to work areas, but in a place where you're not tripping over things all the time.

Remember that not all materials will conveniently fit through available openings. Drywall comes to mind. Watch for U-shaped staircases and undersized, second-floor windows.

If window or door jambs need to be disassembled to accommodate drywall, include the labor costs in your estimate. You may even have to reduce the lengths of the sheets, with a possible increase in hanging and taping time.

Similarly, one-piece tub or shower units are difficult, if not impossible, to get in or out of a house. If you can't find a model that fits and you have to take down a wall to get it into the room, make sure you include this additional cost.

Work area. A good work space, or "site shop," is important to any renovation. Ideally, this should be indoors and close to the storage area and the areas you are working on.

Most of the time, however, the space available for the shop is cluttered, cramped, and poorly lighted. Productivity will suffer in proportion to the distance from the shop to areas where work is being done. If this means traveling up and down stairs, in and out of doors, or around the owners and their possessions, labor costs will be higher.

A shop set up outside is the worst possible situation. Not only will you have to drag your table saw in and out every day, but bad weather can shut you down.

Scheduling. Scheduling can affect costs in several ways. Sometimes, especially in commercial work, the job has to be done during off-hours, which leads to overtime charges for employees and subcontractors. But even a late starting time can disrupt a crew's work routine, and half an hour can be lost each morning while the workers tiptoe around late-rising clients. Remember that special circumstances command a special price.

Even delays that seem to carry no immediate cost will cost you in the long run. First, the client may become disenchanted as the job drags on, and may make you wait for your money. Second, future job starts will have to be pushed back, often jeopardizing profits — for example, by forcing you to postpone a job until late in the season, when bad weather will increase your costs.

And then there's the "handyman syndrome." Every homeowner has a list of little things they've been meaning to get to and will ask you to "take care of it" while your crew is at the house anyway. All of this work should be charged at a premium because it delays starting other jobs that may have a greater profit potential.

Do Your Homework

These are just a few of the multitude of circumstances arising out of renovation that can result in hidden costs to you. The point is, it's too late to do much about these things once the work is under way and a price has been agreed upon.

Do your homework during the estimating stage, and you'll be better prepared to negotiate a price, write a contract, and make a profit. ■

Sal Alfano, formerly a general contractor in East Calais, Vt., is an associate editor with The Journal of Light Construction.

Chapter 6 • Bidding Strategies

Competitive Bidding

by Sal Alfano

Before bidding, assess the drawings, the specs, the schedule, and the competition. The more systematic your approach, the better your chances for success.

Typically, you get two weeks to prepare a bid on an architecturally designed job, and it never seems like enough. Before spending all those hours under that kind of pressure, make sure you *want* the job by carefully considering each of the following:

- Bid form and due date;
- Owners and architects;
- Project scope, including type of work and its cost;
- Labor and time requirements;
- Project schedule and location;
- Other bidders;
- Type of contract.

Bid form. Check the date the bid is due. First, ask yourself: *Can I do a thorough job on the bid in the time allotted?* If you don't have time to give it your best effort, it's probably not worth getting started. Next ask: *Does the bid require a bid bond or a performance bond?* If you've never been bonded for performance, you probably don't have the time to do it before the bid is due. And a bid bond requires you to submit a sum of money with your bid — sometimes as much as 5% of the total price — which is held in escrow until a contract is signed. If you don't have that kind of cash lying around, or a way to raise it, you're finished before you start.

Owners and architects. Check out the owners next. *How will their reputation as sticklers for detail, for example, affect your price and scheduling?* Find out what you can about their prior experience with builders. If you'll be dealing with someone who has a basic distrust of builders, you might want to pass the job up.

What do you know about the architects? If you've had trouble working with them in the past, think twice before going through it again. If you've never heard of them, try to discover how complete their plans are, and how they handle design problems. Ask about their contract with the owners. If it's not a full-service contract, you probably won't get much help from the architects during construction, and there'll be no one around to sign off on the problems that will inevitably crop up.

Project scope. *Is this project similar to jobs you've done before?* If you're accustomed to building single-family residences, think hard before jumping into a multi-unit condo complex. Also be wary of exotic materials or specialized work that requires pricey, hard-to-find tradesmen. These elements can mean more opportunity for things to go awry — and more potential imperfections for the owners and architects to fuss about.

Check your schedule. *Do you have enough time and enough skilled personnel to complete the job?* Timely completion is especially important if penalties will be assessed for finishing late. Consider the job location, too. If it's outside your normal traveling range, it could mean extra costs to transport your crews and deal with unfamiliar suppliers. And administration is always harder when the job is far from home. Find out about your competitors' schedules. If they're full, they may end up dropping out of the bid or submitting a high price.

The competition. *Who are you bidding against?* A list of prime bidders is generally available from the architect's office. If you recognize someone whose bid is always ridiculously low, you can save yourself some trouble by not competing. On the other hand, try to find out about bidders you don't know. Almost every bid has at least one ringer — often an acquaintance of the owner's — whose price will be either wildly high or dangerously low. See if you can learn something about their bidding history from the architect or other builders.

Schedule Your Time

Once you decide to do the bid, treat it like a job and draw up a schedule to budget your time during the bid period. Work backward from the bid due date, noting

Smith Residence
Anytown, Vermont

BID PROPOSAL FORM
January 1, 1992

Bidder's Name: ABC CONSTRUCTION
Address: 421 MAIN STREET
ANYTOWN, VT 11111
Phone #: 802/555-4433

GENERAL:
1. The Contractor acknowledges that he or she has inspected the site and is aware of existing conditions under which the work will be done.
2. The Owner reserves the right to reject any or all bids and to waive any formalities in the bidding.
3. The bid shall be good for 60 days.

I have received the bid documents entitled SMITH RESIDENCE. I have received Addenda numbered and dated as follows:
ONE (1) JANUARY 18, 1992
TWO (2) JANUARY 20, 1992

Base Bid: Lump Sum Base Bid per Drawings and Specifications prepared by XYZ Architects, dated January 1, 1992 including allowances.

Base Bid: FIFTY-SEVEN THOUSAND + NO/100 dollars ($57,000.—)

Alternate 1 - Basement Slab: Provide 4-inch-thick concrete slab reinforced with 6x6 #10 WWM or approved fiberglass reinforcement throughout basement. Slab shall be over 6-mil poly over 4-inch min. well-compacted clean granular fill. Slope slab to floor drain (which is included in base bid). If alternate is accepted, slabs for boiler and pressure tank will not be required.

Alternate 1: NINETEEN HUNDRED FIFTY + NO/100 dollars ($1,950.—)

Alternate 2 - Interior Painting: Omit interior painting and staining of walls, ceiling, doors, and interior trim. Contractor will remain responsible for taping, patching, and smooth plaster finish ready for painting. Contractor will also remain responsible for finish on floors including urethane on hardwood.

Alternate 2: Subtract ELEVEN HUNDRED + NO/100 dollars ($1,100.—)

ADDITIONAL WORK:

Blasting: For additional blasting and removal of ledge, please indicate General Contractor's markup for subcontractor's work, if any.

Blasting: FIFTEEN percent (15 %) of subcontractor invoice.

Starting Date: FEBRUARY 17 1992
Completion Date: APRIL 24, 1992
Signature: [signature] **Date:** 2/3/92
Company Name: ABC CONSTRUCTION

Bids are due at Architect's office by 3:00 p.m., Monday, February 3, 1992.

critical dates and estimating tasks, such as the site inspection, and set a target for when you expect to complete each one. If your schedule is tight, you may have to shuffle your appointments for the next two weeks, or plan on burning some midnight oil.

Make sure your schedule takes your subs into account. Because their work is specialized and often in high demand, subcontractors need a lot of lead time to prepare their quotes. Get the plans and specs to them as early as possible, and make sure they know when the bid is due.

If you're going to solicit competitive bids from several subs in the same trade, decide in advance who they are, and list them by trade. Create a column next to each name for their price, and leave space to record any notes you may want to make about their estimates. Also leave room for unsolicited subcontract bids as well, since some will probably arrive at the eleventh hour.

As you go through the estimate, write any questions you have for the subs on the subcontract-bid list, so that when you get their phone calls you'll be ready. Familiarize yourself with the scope of each trade's work so you can ask intelligent questions. In renovations, for example, the mechanical specs sometimes require demolition of existing equipment or portions of the structure that obstruct the new work. You have to know whether or not the subs have included it in their price. Once the contracts are signed, it's too late to argue about who pays to patch the walls.

Read The Specs

The spec book — or "bible" as it is sometimes called — is, along with the plans, the focus of any bid, and you should read it cover to cover.

The spec book contains three kinds of information. The first describes the kind and quality of materials to be used: the size of the studs, the kind of tile, the thickness of the underlayment.

A second kind of information describes procedures, like how to set the tile or when to use cold-weather concrete protection. Special procedures can be costly, and what you charge for them should reflect both their actual cost as well as how much you can get away with and remain competitive. In a tight bid, you may omit some of these costs, but it should be an informed decision, not an oversight.

The third kind of information deals with project administration, including the bid process itself, and should be scrutinized carefully. Two bid requirements that often cause last-minute panic are documents that need to be notarized, such as noncollusion affidavits, and project schedules. The former are easy to handle as long as you notice them early enough. Project schedules, on the other hand, are often incorporated directly into the contract documents, and can have serious consequences down the road. For example, many contracts include penalties for failure to complete on time. A bid that includes a hastily thrown together schedule is a bid you may live to regret.

Stipulations in the specs may also require you to change your normal routine. Minimum liability insurance coverage, for example, may exceed your current policy limits, and you'll need to include the increased premiums in your overhead for the job. Or you may be required to obtain and pay for special permits and licenses, or prepare elaborate shop drawings and operating manuals for equipment. All of these generate expenses that should be included in your bid price.

The Drawings

Like the specs, the drawings ensure that everyone bids on the same building. An incomplete set of drawings is the hardest to handle. It's difficult enough to account for information that's plainly included on the drawings, but it's harder still to anticipate costs for work that is hidden or not explicitly stated. For example, the site plan may not indicate that there is not enough space on the site to store construction materials, park cars and trucks, and still give the excavator room to operate heavy equipment. This may increase the labor needed to handle materials.

Incomplete plans may not be fully dimensioned, so your crews will be scratching their heads on company time while trying to figure out how long the studs are. Detail cut sheets may make a particular construction detail appear simpler than it really is, or not fully describe multiple levels or complex intersections. If the architects can't give you the answers you need during the bid period, your profit may disappear while you chase them down during the construction.

Often there are discrepancies between drawings: the plan view shows a window in a certain location, but the elevation shows it in a different location. Generally, contradictions are resolved by following the spec book over the plans, and the larger-scale drawings over the smaller-scale drawings. But don't guess: Clarify ambiguities with the architect.

Addenda

If the problem is important enough, an addendum that resolves the matter will be issued to all bidders. Receipt of addenda must be acknowledged in the space provided on the bid form. Pay attention to the addenda, which are usually delivered by mail, and make a checklist of the items they contain. Remember, you aren't the only one bringing questions to the architects, and your bid is assumed to include all of the information the various addenda provide.

Slippery Language

The specs and drawings will probably include a phrase like "provide Schlage A-Series locksets or equal." Watch out. What is "equal" is usually determined by the architects — and the architects only. Find out before you submit your bid if your substitution is acceptable, and look for other approved substitutions in the addenda.

Some drawings are sprinkled with little notes describing various portions of the building as "typical." This means the architect has thought through this particular detail once, then generalized it to include similar details throughout the building. When you see this, scour the plans for every possible occurrence of the detail and think each one through. You can't afford to assume anything.

Two other phrases will bite you if you aren't careful: "as needed" and "match existing." Both are found primarily in projects that require some renovation of an existing structure, which usually means a site inspection is needed. Each occurrence of these phrases, along with the others already mentioned, should be put on a separate list and checked off one by one during the site visit. This is usually the only opportunity you'll get to inspect existing conditions, so make the most of it.

To "match existing" conditions, you have to know what they are and determine if it's *possible* to match them. If existing trim must be duplicated, it may require special milling and extra labor. If you need to match materials like roofing, which varies from bundle to bundle and may be discontinued by the manufacturer, find out how close the match must be, and what alternatives are available if an acceptable match isn't possible. Again, the answer may turn up in an addendum.

"As needed" is perhaps the most insidious phrase of all, since it's entirely a judgment call. I've looked at plans that said "patch roof shingles as needed" when there were two feet of snow on the roof and no way to discover how much patching was needed.

Allowances And Alternates

In this particular instance, I asked for an allowance for roof patching. Allowances are commonly used to account for unknowns (like subterranean ledge during excavation) or as a means of obtaining prices for work that the owner isn't sure about, like floor covering or kitchen cabinets. Alternates are similar and are usually used to find out which of two options is less expensive. Both are usually called out on the bid form and often play an important part in deciding who gets the job. You may have the lowest base bid, but unless you're still lowest after the allowances and alternates are totalled, you'll be sitting this one out.

The last few days of preparing the bid are pretty hectic. Up to this point, you've been preoccupied with details, but now you should focus on the big picture and tying up loose ends. The things you do or fail to do from here on out can make or break your bid.

First, close out the bid well in advance of the time it is due, usually the end of the day before. This doesn't mean you won't be accepting any more prices, but simply that you ought to organize the information you already have. For this you need a summary sheet which lists all of your worksheets, page by page, and the totals from each. Totalling each page is safer than carrying subtotals through to the end. Headings for the page should read "Description," "Amount," "Revised Amount," and "Bid Amount."

On each line, fill in the column labelled "Amount" *whether you have a hard price or not*. This allows you to play around with the total bid price before you actually have all your hard prices. Devise some system for indicating which prices are hard and which may still be revised (different colors, circled numbers). Total up the summary sheet and add your overhead and profit. If your intuition tells you that this preliminary total is too high to win the bid, now is the time to start looking for places to cut the price, remembering, of course, that some of your soft prices may change everything.

Where To Cut

There's no hard and fast rule about where to make your cuts. You are least likely to find savings in material prices, unless you discover an outright mistake, like calculating subflooring twice, or accidentally doubling the quantity of siding. But suppliers often differ substantially in their price for windows and doors, and this is a good place to shop for price.

Labor costs hold the most promise, but also the greatest uncertainty. Other than outright errors, you can adjust several things. Obviously, you can rethink certain tasks and reduce the total number of hours to complete them. You can also plan to use less skilled and therefore less expensive labor for certain kinds of work, but this may compromise quality and won't always save money in the end. Another possibility is to reduce the number of workers on the site. This will probably increase efficiency, but will extend the schedule.

Subcontractor prices often vary by wide margins and offer another place to save. Some builders play subs against each other to get them to reduce their price, but my experience is that this will catch up to you sooner or later. A better approach is to either negotiate price reductions with a sub you've worked with before, or take on some of the sub's responsibilities inhouse (demolition and prep work, for example). The other option — accepting a low price from an unknown — is riskier, and is strictly a judgment call.

Bid Day

Before you call it a night, organize your materials for the next morning when things will get hectic. Make a list of any phone calls you will need to make to firm up prices. The list of subs you compiled last week now moves to the top of the pile because tomorrow they're all going to call you and you'd better be ready.

When the subs finally phone in their prices, ask the same questions of everyone, and take notes on their answers. Be sure you know what each quote includes and what it leaves out. This is particularly important where trades overlap, as when the heating sub assumes the electrical sub will wire up the furnace.

Sometimes you'll get a "split quote," where a single sub will give two prices, one if he does the whole job, and another for each of the individual parts. This is not uncommon for things like large heating or air-handling systems. Your lowest price might be some combination of quotes from three or four bidders for separate parts of the job, but this isn't always the best route. Do you really want to deal with three or four subs on work that one could do, even if it is at a slightly higher cost?

The Opening

Most bids are delivered in person, and this can present problems. Subs often wait until the last minute to phone in their quotes (giving you virtually no time to negotiate), so you may not have all of their prices until an hour before the bids are due to be opened. If the architect's office is forty-five minutes away, that's cutting it close. One solution is to send somebody ahead with the bid form in an unsealed envelope, and with all of the important numbers left out. You stay at the office to take phone calls from subs, and when you've made your choices, call your accomplice at the other end. Another possibility is to stop on the way to the bid opening to make a few last-minute calls. In either case, don't fill in the prices until the last moment.

Stay for the opening if you can, and record the other bidders' prices as the envelopes are opened. The information may come in handy in future contests against the same companies. Another reason to stay is to make yourself available to answer questions or maybe ask a few of your own. Your presence *may* give you an opportunity to affect the outcome.

For example, the bids may be so close that the owner will, in effect, be able to take his pick. Some owners think they have to take the lowest bid, no matter what, and that the bidding is over once the envelopes are opened. In many cases, neither is true. You might point out that price isn't everything, and suggest that the owner consider other elements, such as the construction schedule. If the bids are far apart, you may put a bug in the architect's ear about checking to see who followed the specs and who didn't.

But it's worth staying just to find out who won. You've put in all that work; why leave at the finale?

Even if you lose the bid, it's never a complete waste. You've learned something about your business and your competitors that you can use to your advantage. There's always next time. ■

Sal Alfano, formerly a general contractor in East Calais, Vt., is an associate editor with The Journal of Light Construction.

Using Allowances To Cover Your Costs

Include allowances in your bid to reduce the risk of "hidden costs" and to help arrive at a more accurate price and schedule

by Sal Alfano

Often you have to begin construction before you know exactly what it is you're building. In situations like this, pricing certain items "on allowance" can save the day, as long as all parties properly understand the concept.

An allowance is useful when pricing work whose extent remains unknown. Soil tests may have determined the existence of concealed ledge, for example, but how much ledge is there and how it will interfere with the foundation may remain unknown until you actually start digging.

Another common situation develops when certain items are included in the drawings and the contract, but not in the specifications. For instance, the owner may have decided on the location and quantity of kitchen cabinets, but not on what kind of cabinet to use. The same is often true of lighting fixtures: they're in the drawings, but no one knows what kind of fixture goes where.

In almost every case, using allowances is preferable to simply leaving out the items that are still undetermined. This lets construction begin before all of the details are nailed down.

An allowance gives the owner a fairly accurate idea of the actual cost of the project, thus reducing the magnitude of "hidden costs." And it gives the contractor a more complete idea of the scope of the job, permitting accurate pricing and scheduling.

What Type of Allowance?

Allowances come in several flavors. A *material allowance* affects only material cost. The labor to install, as well as overhead and profit, are figured into the base price agreed upon in the contract. This kind of allowance is commonly used when the class of material is known, but the type is not yet determined.

The owner may, for example, know that the living room will be carpeted, but hasn't chosen a carpet. A material allowance of, say, $15 per square yard is included in the contract, subject to change when the specific carpet is chosen. Since carpet is carpet, installation costs are fixed, and are included as part of the base price.

A *labor allowance* is similar: only labor is affected. In the case of concealed ledge, it may be impossible to know how many cubic yards of ledge will have to be removed, but it is possible to calculate the cost of removing a single cubic yard. Consequently, an allowance price of, for instance, $60 per cubic yard can be included in the contract.

The final two types of allowance prices are a combination of the first two. The *material and installation allowance* includes both labor and material, leaving only overhead and profit in the base price. The *total-cost allowance* includes everything — labor, material, overhead and profit — and nothing is carried in the base price.

Every allowance price eventually is transformed into a change order. When the cabinets and light fixtures and carpet finally are chosen, and the backhoe crunches into solid rock, the allowance price sheds its skin and changes into a hard price.

This price change has important consequences. A material allowance will allow change only with respect to the material costs. If the carpet costs $10 per square yard instead of the $15 per square yard allowed, a change order showing a net deduction of $5 per square yard will be written. Installation costs will remain the same.

Overhead and Profit

Which type of allowance is best for the builder? The total-cost allowance is best, and here's why.

In the other three types of allowances, your overhead and profit are fixed and are unaffected by the change order. Since your overhead and profit cover, among other things, certain kinds of losses that can occur on any job, your exposure to these losses can increase dramatically if the material choices made by the owner are priced well above what was allowed. And experience teaches that in most cases allowances are low: change orders show an *increase* in costs, not a decrease.

Take kitchen cabinets again. Several years ago, I entered into a contract that had a $5,000 total cost allowance for all kitchen cabinets and bathroom vanities. The cabinets that the owner ultimately chose actually cost more than $15,000, not because there were more cabinets, but because each one was more expensive.

The problem is that it's not uncommon to damage a cabinet during installation, and the only place to recoup the loss is in overhead and profit. Unless overhead and profit are tied to the actual cost of the allowance items, you are taking a big risk. Had I been operating under a simple material allowance, I would have been three times more vulnerable than I thought I was when I signed the contract.

Liability and property-damage insurance are also affected. Since your work product is almost always excluded from any claims adjustment, you will have to repair or replace it out-of-pocket if damage occurs through your negligence. Again, overhead and profit are your chief resources. If the use of allowance prices is increasing your gross receipts without a proportional increase in overhead and profit, you are selling yourself short.

The loss need not occur on the job where a particular allowance is used, because your overhead and profit are cumulative, job to job, season to season. It doesn't matter when the loss occurs or what your volume is for a particular year: you still are liable for all completed work, year after year.

An allowance price that fixes overhead and profit before actual costs are determined also can cost you money if the work covered by the allowance causes scheduling delays. Delays make your crews less productive and increase your management time while you shuffle schedules. And delays can carry over from the current project to all future projects that season. A modest allowance can blossom into a high-priced change order, and you should be compen-

Typical Allowance Clauses

Material Only

"***Carpeting:*** Carpeted areas are shown on drawings. Allow $25.00 per yard for carpet and pad material only. Include installation costs in the base bid."

Labor Only

"***Appliances: Cooktop with Downdraft Venting:*** To be provided by owner. Contractor allow $200.00 for installation and hook-up."

Material and Labor

"***Cabinetry:*** To be selected by owner as indicated on drawings. Allow $3,000.00 for materials and installation of kitchen cabinets, and $1,250.00 for materials and installation of bathroom cabinets."

Total Cost

"***Fuel System:*** Provide an allowance of $3,000.00 for installation of the underground fuel tank by the appropriate utility. Allowance shall include material, labor, and all overhead and profit."

sated for any extra time it takes to complete the work.

A good example of this is a staircase that I once built on the basis of an extremely low total-cost allowance. It was low because we signed the contract before the architect and the owner had any idea of what they wanted. The final design and detailing approved by the owner quadrupled the allowance price. The stair itself required three subcontractors, as well as labor from my own crew, and it took a great deal longer than expected, forcing us to revise the schedule. Had the allowance fixed my overhead and profit, the change order never would have covered these extra costs.

Finally, adjusting the allowance price always involves extra management time for the contractor. After all, the reason the carpet wasn't chosen was that nobody took the time to show samples. That job will invariably fall to you, and you ought to be able to include those charges in the change order.

The allowance is a necessary and practical solution to common problems in contract specification. But, depending on the nature of the work affected, the type of allowance used can have serious consequences for the builder.

Be careful to include in the base price all costs that are excluded from the allowance, and negotiate to change the type or amount of the allowance if you foresee any problems. ■

Sal Alfano, formerly a general contractor in East Calais, Vt., is an associate editor with The Journal of Light Construction.

Working For Costs Plus a Fixed Fee

The cost-plus-fixed-fee contract fairly divides risks and rewards between you and your client

by John Sylvestre

Every contract divides the risks and benefits of a transaction between the two parties. The trick to a fair contract is finding a way to divide both risks and benefits evenly. Some building and remodeling contracts come closer to this ideal than others.

The lump sum contract, for instance, in which the contractor agrees to do the job for a given price, places all of the risk on the contractor. It is the contractor who takes the hit if costs or schedules get out of whack, even if factors are beyond his or her control.

The most common alternative to the lump sum contract is the time-and-materials contract, also called the cost-plus contract, since it includes charges for time and materials, plus a percentage markup for overhead and profit. As costs increase, so do the contractor's profits. Seen from the client's point of view, this shifts the risk too far the other way: If the job takes longer or requires more materials than expected, for whatever reason, the client reaches deeper, while the contractor profits the more.

A Sane Alternative

It's for these reasons that many builders and remodelers prefer a third type of contract, the cost-plus-fixed-fee contract. This approach divides the risks more evenly between contractor and client. Basically, it works like this:

The owner is charged for all direct costs — material, the contractor's hourly labor costs, and subcontractor and other fees — at actual invoice cost. The contractor's profit comes from a fixed fee, which is negotiated as part of the contract.

This fee represents the contractor's sole source of profit. It is therefore in the contractor's best interest to finish the project as soon as possible at the least direct cost. A speedy finish produces the maximum profit for the time spent, and allows the contractor to get on to the next job. At the same time, the owner bears less risk, paying only for what is actually used, with a contractor's fee that does not go up if material or labor costs increase.

Eliminating fudge factors. Perhaps the nicest thing about these contracts is that they let both parties benefit from the elimination of the protective fudge factors that most contractors include in bids (or omit at their peril). For starters, you can safely negotiate a fee that is smaller than you would ordinarily accept for a given job, because your fee is guaranteed and can't be eroded by cost overruns.

In addition, you can omit from the bid many of the markups you include to protect yourself. For instance, for ordinary lump sum bids, I usually include a 10% to 15% labor contingency, as well as a similar lumber contingency, to guard against price fluctuations or overruns.

I also generally include a percentage, again varying with job complexity, to cover subcontractor fee overruns. This pays for the little tasks that subcontractors might have overlooked when making up their bids, and for which they may want additional payment at invoice time — things such as the drywall crack in the back bedroom that needs to be patched, mudded, and painted, or the extra three-way switch in the hallway. I don't want to go back to the owner for these near the end of the job, so I'll build them into a lump-sum bid.

In lump-sum bids, I need these markups to protect me from unexpected increases in labor or material costs. But in a cost-plus-fixed-fee bid, I don't need or include any of these percentages, since I know that such increases will be covered by the owner. My motivation for preventing overruns, on the other hand, is customer satisfaction and my own interest in seeing the job completed on schedule and within budget.

Trust me. These contracts work best when you are able to establish a trusting relationship with the client — that is, when the client is one who can see that you have nothing to gain from time and cost overruns, and therefore trusts you to keep them in check. Other independent business people, for instance, tend to understand that you are honestly trying to fairly balance the risks and make a reasonable profit.

Some clients, on the other hand, simply aren't ready to share the risks. Scared by horror stories about shady contractors, they feel safe only if the contractor carries all the risks, even if they have to pay a little more to not carry any themselves. With these folks, I usually end up giving a lump sum bid.

Change Orders

The cost-plus-fixed-fee contract also creates a fair and orderly way to deal with change orders. We include a 110% change order allowance in our contracts — that is, the owners can change or add orders that add up to 10% of the original estimated cost of the project before we assess further fees. (The owner, of course, picks up the actual increase in costs.) Once costs exceed 110% of the original estimate, we collect a 25% markup on the excess.

This gives the owners reasonable leeway to change their laminate to Corian or choose a nicer baseboard, but compensates us if their changes keep us on the job much longer than expected. It's important, of course, to remind the owners that every change order does increase costs, and to let them know if they are nearing their 110% mark.

Contract Essentials

In general, cost-plus-fixed-fee contracts must carry all the safeguards that ordinary lump sum contracts do. These include:

- a detailed list of the scope of work contracted;
- detailed material and design specifications;
- a detailed list of any materials the owner has agreed to supply — appliances, for instance, or wallpaper, or that great sink they saw in Kankakee.

In addition, there are some essentials, peculiar to the cost-plus-fixed-fee contract, that we always include:

- a detailed list of the expected costs, both materials and labor;
- an exact fee for our services, to be charged on top of the reimbursement for our labor, material, subcontractor, and other costs.

As a courtesy, we update the cost estimate with each payment, letting the owner know exactly how the costs are progressing. Sharing the risks means sharing the facts.

Cost-plus-fixed-fee contracts address the interests of both parties: The client gets the best job at the best price, and the contractor makes a safe and reasonable profit for doing a good job. This sense of mutual benefit can help create that rare sense of teamwork between contractor and client — when both feel they are working not as adversaries, but together toward a common goal. ■

John Sylvestre runs Sylvestre Construction, Inc., a design/build remodeling firm in Minneapolis, Minn.

Chapter 7 • Contracts

A Plain-Language Construction Contract

A simplified contract can protect you from pitfalls on small jobs

by Thomas O'Brien and Paul Barbahen

"Get it in writing" is a pretty good rule of business. For a variety of reasons, however, contractors often ignore this adage. The job is too small, or you know the client; or, maybe you think it's not worth the time or money to pay an attorney to draft a proper contract. But informal documents, verbal agreements, and handshakes can break down and lead to conflicts or courtroom battles.

A written contract that anticipates potential areas of conflict helps prevent later misunderstandings. On smaller jobs, the contract needn't be pages thick with dense legal lingo. A simple, "plain-language" contract will serve just as well. In fact, the clearer the terms are, the more likely it is that you and your client are starting out with a common set of expectations.

For this purpose, we've developed a sample "plain-language" contract ("Sample Construction Contract,"pages 89-92). With minor changes, the sample contract can be used for new or existing construction. The contract is designed for use in Illinois, where we practice construction law. Go over the suggestions in this article with an attorney licensed in your own state: Each state's construction laws are a little different.

The Construction Contract

In any project beyond handyman dollars, you'll want your contract to have a general description of the work, right in the body of the contract, even if plans are attached. The payment terms should be clear. The insurance provisions and completion dates should be clear as well.

1. ***General.*** The contract must provide an accurate written description of the work, with plans, specifications, and a visual representation of the work to be done. The specifications should describe the quality of the materials (for example, 1/2-inch, CDX ply sheathing; 3-tab fiberglass shingles). Drawings should be clear. Simple drawings are better than none because they reduce possible misunderstandings.

If you are using a third party's documents, such as an architect's plans, you have a duty to look them over carefully when you bid. The courts frequently side with the owner if you err in your bid.

Be sure to have adequate language regarding change orders and extras, and simple language alerting the owner to unforeseen conditions.

2. ***Price.*** The price should be crystal clear and set out on the very first page of the document.

3. ***Starting and completion.*** For new construction, you should spell out starting and target completion dates, along with permitted reasons for delay. Contractors can still get owners to sign contracts without dates, but this frequently cuts two ways. The owners say X was a reasonable date, and the contractor says X plus 30. Since the law presumes a "reasonable time" where no precise time is set out, failing to use dates only invites litigation, unless there truly is an understanding that the project may proceed at its own pace.

Small companies frequently don't have as much experience scheduling and coordinating jobs as they do with the construction itself. But even a small company can meet completion dates by agreeing only to realistic time frames, making sure the crew size is adequate, making allowances for weather delays, and most important, communicating with the owners about the many legitimate reasons for delays. Owners will usually accept that a rough-in plumbing inspection is holding things up, if they are told. Owners also have to accept responsibility if they cause delays by not making materials' choices on time.

Related to completion-date terms are penalty clauses, usually in dollars per day, for late completion, and bonus/incentive clauses, also in dollars per day, for early completion. While these are primarily used in bigger projects, they are often asked about by owners anxious to establish a move-in date.

If possible, avoid penalty clauses, since they create an area of dispute, and few of us want to admit fault in a late completion case. You might try leaving in a bonus/incentive clause only for early completion: it will probably be received well by owners who have heard their share of construction horror stories.

4. Permits and codes; compliance with local law. A clause reflecting the laws of your state in this regard may be required by statute and is always appropriate. Obviously you have to do work according to codes. Some states, such as Illinois, still give unpaid material suppliers and subs the right to place a lien on property and be paid, even if the general contractor has been paid. It also is a sign of professionalism and invites a discussion with the owners about how the payouts they make will be applied. You should review not only the building codes, but the lien laws as well, in order to protect yourself and prevent possible problems.

5. Specific requirements for materials and workmanship. We have put enough space in the sample form to provide specific details. This makes the form more flexible, since not every project calls for separate specifications. Also, many owners can describe in a few words what they want, based on work they've seen on a nearby project. Even when the particulars are repeated on a plan or spec sheet, a few extra words are better than a misunderstanding.

Allowance provisions are common in new home construction. Fixed dollar allowances permit the owner to select appliances, colors, carpet, or light fixtures. Allowances work well unless they are unrealistically low. Large developers usually limit the buyer to a fixed number of choices.

6. Payment. Owners expect to make a reasonable down payment, often as much as 25% of the project's total price. In at least one state, consumer laws limit front money by law to 25%. You must know any such law in your area and incorporate its terms in your documents.

Most owners will willingly sign a contract with payments keyed to developments they can see, such as a portion after excavation and foundation work, a portion after framing, etc. But problems often arise when the owners refuse to make a final payout, based on a minor punch list. Key your entitlement to the last payment to a Certificate of Occupancy, or some other objective standard, rather than to the owner's satisfaction.

A key phrase in this section is "time is of the essence." This time-tested legal phrase permits the contractor to stop work without being in default if the payment due is not made when due.

7. Signature. Sometimes work starts without a signed contract. Since your rights under your local lien laws may be limited to the "reasonable value" of the work unless there is a contract with the value agreed upon, an unsigned contract is worthless.

General Conditions

The second part of most construction contracts is called the "General Conditions." These are the things that stay the same from job to job. As in the main body of the contract, there may be differences from state to state. You can have the "General Conditions" reduced to illegible "fine print," but that is a mistake. Go over these provisions with your client.

1. Contractor's duties. This section spells out who is going to be responsible for what. Two of the most important points are

continued on page 93

Sample Construction Contract

This contract, dated_______________, is by and between the following owner and contractor:
Owner:________________Property address____
City, State, Zip ___________________________
Telephones (days) ____________(eves)____________Mailing address (if different)____________________
____
Contractor:__________________________Representative:____
Street___________________________City, State, Zip___________________Telephone___________________

1. GENERAL
This contract is for the following work and materials to be performed by the contractor on the property address above. This project is for new construction. (Strike if not applicable.) The project is generally described as follows:
____
____

The contract consists of this document, any plans or specifications or exhibits referenced herein, and the General Conditions following the signature page. (Identify here any plans or drawings, with enough specificity to show which attachments are included in the contract.)

Change orders and modifications shall be in writing and shall become part of this contract.

2. PRICE
The total price for the work agreed upon is $_________. Payment terms are set out below in Paragraph 6.

3. STARTING AND COMPLETION PROVISIONS
The work will begin on ___________________________ , and will be completed, absent unusual circumstances, on ____.
(Optional) Owner agrees that if the work is completed earlier than the above date, a bonus of $______ per day will be paid. A penalty of $___________ per day will be deducted from the final payment for each day beyond the completion date that the work remains incomplete.

4. PERMITS AND APPLICABLE CODES; COMPLIANCE WITH LOCAL LAW
A. All work to be done under this contract will be in accordance with the building codes presently in force in the City of ____________________, ________________ County, State.
Contractor shall obtain all necessary permits and pay all required permit and plan fees from the contract sum. Unless otherwise agreed, all plan and permit fees shall be paid by Contractor from the down payment paid by Owner.
B. Contractor shall at all times comply with the laws of this state regarding mechanic's liens.

5. SPECIFIC REQUIREMENTS FOR MATERIALS AND WORKMANSHIP
A. This contract will be completed by Contractor in a good and workmanlike manner, using good quality materials. The parties agree upon the following materials specifications and work description, together with any plans or specifications incorporated herein:
windows ___________________________________ wall construction ______________________________
roof detail ___________________________________ cabinets ___________________________________
fixtures____
other as applicable____
B. If applicable, the contract price includes the following allowances:
kitchen cabinets ____
bathroom tub, toilet, and vanity____
other as applicable ____

6. PAYMENT
A. Timely payment by Owner of all sums due under this contract is of the essence to this contract. The parties agree to the following schedule of payments:
1. Initial payment: ____

2. Progress payments:

stage of work	date expected	amount
a. ______________	______________	______________
b. ______________	______________	______________
c. ______________	______________	______________

Contractor shall provide Owner with his own waiver or cumulative subcontractors' waivers equal to the amount paid for any progress payment.

B. Contractor may cease operations if any progress payment is not made by Owner as required herein, and proceed to collect any balance due with any legal remedy. Alternatively, Contractor may continue operations, as set forth in the attached General Conditions.

7. SIGNATURE

Attached hereto are General Conditions governing the rights and obligations of the parties to this contract. The parties are further subject to the laws of this state governing contracts and mechanics' liens.

IN WITNESS WHEREOF, we have hereunto set out hands and seals this ____________ day of _________, 1992.

______________________________ ______________________________

Owner Contractor

GENERAL CONDITIONS

These General Conditions are part of the contract between ______________________ ("Owner") and ______________________ ("Contractor") for work at ____________________ _____________, ______________________, (dated) ______________________.

1. CONTRACTOR'S DUTIES — GENERAL

A. To direct and control the work contracted for in accordance with the terms of this contract and all applicable codes, laws, and regulations, and as the building permits, if any, issued for this project require.

B. To inspect the site, examine the plans and specifications, if any, and supervise all of Contractor's employees, and to direct the work of all subcontractors selected by Contractor.

C. To maintain the work site in a safe and clean condition, to the extent consistent with the contract.

D. To advise Owner promptly if concealed conditions are ascertained which require additional or different work, and to proceed in such event in accordance with this agreement.

E. To provide locked storage for any equipment, tools, or other property used in the performance of this contract, unless otherwise agreed upon in writing.

2. OWNER'S DUTIES — GENERAL

A. To provide adequate utilities for the work agreed upon.

B. To advise Contractor of any condition of the property which affects Contractor's ability to perform.

C. To provide secure storage areas for materials delivered to the work site.

D. To execute in a timely manner all permit applications and other documents necessary for the work to proceed.

E. To perform no work on the project without a written agreement with Contractor.

F. To avoid interfering with workers.

G. To make no agreements with any tradesperson, subcontractor, or Contractor's employee outside the scope of this contract without the written consent of Contractor.

H. Owner shall be entitled to make periodic inspections of the work site when accompanied by a representative of Contractor, provided such inspections do not interfere with the work and can, in the sole judgment of Contractor, be made safely. Any other entry onto the construction site shall be at Owner's risk.

I. Owner shall notify his insurance agent of the execution of this Agreement and obtain any necessary Riders to his current coverage or any locally customary forms of coverage, such as Builder's Risk, to cover Owner's interests and liabilities during the construction process.

3. CLOSING (new construction/with sale only)

This transaction shall close within 10 days of the contractor's notification to the purchaser that the property is ready for occupancy. The closing shall take place at a location agreed to by the parties or at the office of the title company. The parties agree that the home shall be ready for occupancy at such time as a Certificate of Occupancy is issued by the appropriate government agency.

As to those minor finish matters which may not be completed at closing, a pre-closing inspection shall take place and any such unfinished matters noted. Contractor shall complete any such items within _____ days of closing: as to such matters, Owner's agreement to close does not constitute acceptance of such unfinished work which has been identified.

4. DEED (new construction/with sale only)
The Contractor shall deliver his stamped Warranty or Trustee's Deed conveying the subject property to the Purchaser. This Deed shall be subject only to real estate taxes for the current year that are not yet due and owing, easements and restrictions of record, building and zoning laws, highways, and rights of way.

5. TITLE INSURANCE (new construction/with sale only)
At least three business days prior to closing, Contractor shall deliver to Purchaser, at Contractor's expense, a commitment for ALTA Extended Coverage Owners Title Insurance Policy in an amount covering the total purchase price of the property, including extras, but excluding any personal property. Said commitment may be subject to matters to which the conveyance will be subject.

6. MATERIAL SUBSTITUTION
Contractor reserves the right to substitute other materials, products and/or labor, of similar, equal or superior quality, utility, or color. Contractor reserves the right to make alterations to the heating and/or cooling system, provided any such substitution or alteration has comparable durability and performance characteristics. In the event of the substitution of any appliance or heating equipment, the warranty terms of the substituted materials shall be equal to those originally specified unless Owner otherwise agrees in writing.

7. DELAY
Contractor shall not be responsible for delays caused by events beyond the control of Contractor, including but not limited to: strikes, war, acts of God, riots, governmental regulations and restrictions. Delays caused by Owner's failure to make allowance materials' selections or caused by the performance by Contractor of extras or necessary work (as described in Paragraph 9) shall likewise be excusable delays.

8. INSURANCE
Contractor agrees to maintain all necessary forms of insurance to protect Owner from liability for any occurrence arising from the performance of this contract. Contractor agrees that he shall cover his own employees for work er's compensation and carry general liability, and that all forms of insurance carried hereunder shall be with reputable companies licensed to do business in this state.
Owner agrees to carry full coverage on the subject property covering Owner's risk of loss during the construction period, together with all special forms required by reason of the performance of this contract. Specifically, Owner shall contact Owner's insurance agent and secure any necessary Builder's Risk coverage prior to the commencement of the work.

9. HIDDEN, CONCEALED and UNFORESEEABLE CONDITIONS
The parties agree that in the event Contractor discovers a condition requiring an extra cost that they shall proceed as follows: Contractor shall notify Owner verbally at once to expedite agreement as to the charge to correct or cure such condition, and provide a written estimate as soon as practicable. The parties must agree to such extra charges, or agree to a resolution method, or this contract may be canceled by either of them.
For purposes of this section, a "hidden, concealed and unforeseeable condition" shall mean a condition not readily observable to a prudent contractor inspecting the subject property for the purpose of performing this contract.

10. EXTRAS
Any extra work or materials desired by Owner shall be agreed upon in writing and such extras shall become a part of this contract. Unless otherwise agreed, extras shall be paid for as performed. Failure of Owner to sign an extras order shall not preclude recovery for same by Contractor, and acceptance of said extra work or materials shall be presumed, unless there is written notice to the contrary.
Contractor shall advise Owner at the time of agreement on an extra as to any additional time required to perform this contract.

11. SUBCONTRACTORS
A. Contractor shall select subcontractors as required to complete this contract. Owner acknowledges that various portions of the work will be done by subcontractors. Any subcontractor selected by Contractor shall have all requisite licenses for the work to be done by such subcontractor, and Contractor shall issue subcontracts in

writing whose specifications are consistent with this agreement.
B. It shall be the duty of Contractor to use reasonable care in the selection of subcontractors. Absent objectionable performance by any subcontractor, the selection of subcontractors shall be with the contractor exclusively. Contractor shall require all subcontractors to have such types of insurance in force as are required to hold harmless and indemnify Owner from any claim for injuries or property damage by any agent or employee of any subcontractor.
C. Contractor shall pay subcontractors on a timely basis and obtain from subcontractors any necessary documentation required to release their lien rights, if any, as the work proceeds.
D. Contractor shall exercise reasonable care in the selection of materials used by subcontractors, but shall not be responsible for later discovered materials' defects or damages from installation methods, not reasonably ascertainable at the time of installation.

12. ARBITRATION

This contract shall (not) be subject to binding arbitration by the ____.
Such arbitration shall be (non) binding on the parties hereto.

13. TERMINATION and CANCELLATION

Contractor may terminate and cancel this contract if any payment called for hereunder is not received as scheduled, provided that notice is given to Owner as provided below. Upon such termination, Contractor shall have all remedies provided by law, including such lien rights as then apply.
Owner may terminate this contract upon the following conditions:

A. Failure of Contractor, or his subcontractors, to pursue the work contracted for, absent excusable delay, as provided in Paragraph 7 above, for a continuous period of ____ days, without a written agreement permitting same, which may be satisfied by a simple notation to this agreement.
B. Failure of Contractor to rectify any condition regarding which building code enforcement authority has issued a citation or violation notice, within ____ days' notice of such violation, unless Owner and Contractor otherwise agree.
C. Any other failure to perform this contract required by the terms of this contract.
D. No termination shall be effective unless 10 days notice of Owner's intent are given as provided below, during which time the default may be cured by Contractor.

14. WARRANTIES

A. The work of Contractor, including materials and labor, shall be guaranteed for a period of ____ years, during which period Contractor shall at his own expense correct any defect arising from his work unless Paragraph 11(D) of these General Conditions applies. This provision is in lieu of all other warranties, express or implied, and Owner has no action at law or in equity against Contractor after said date.
B. Any and all warranties for appliances or mechanical systems shall be delivered to Owner when Contractor's final payment is received.
C. Notwithstanding any manufacturer's warranty of any component, appliance, or system, no action may be brought against Contractor on this contract, for the performance of the work, except as provided above.

15. NOTICES

Notices may be sent to either party at the addresses shown above, or mailed by certified or registered mail. Any mailed notice shall be deemed given as of the date of mailing.

16. SEVERABILITY

If any portion of this agreement is found invalid or unenforceable by any court, the remaining provisions shall remain in force between the parties.

17. ENTIRE AGREEMENT

This contract consists of the documents defined above, and constitutes the entire agreement of the parties. It can be modified only by a written document.

IN WITNESS HEREOF, we have hereunto set out hands and seals this ______________________ day of ______________________, 1992, at ______________________________, State.

______________________________ ______________________________
Owner Contractor

continued from page 88
D) to notify the owners about unforeseen conditions and E) to provide locked storage. You don't want the owners to worry about their child getting into your tools.

2. *Owners' duties.* The model contract contains self-explanatory owners' duties. While the owners may be under the impression that they have no obligations beyond the payment of the construction bill, this is incorrect. Owners are involved in selection of materials, negotiation of extras, and final approval of the project.

Owners today frequently want to participate in construction. This often sounds okay until you consider your liability. Formalize any arrangement about client labor, in writing. Any claims against you should be expressly waived by the owners.

3. *Closing (new construction only).* You are entitled to close on a spec house as soon as you finish construction and receive a certificate of occupancy. If a "perfect home" standard is specified, you may never get agreement that the house is perfect. Most owners, however, will sign a clause that they are obligated to close when the building inspector says the structure is habitable.

4. *Deed (new construction only).* If you are developing and selling new homes, your local title insurance company or escrow service probably handles the legal work on the deed. Select a deed clause that is customary in your area.

5. *Title insurance (new construction only).* Any sale requires title insurance acceptable to the buyer's lender. Most title insurers are members of the American Land Title Association and use a standard title insurance policy. Loans are often resold to investors who want to see a standard title insurance, not an abstract. If you build spec houses, you will get a builder's rate from your local title company and a great deal of valuable assistance processing lien waivers and releases.

6. *Material substitution.* The "substitution clause" allows a contractor to substitute materials that are in substantial conformance with the materials set forth in the contract. That way, if you get a price break from a supplier of a comparable brand, you can switch.

Be careful, however, about switching brands of major components on an owner without notice. If an architect has specified a particular Lennox air conditioning unit and you think Carrier is better, the prudent course of action is to get the buyer's consent, regardless of equivalency.

7. *Delay.* The "acts of God clause" protects you from liability for acts or events that are beyond your control. These include war, riot, lightning, earthquake, tornado, strikes, and the like.

Owners must be conscious that extras and changes will cause delays, as will their failure to select materials on schedule. Some General Conditions expressly include "owner-originated" delays.

8. *Insurance.* Insurance is a hot topic among contractors. It costs too much and covers too little. Buy whatever you need for the work you do. Not paying for insurance is the world's worst idea for saving money.

You need to remind owners to get "builder's risk" or owners' risk policies. These cover losses to buildings undergoing substantial improvements. If a fire burns down the building, and the owner hasn't insured the cost of improvements, the owner may not be able to pay for your work.

9. *Hidden, concealed, and unforeseeable conditions.* These clauses alert the owner to necessary cost increases if concealed conditions require the owner to spend more money than the original contract called for. Common remodeling surprises include asbestos, radon, structural defects, and lead plumbing. Owners justifiably react with alarm when told that the new bathroom will cost $1,000 more because the contractor has found (now that the walls are open) that there's no vent stack. You'll have a particularly tough time persuading an adamant owner that you're getting a raw deal if you don't have this provision in the contract. Once the walls are open, try to figure out what's reasonable, and put the resolution in writing.

Remember that the owner thinks you are the expert and should have known what was behind the walls. Take the time to educate the owner up front about this important clause.

10. *Extras.* Most contractors want extras paid for in cash, when ordered. Problems can arise when owners change their minds after giving verbal approval. But stick by your guns on any extras. It's best to get all orders in writing, and get paid before you order.

11. *Subcontractors.* Contractors need to

develop written subcontract forms to maintain orderly scheduling and improve the efficiency of their subs. Too many subcontracts are verbal, and on some smaller jobs, the subs don't see the prime contract and may not have all the information they need to perform their subcontracts properly.

Do not allow owners to pay subs directly, or you may lose control of your job. Be prompt about your payments, however. In those states where subs can file liens, the owner can be directly affected by your failure to pay your subs as required.

12. Arbitration. Arbitration sounds good, but don't agree to it unless you find that it really works in your area. You and the owner can select a dispute resolution mechanism and put it in your contract. However, many contractors feel that the "Better Business Bureau" type of arbitration may be partial to owners. Sometimes the courts provide a more equitable decision.

13. Termination and cancellation. The courts interpret termination and cancellation clauses on a case-by-case basis. The contractor should have a right to stop work if a payment is not made within a specified grace period, and owners have rights when contractors don't show up for days at a time. The law also excuses performance when property is destroyed.

It is hard to draft a clause that adequately compensates a contractor for the owner's default; if the owner isn't paying, chances are the dispute is headed for court. Problems also arise when the owner has reason to fire the contractor, such as for not showing up. It's hard to determine the fair value of the work in place, especially when a second contractor cannot finish the work for the unpaid balance. A contractor would not, however, be in default if the job site flooded.

14. Warranties. By all means use something in writing, or you may find that your state law dictates an implied warranty far longer than you thought. New home contractors often have warranty provisions separate from their contracts. We suggest that you give a reasonable warranty on materials and labor for one year. Thereafter there should be no warranty or a limited warranty. Such limited warranty should indicate that the contractor is not responsible for any labor costs to replace defective materials. Further, the contractor should not give any warranty for materials greater than that given by the manufacturer. Many new materials are installed by small contractors who can least afford the labor costs of replacing them if the materials prove inadequate through no fault of the contractor. If you have doubts about a novel building product, remember that you are on the hook unless an architect specified it.

15. Notices. There should be a notice clause so that it is clear whenever you have properly notified the owner. Sometimes a notice of termination of the contract is not effective because it can't be properly delivered.

16. Severability. These clauses protect you from having an entire contract declared invalid, just because one part of it is invalid. If a lawyer in your state says to use this clause, do so.

17. Entire agreement. These clauses protect you from having some scrap of paper "incorporated" into your contract by a court. Additional provisions to a contract should be formal enough to be recognized as part of the original contract.

Conclusions

A good contract is something like a good car; no single model captures all the good features. Not every clause in our model contract applies to every project, nor does every project require every clause.

Analyze your contracts and contract procedures just as carefully as you do your business plan and your hands-on work. Develop your own forms to eliminate future problems. Bear in mind that each year contractors' lawyers lose cases they should win because the contract is not as clear as it could be, the attached plans and specifications are vague, or the contractor fails to document a change order or an understanding between himself and the owner with a cover letter. While there seems to be too much paper in the world, the one piece you need to win a case is often missing due to a lack of attention to the most basic aspect of contracting — the contract between the parties. ■

Thomas O'Brien and Paul Barbahen are partners in the law firm of O'Brien and Barbahen, in Chicago, Ill., which specializes in construction law.

Contracts: Reading the Red-Flag Clauses

Read and understand these key clauses or proceed at your own peril

by Sam Starobin

How often back in school did we race to give the answer before fully reading the problem? We paid the price in grade points.

Well, we are older now and foolish things like not reading the problem are likely to cost us money. In this case, the problem is the construction contract, and too many contractors accept work under contracts they don't fully understand. When a contractual problem arises, they learn the contract the hard and costly way.

As a former government contracting officer who has administered hundreds of contracts, I have observed a scenario like the following all too often. Contractor X has completed his contract, added up all his costs, and to his great surprise and sorrow, found that he has lost money on the contract. What to do? He assembles his project superintendent and his engineer and reviews the job to see where he can squeeze out the money to make himself whole.

The project superintendent speaks up: "Remember when we started excavating? We bid on clean dirt but we ran into that trash and we had to overexcavate and bring in clean fill." "Great," says Contractor X, "that cost us a bundle. We'll write it up as a claim."

Then the project engineer chimes in. "Remember when we submitted those shop drawings for rebar? We were supposed to get them back in 15 days but it took 45 days, and we had to hold up the job for a week because of it. That cost us money."

The group throws together stories for four or five such episodes, cooks up some costs (since they have kept no accurate records), adds up the total, and breathes a sigh of relief. "We're saved. This makes up our loss."

When their claim finally reached my desk, I had a sad duty to perform. In spite of

Where to Find Red-Flag Clauses

	CONTRACT FORMS			
CLAUSES	**Federal Standard Form 23A**	**Massachussets Div. of Capital Planning and Operations**	**American Institute of Architects (AIA)**	**Engineers Joint Contract Document Committee (EJCDC)**
CHANGES	GP 3	ART VI	ART 12.1 ART 12.4	ART 10 ART 9.5 ART 11.2
DIFFERING SITE CONDITIONS	GP 4	ART VI 5a	ART 12.2	ART 4.3.2 ART 4.2.3
DISPUTES	GP 6	ART VI 3 ART VI 5b ART VI 5c	ART 2.2.9 ART 2.2.12 ART 7.4.1 ART 12.3 ART 7.9	ART 9.11 ART 9.12 ART 6.29 ART 16
SUSPENSION OF WORK	GP 17	ART V 3	ART 3.3 ART 14	ART 15
DELAY, TIME EXTENSION, LIQUIDATED DAMAGES	GP 5	ART V 1c ART V 2c	ART 8.3	ART 12
EXCULPATORY	GP 10 GP 13	ART IV 7 ART V 3	ART 6.2.2	ART 4.3.1.1 ART 4.3.1.2

Use this table to locate the clauses listed at left in the major construction contracts. For example, Differing Site Conditions is covered in Article 12.2 of the AIA contract.

my sympathy for their situation, I could not expend public money on unsubstantiated claims. First, the contract set time limits for submitting claims for various causes, and missing these dates waived the right to compensation. Also, a contractor must be able to substantiate his claim. It is not enough to have a legitimate claim; the contractor must be able to prove it with documentation.

There are ways to avoid the problems faced by Contractor X, however. Assuming that you are working on fixed-price contracts, you are familiar with the General Agreement, the document you sign that sets out the price, the time, and the product of the contract. You have studied the drawings, and read the specifications (hopefully). But do you really understand the General Conditions, that section of dry legalese? This is where many contractors, like Contractor X, trip up.

The General Conditions define the basic relationship of the parties and establish the procedures for dealing with various situations that arise in a project. The provisions of the General Conditions that deal with particularly troublesome areas such as changes, delays, differing site conditions, disputes, and suspension of work, have been called the "red flag clauses." It is vital that you understand these clauses. This article will discuss their content in four basic contract documents, the Federal contract form (GSA Standard Form 23A), the Massachusetts Division of Capital Planning and Operations contract, the American Institute of Architects (AIA) contract, and the Engineer's Joint Contract Documents Committee (EJCDC) contract form, which is the one endorsed by the Association of General Contractors. These four should give you a basis for reading and interpreting other contract documents. The accompanying table shows where these clauses are found in the various contract forms.

The Red Flag Clauses

Changes. Changes are an inevitable part of the construction process and all contract forms establish rules for handling them. In general, they establish the right of the owner to direct changes *within the general scope of the contract*, and to require the contractor to perform those changes. The contractor is required to submit his request for adjustment of time and money within a reasonable time (30 days in the case of the Federal contract). A change request that is beyond the scope of the contract because of its size or scope may constitute a *cardinal change* and call for renegotiation of the contract.

> ***Changes***
>
> "The Contracting Officer may, at any time, without notice to the sureties, by written order designated or indicated to be a change order, make any change in the work within the general scope of the contract..."*
>
> — GSA Standard Form 23A

The tricky part of the changes clauses has to do with the so-called "constructive change": a change in the scope or character of the work that has not been declared by the client or the architect. Let's say you have submitted a shop drawing for a detail of the building structure. The shop drawing comes back from the architect with a note saying that additional structural members must be added in order for the drawing to be approved. It is obvious that the architect has been busy redesigning the structure, which in turn throws off the bid. This is a constructive change. Or you show up on the site and find that a pile of trash that was to have been removed by others is still there and you must remove it in order to start work. This too is a constructive change.

If the contractor feels that a change has been imposed on him, he must serve notice on the owner in a timely manner (within 20 days in the Federal contract), citing the circumstance and requesting a change order. Failure to meet the reporting requirements may forfeit your right to claim compensation. In the case of disputes, the contractor may submit a claim under the disputes provisions of the contract, but must perform the work, unless he believes he is faced with a cardinal change. If you think you are faced with a cardinal change, seek legal advice. This is a complex area of contract interpretation.

An important point: Your right to a

change order does not depend on the financial status of your project. Like Contractor X, some contractors try to make up for losses by submitting a flurry of claims at the end of a project. This rarely works. If you have a right to a change, act on it, but at the proper time. Late may mean never!

Differing Site Conditions

I'm sure you have been there. You start excavating in an area where the boring data promised easy digging and run into a rock outcropping. Or you start renovation of a building and find the piping badly corroded though the contract documents indicate good piping. What are your rights? What do you do to protect them?

All four contract documents used in this discussion make specific provision for adjusting the contract in such cases. All emphasize two requirements; timeliness and documentation. The owner must be notified as soon as possible after the differing site condition is discovered, the notification must be in writing, and any claim for adjustment of contract time or price must be submitted in a timely fashion.

> ***Differing Site Conditions***
>
> "The Contractor shall promptly, and before such conditions are disturbed, notify the Contracting Officer in writing of: (1) Subsurface or latent physical conditions at the site differing materially from those indicated in this contract, or (2) unknown physical conditions at the site, of an unusual nature, differing materially from those ordinarily encountered..."
>
> —GSA Standard Form 23A

The Federal and EJCDC contract forms are more specific. They require the contractor to inform the owner in writing as soon as the condition is discovered. Further, he must not modify those conditions until the owner has had a chance to examine them and determine whether a contract modification is justified. If you change the condition before the owner can observe it, you may lose your right for compensation. In any case, do all you can to document the case. Take lots of photographs. Good documentation will help you to establish your case with the owner, or in a disputes procedure if it comes to that.

Disputes. Construction is a field that provides plenty of opportunity for disagreement. If you are fortunate and have a reasonable owner to deal with, and if you have worked to establish an atmosphere of cooperation, most disagreements will be resolved by negotiation. For those that are not, the contract documents provide for a dispute mechanism.

> ***Disputes***
>
> "Except as otherwise provided in this contract, any dispute concerning a question of fact arising under this contract which is not disposed of by agreement shall be decided by the Contracting Officer..."
>
> —GSA Standard Form 23A

The decision to trigger the dispute mechanism is up to you. The four contract documents used here vary in details but agree in substance. If you feel your attempts to resolve a disagreement have come to a dead end, you must then submit your claim in writing to the owner or his designated agent. He must respond with a decision in writing within a set time. This decision is final, unless you decide to appeal the decision to a higher level. In the Federal system, this higher level is generally a Contracts Review Board. Under the Massachusetts contract, it is a hearings officer for small disputes and the court system for larger disputes. The AIA and EJCDC contract forms call for arbitration.

If you have come to this step, remember that documentation is the key to success. Obtaining justice depends not only on being in the right, it depends even more on being able to prove it.

Suspension of work. The owner has decided on a change in the structure and you have been given an order to suspend work until the redesign is accomplished. Or, you have submitted shop drawings for the next phase of work. They have not been returned and you can't proceed. In either case, there you sit, the money meter running and unable to move. Can they do this to you and what are your rights?

> ***Suspension of Work***
>
> "The Contracting Officer may order the Contractor in writing to suspend, delay, or interrupt all or any part of the work for such period of time as he may determine to be appropriate..."
>
> —GSA Standard Form 23A

All the contract documents authorize the owner to suspend work for various reasons but they provide for compensation for suspensions beyond a certain length. The Federal clause states compensation may be claimed for an *unreasonable* period of delay, whatever that is. Massachusetts is more specific. You can claim compensation for delays beyond 15 days.

You may be forced into a work suspension by an act or failure to act on the part of the owner. This constitutes a constructive suspension. Since the owner has not explicitly directed a suspension, you must inform him that his action or failure to take action has forced you into a suspension. Do this, in writing, as soon as possible. Both the Federal and the Massachusetts contract forms state that the contractor must submit this written justification within 20 days after the act or failure to act that caused the delay. Claims for damage due to the delay must be submitted as soon as possible after the end of the delay.

Delays, time extensions, liquidated damages. Have you ever overrun the contract time? You are a rare contractor if you haven't. Overrunning the set completion date exposes you to financial penalties. Both the Federal and Massachusetts contract forms specify *liquidated damage* as a penalty for overruns. However, the absence of a specified amount for liquidated damages does not remove the threat of financial penalties. There is always the possibility of actual damages being assessed against you.

What protections do you have? All four contract forms recognize the possibility of excusable delay. In general, acts beyond your control may constitute such a delay: strikes, floods, unusually severe weather, acts of God, etc. The important thing in claiming a time extension for such a cause is timeliness. The claim must be submitted soon (ten days in the Federal contract) after the start of the delay.

> ***Delays***
>
> "The contractor's right to proceed shall not be so terminated nor the contractor charged with resulting damage if:
>
> (1) The delay in the completion of the work arises from unforeseeable causes beyond the control and without the fault or negligence of the contractor, including but not restricted to, acts of God, acts of the public enemy, acts of the Government..."
>
> —GSA Standard Form 23A

Exculpatory clauses. This is a fancy term that means the owner shrugs off responsibility for certain situations. A typical exculpatory clause is one that denies the contractor either time extensions or additional compensation regardless of the cause. The contract documents discussed here have no significant clauses of this type since they were drafted with a view to fairness. Many private owners, however, have no such concern for fairness, and strengthen their position with exculpatory clauses. Watch for them and consult with your lawyer as to their validity. Exculpatory clauses have not always held up in court, but you can never be sure. Don't let them sneak up on you unawares.

This discussion should serve as an introductory guide to understanding your contractual position. Contracting is a tough game. It is hard enough to be successful when you know the rules; it may be impossible if you don't. If you are dealing with an unfamiliar contract, make sure you understand the General Conditions. Get a lawyer to help if you have trouble plowing through the legal verbiage. The time and money spent may save you considerable time and money later — and you will sleep much better. ■

Sam Starobin is an engineering and construction manager with Stone & Webster Engineering Corp., in Boston, Mass.

A Practical Approach To Change Orders

by Michael McCutcheon

Avoid change orders whenever possible. When they are necessary, handle them with clear policies and procedures.

Change orders are one of the messiest areas in all of construction. They can be financial and scheduling disasters, and are also a common area of litigation between builders and owners. I've learned from hard experience that without some kind of change-order system, you'll end up working for free. Every builder should develop a system to overcome the obstacles posed by change orders and turn them into profit and referral-making opportunities.

I'm a design/builder, and I've refined my change-order system to fit the way I do business. The guiding principles are: minimize the number of change orders by planning carefully in the design stage, and then handle any changes that do come up with standardized procedures and forms.

Plan Ahead

In general, my advice is to try to avoid change orders if you can. For one thing, they take a lot of time to estimate. You have to talk to the owner and architect, as well as to your foreman, suppliers, and subs. Then you must find the time to calculate the costs, write the change order, check it, give it to the owners, get their reaction, and make any revisions. If it's accepted, you must communicate it carefully to your field people, order any required materials, and reschedule your subs. All in all, it's a big headache with a small financial reward.

Change orders also goof up your schedule. And even the smallest change becomes critical because of the increased attention from the owners. It's important to perform change-order work carefully and on time because it will probably receive more scrutiny than the rest of the work.

Consequently, the best case is to have no change orders at all, and my company does several things to avoid them. First of all, we work on a design/build basis whenever possible. This allows us to prepare our own plans using either an in-house designer or an independent architect. In design/build work, we get to go over the job carefully with the owner during the design phase to make sure that we anticipate as much as possible in the original plan.

Even when the owners want to use their own architect, we get involved in the design process as early as possible. Often, we prepare a budget in conjunction with the architect before plans are drawn, participate in design meetings, and estimate schematic drawings.

But regardless of who develops the design, we insist that the owners purchase complete plans and specs. There is no excuse for a lack of detail in the construction documents, even on design/build projects. The more detail there is in the final drawings and specs, the better we can streamline communication for the whole course of the job. The owners, our employees, and our subs all have a much better idea of exactly what we've agreed to build. And if there is a problem later, we can refer to the plans to decide whether or not the resolution requires a change order.

Another way we minimize change orders is by holding a preconstruction conference with the job foreman, the designer, and the owners. This is typically done the week before the job is scheduled to start. The meeting gives us a chance to go over the scope of the project and to identify any gross misunderstandings before work begins. Sometimes the owners will even tell us at this meeting about extras they are already contemplating.

Finally, we have found that the jobs with the fewest change orders have been those during which the owner was out of town. Owners virtually never make changes when they are out of town, although there are exceptions. On a recent job, the owners called from Europe because they realized in the middle of their trip that they wanted an extra hose bibb on their new deck. I can only imagine what it would have been like had they been at home.

Of course, sometimes changes are inevitable, such as unanticipated termite work or demands made by the inspector. That's why communication between the job site and our office is important. Our foremen constantly monitor the job for possible change orders,

CHANGE ORDER

Date: 12/9/91

Description of Change:

Provide a Velux TPS-4 skylight in the Master Bedroom. This includes framing the opening, patching the roofing and finish materials to match existing, and all labor to supervise and install.

Contract Price Change:

Principal	1.0	hours @	$ 60	$60
Project Manager	1.0	hours @	48	48
Administration	0.5	hours @	30	15
Foreman	12.0	hours @	39	468
Carpenter	0.0	hours @	35	0
Laborer	4.0	hours @	25	100
Velux TPS-4, type-L flashing				$495
Roofing, drywall, paint				50
			Subtotal	$1,275
			Contingency (5%)	$64
			Adjustment to Target	$1,339
			Adjustment to Maximum (+10%)	$1,473

Contract Time Change

Current Completion Date:	2/28/92
Addition of calendar days caused by change:	2
New Completion Date:	3/3/92

Schedule for execution of the Change Order:

This work will begin as soon as approved by the Owner

Please notify our office immediately if you have any questions about this change order.

________________________ ________________________

Michael McCutcheon (Owner's signature)
McCutcheon Construction, Inc.
1280 Sixth Street ____________________
Berkeley, CA 94710 (Date of acceptance)

Figure 1. *A standard change-order form speeds the process and keeps the job moving. A detailed breakdown of costs for the changes will minimize the owner's questions and promote good will.*

even though they don't make changes without office approval. To help them do this, we make sure they have not only a set of plans, but a copy of the contract and final budget, so they can determine when any legitimate changes arise.

Let Owners Lead the Way

It is tempting to use change orders to try to make up for a low initial bid. One local contractor has the reputation of insisting that his foremen come up with change orders amounting to at least 10% of the bid price so he can recover profit given away in his lowball bid. In my view, change orders that are not initiated by the owners can very quickly become annoying to them.

There's usually no controversy when the owners initiate a change. For this reason, we try not to approach the owners for those "nickel and dime" changes that occasionally crop up over the normal course of a job. Why bother them for small changes that they haven't requested? This may sound risky, but it works for us because, on most of our jobs, we bill for time and materials up to a guaranteed maximum. In return for the security of this cap, the owner agrees to pay us half of any savings below a "target" price, which is about 10% below the maximum. That way, when a job goes very well, we get a bonus. This happens on about one-third of our jobs and helps balance the few that go over the maximum. We prefer to work this way because it provides the owners with a cap, but allows for the uncertainties involved in remodeling. Since the target price includes a 5% contingency on top of our tight estimated price, we have built-in flexibility that allows us to deal with minor surprises without hounding the owner for a change order.

Consequently, most change orders we write are for work requested by the owners that changes the scope of the project. For example, people often decide in the middle of a remodeling job to add a skylight. This is substantial enough to warrant a change order, which, when handled properly, can be lucrative (Figure 1).

When the owner asks for a major increase in the scope of the job, we prefer to write a separate contract rather than a change order. For example, we currently have a $45,000 project to which the owner wants to add over $20,000 worth of work. We have told him we will do the additional work only with detailed plans and a separate budget, contract, and schedule. This protects us both from getting in over our heads.

Figuring A Change

We price change orders fairly, and apply the same markup to them as to the rest of the job. We figure them as quickly as possible so things keep rolling. As soon as one of our foremen discovers a need to change the work, he notifies the office. Then we review the contract and estimate to verify that this work is indeed a change. If it is, we find out from the foreman how many man-hours and what materials the change requires, and get prices from any subs involved. Finally, we enter all of this information on a template in the computer.

To avoid problems, all changes should be authorized in writing, but sometimes it is just not possible to have a piece of paper prepared and signed before the change must be made. Recently, we had an owner who wanted to add some lights, but did not decide this until the electrician was at the site and almost finished with the rough wiring. In cases like this, I will accept a verbal agreement after I have worked it out on paper. This helps keep the job on schedule. I still confirm the change in writing as soon as possible.

Change-Order Form

Usually change orders involve an increase in price, but sometimes they involve a credit, as when the owners decide to take carpeting out of the contract and handle it themselves. Either way, we use the same form.

Our change orders include:

- a detailed description of the change;
- the estimate breakdown;
- markup;
- the effect on the schedule;
- the date the change order will take effect;
- a request for the owners to contact us immediately with questions.

Our estimate breakdown includes labor,

Learning the Hard Way

In 1983, I did my first large job, an extensive remodel and addition in Oakland for very wealthy clients. Because they liked to fly into town (in their own jet) to spend an occasional day or two in San Francisco, they wanted to upgrade their Oakland home.

Naturally, I was excited by the job and the prospect of working with people for whom cost would be no object. Well, I learned that some people have more savvy about money than I do. I got a big-league education from these people about change orders.

They started off wanting the addition and remodel — perhaps $50,000 worth of work at the time. But soon they started asking me for "favors." Would I mind coordinating the installation of the carpet they were picking out at the design center in San Francisco? And could I help them replace the roofing on the existing house? After all, we already had a roofer at the site. By the time all was said and done, we had performed over $20,000 of extras, for which I never charged them any fee. We did it all at cost.

When I added up the damage at the end of the job and it finally hit me what had happened, I vowed never again to let being a "nice guy" prevent me from charging a reasonable fee for extras. The kicker was that these people manipulated me into saving them thousands of dollars in legitimate supervision fees, and never even said thank you. — M. M.

material, and subcontractor costs. And because change orders often require preparation time in the office and additional supervision in the field, we also include any extra administrative time.

Our markup for change orders is the same as for the rest of the job, and we make sure the owner understands this from the outset. It promotes careful planning, discourages frivolous change requests, and ensures that the effort required to make changes is well-compensated. The markup is included for both decreases and increases in the cost of the job as a result of the change.

It's particularly important to remember that a change order affects not only the price of the job, but also the time to complete. Your change-order form must have a provision for adjusting the contract time.

The Bad News

This system has some shortcomings. For one thing, it doesn't work as well with fixed price contracts, which aren't as flexible as cost-plus agreements. Nor is it well-suited to competitive bidding, where the builder has virtually no participation in or control over the design.

Another problem, which is not unique to this system, is that it assumes the owner is honest, reasonable, and fair. Unfortunately, as we all know, not everyone has these attributes. To guard against this, we include specific language in our contracts to deal with change orders (Figure 2). For example, the contract specifies that any change required by a building inspector constitutes an automatic change order. Of course, we try to minimize this sort of thing in the design stage, but when something slips through the cracks, we're covered.

Occasionally, we work with customers who nitpick and won't allow even an obvious change order. One customer was so tough that he would not allow me to write a change order for work demanded by the inspector. He refused to pay, even though it was clearly an automatic change order in our contract. This left us with the difficult choice of either completing the job under unfair circumstances or leaving it halfway. In such cases, I make peace with myself. I know the only way I can be happy is to do the "right" thing even if it costs me some money. So I complete the job and collect as much money as I can.

The Good News

There are several advantages to my approach to change orders. I believe the system is partly responsible for the fact that, while completing over $5 million of work in the last 11 years, I have never had a lawsuit, and have incurred only $3,000 of bad debt. My system allows me to execute required changes quickly, and get on with the job. And the equitable handling of changes relaxes the client, shortens the punchlist, and reduces callbacks.

It also encourages referrals and repeat business. We recently finished a major second-story addition for some people who had heard so many horror stories about remodeling that they dreaded what might happen to them. We worked out a design/build agreement, including an independent architect, and came in ahead of time and under budget. Not only were they delighted with the job, they added about $20,000 in upgrades to enhance the project. This was a case where owner-initiated change orders made the owners happy and allowed us a fair profit.

Changes from the original contract are an opportunity to demonstrate your business ethics, impress your clients with your thoughtfulness and concern, keep your jobs rolling, and make some money in the bargain. Develop a system that works for you and continue to refine it, and you will have solved one of the toughest problems in construction management. ■

Michael McCutcheon is a design/build contractor in Berkeley, Calif. He does mostly remodeling work, with some new construction.

ARTICLE 10 — CHANGES IN THE WORK

10.1 Changes in the Work must be agreed to in writing by both parties. The amount for such extra work shall be determined in advance, if possible, with the Contract Time, Project Target, and Project Maximum adjusted accordingly.

10.2 Any changes, alterations, additions to or omissions from the drawings and specifications that may be required by any public body, utility, or inspector shall constitute a change in the Work and shall be paid for in the same manner as any other change in the Work.

10.3 Full payment for changes in the Work approved by the Owner shall be due and payable upon substantial completion of that particular change in the Work.

Figure 2. *Specific provisions for change orders should be included in every contract. Be sure that your attorney prepares or reviews the language before you use it.*

Who Pays for Job Surprises?

A differing-site-conditions clause can help you avoid paying for unforseen problems

by Michael C. Loulakis and Jeffrey G. Gilmore

Renovation contracts pose a unique problem for a building contractor: In bidding, a contractor cannot tell precisely what hidden conditions may increase the job's cost. But it is possible to ensure that you get compensated for such unexpected work. The best vehicle for this is the inclusion of a *differing site conditions* clause in the contract.

This clause allows the contractor to recover costs if he or she encounters either of two conditions while performing the work. The most common condition, called the "Type I differing site condition," is a latent physical site condition that differs materially from the conditions indicated in the contract documents. The second condition, called the "Type II differing site condition," is an unusual and unknown physical site condition that differs from those generally recognized as inherent in the type of work specified in the contract.

A case that came before the General Services Board of Contract Appeals some years ago, *Alart Plumbing Company, Inc.*, GSBCA Nos. 6487, 6690, 6923, 7123 84-1, BCA Para. 17229 (1984), illustrates how a differing-site-conditions clause can protect a renovation contractor. In the Alart case, the contractor agreed to install a sprinkler system in an existing government building. This called for a main feedline to be installed in a crawlspace above the ceiling.

The drawings indicated the height of the crawlspace above the ceiling, but gave no indication of what, if anything, was in the crawlspace. The contract documents also contained the following clause:

> "The drawings indicate the general arrangement of piping and the location of the sprinkler heads, but are not intended to establish or show actual or relative elevation of piping with respect to other piping structure or ducts, nor are the drawings intended to find exact locations of cross mains and branch lines. The drawings are not intended to relieve the contractor of any responsibility for avoiding conflicts or obstructions or installing the number of sprinklers and supply piping thereto as are required."

The contractor visited the site to inspect the work before bidding, but did not look into the crawlspace because doing so would have required him to cut at least one access hole in a wall or ceiling. The contractor determined from the drawings that one worker could place the main feedline in the crawlspace by moving through it from one end to the other.

Unfortunately, when the contractor started work, he found that the crawlspace contained a series of beams blocking access to the space. The beams prevented a worker from entering the space and running the feedline as planned, so the contractor had to employ a much more costly method of running the lines. Alart sought recovery under the differing site-conditions clause, alleging that the contract documents failed to depict the true conditions of the area above the crawlspace. Despite the government's arguments to the contrary, the General Services Board found that the contractor was entitled to recovery and awarded him the additional costs.

The government's prime defense was that the drawings were silent on what, if anything, was in the crawlspace. Remember that the Type I clause applies when site conditions differ from those "indicated in the contract." Thus the government argued that because the plans were silent about what was in the crawlspace, the Type I clause could not be used.

But the board ruled that this silence indicated that nothing in the crawlspace would preclude the contractor from working as planned. The board also noted that the specifications and drawings "bristled with all sorts of details." Notes were contained on all drawings that alerted the contractor to many local conditions. They even gave the exact height of the crawlspace. Given such exhaustive contract information, the board found that the government's silence about what was in the crawlspace "was as good as a warranty that nothing was there."

The government, however, argued that if the contractor had performed a reasonable site investigation, he would have known that the beams existed. The board rejected this defense, stating that because the building was in use during the prebid inspection, it would have been "unimaginable" for the contractor to have poked holes in the ceiling.

The purpose of a differing-site-conditions clause is to avoid inflating a bid to account for contingencies that may or may not arise. By giving the contractor a means to recover the actual costs of such contingencies, owners benefit from lower, and more realistic, bids. Without such a clause in the contract, renovation contractors should include sufficient contingency factors in their bids to allow for the unexpected.

A further point to remember is that even with a differing-site-conditions clause, a contractor on a renovation project is charged with reasonably inspecting the site and determining the actual conditions. While what constitutes a reasonable site investigation will depend on the facts of a given case, courts look very carefully at the extent of the contractor's site investigation before awarding any recovery based on a differing-site-conditions clause. ■

Michael C. Loulakis and Jeffrey G. Gilmore are lawyers with the firm of Wickwire, Gavin & Gibbs, P.C., specializing in construction and public-contract law.

Chapter 8 • Keeping Your Books

Overhead, Markup, and Profit

by Robert Criner

To make a profit, track your expenses accurately and calculate the correct markup

Ask the average small builder on the street what his net profit is and you might hear something like, "Well, I made $70,000 dollars last year." But, in fact, that number is probably what's left over after paying for materials, labor, and subcontractors. What most builders forget is that the cost of doing business still has to come out of that money. In other words, they forget about their overhead.

But even builders who know what their overhead is often confuse it with the markup they apply to get the final selling price for the jobs they estimate. To know the difference between overhead and markup, you have to understand the relationship between three basic concepts: *gross income, direct costs*, and *indirect costs*.

Cost Categories

Totaling up the dollar amounts for each of these three categories is the first step. You don't have to be a CPA to track these numbers. If you know how to keep a checkbook, you have what it takes to compile the figures you need. But this is precisely where most people fall down. They don't have a bookkeeping system that accurately records their company's expenses, nor the discipline to add up everything at the end of each year and look at the important relationships between income and expenses.

One advantage I have is that my remodeling business is small enough ($300,000 per year) that I can keep an eye on all expenses. But my system for recording costs will also work for companies larger and smaller than mine. The basic idea is to set up specific cost categories and make sure that each expenditure or income transaction is recorded only once, in the proper category.

Gross Income

For small builders like me, income is simple. Most of the money flowing into my business comes from contracts, interest on my checking account, or discounts earned from vendors (Figure 1). The total amount from these and all other sources is called *gross income*.

Direct Costs

Expenses are a bit more complicated and cause the most confusion. There are two kinds of expenses — *direct costs* and *indirect costs* — and you have to keep track of them separately to determine your overhead.

Direct costs are the most obvious. I think of direct costs as *any expense I would not incur if I did not do a specific job.* I keep track of five general categories of direct costs: permit fees, Home Owners Warranty (HOW) premiums, subcontracts, labor, and materials (Figure 2, next page).

Labor includes more than just hourly wages. I also add up all the costs I would *not* incur if I did *not* have field employees. This includes social security tax, federal and state unemployment taxes, and all fringe benefits, like health insurance, vacation pay, and holidays. I also count workers compensation premiums as direct costs because they are linked directly to total payroll. I lump all these labor costs together for

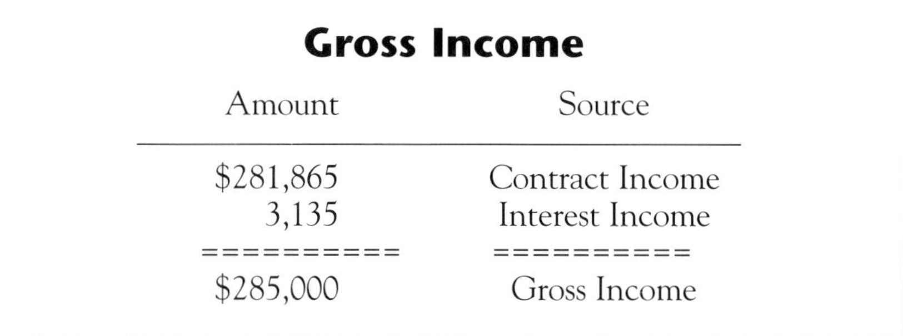

Gross Income

Amount	Source
$281,865	Contract Income
3,135	Interest Income
==========	==========
$285,000	Gross Income

Figure 1. *Gross income is your total income from all sources. If you have more than one kind of business interest—say new construction, remodeling, and a handyman division—you might want to track income and overhead for each separately.*

Where the Money Goes

Indirect Costs (Overhead)

Amount	Percent Gross Income	Category
$1,995	.7%	Advertising
1,425	.5	Dues & Subscriptions
1,425	.5	Education
2,850	1.0	General Liability Insurance
600	.2	Legal & Professional
5,700	2.0	Medical Insurance
11,400	4.0	Office Rent
31,350	11.0	Office Salaries
5,700	2.0	Office Supplies
5,700	2.0	Part-Time Office Help
710	.2	Repairs & Maintenance
400	.1	Shop Expenses
2,850	1.0	Small Tools
2,850	1.0	Taxes & Licenses
1,995	.7	Telephone
1,425	.5	Travel & Entertainment
1,425	.5	Trash Removal
7,695	2.7	Truck Expenses
2,850	1.0	Utilities
855	.3	Work Clothes
$91,200	32.0%	Total Indirect Costs

Net Profit

Amount	Percent Gross Income
$10,545	3.7%

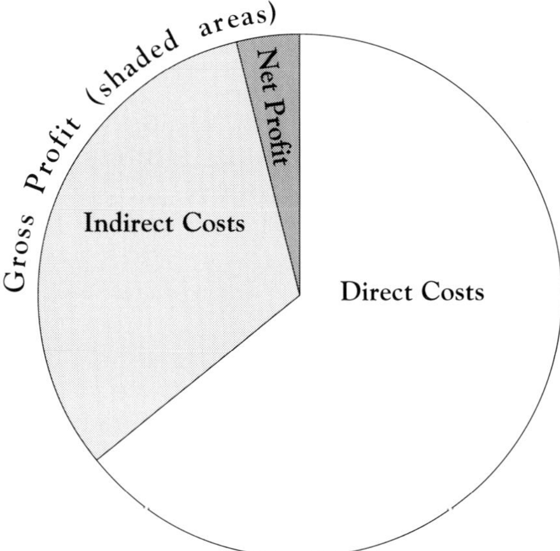

Annual Gross Income

Figure 2. *Your gross receipts fall into one of three categories: direct costs, indirect costs, and profit. Direct costs follow this rule: If you have an expense only because you're doing a particular job, it's a direct cost. Indirect costs, often called overhead, are expenses you incur just by being in business. What's left after all these expenses are accounted for is net profit — what you get to keep or reinvest in the business.*

The formulas are as follows:

Gross income – direct costs = gross profit
Gross profit – indirect costs = net profit

Direct Costs

Amount	Percent Gross Income	Category
$570	.2%	Permit Fees
285	.1	HOW Premiums
68,400	24.0	Materials
57,000	20.0	Labor
57,000	20.0	Subcontracts
$183,255	64.3%	Total Direct Costs

the purposes of job costing and calculating my overhead, but I keep them separate for tax purposes.

Indirect Costs

When you subtract direct costs from gross income, you are left with *gross profit*. Unfortunately, this is the number many builders quote when asked how much money they made last year. But gross profit doesn't go into your pocket. It includes expenses that represent the cost of doing business in general but that don't specifically apply to this or that job. These are called *indirect costs*.

Many of the indirect cost categories I use are self-explanatory and will work for almost any small builder or remodeler. The basic rule I follow to determine my indirect costs is just the opposite of the one I follow for direct costs: *A cost is indirect if I incur it whether or not I do any remodeling.*

For example, my own salary and the hourly labor costs of my part-time secretary are indirect costs because we work strictly in the office. If I swung a hammer part of the time or did direct field supervision, that portion of my time would be counted as a direct cost to the job. But my secretary and I work only in the office, and we work whether or not anyone is working in the field. It's true that we might not be employed very long if there were no remodeling jobs at all. But the point is that these personnel costs are not incurred by a specific project — the Jones or Smith job — but by the fact that Criner Construction exists as a business entity.

I treat general liability insurance premiums the same way because they're based on gross receipts and not tied to a single job. The same is true for office rent and utility bills, telephone charges, advertising, and all the rest.

A couple of indirect cost categories I use may seem puzzling, but they work for me because of the way I run my business. For example, I consider small tools an indirect cost because I supply all hand and power tools to my two full-time field employees. The tools are not consumed during the course of any single job, and the purchase price and maintenance costs are expenses I incur over a long period of time.

I also supply my field employees with company vehicles, and for the same reasons, their operating and maintenance expenses are indirect costs. The purchase price for large capital expenditures like this should be amortized over the expected life of the equipment. Otherwise it will throw your numbers out of whack. For example, if you pay $12,000 cash for a new vehicle and include the total amount as an overhead expense in just one year, your indirect costs will be inflated for that year. Unless you expect to buy a new truck every year, you should divide the $12,000 by the expected useful life of the truck — say five years, at $2,400 each year.

Keeping Track

Once your cost categories are set up, it's important to make sure that no money is spent without being recorded. I have one checkbook and it never leaves the office. I have credit accounts with all my vendors, and I give every employee a charge card. This ensures that I have a canceled check, vendor invoice, or charge card receipt for every expense. On the rare occasions when a credit card isn't acceptable, I reimburse employees by separate check for out-of-pocket expenses and log it into the proper cost category. Once in a while I may have to run out with a check to pay for something in person. But I don't mind doing it because it's easier than losing track of the accounting.

Everything begins with estimating and job costing because you can't calculate your overhead without knowing what your direct costs are. When I do an estimate, I group the direct costs into 25 categories, such as "carpentry," "roofing," and "insulation." As the job progresses, I use the same 25 categories for job costing. I don't use a purchase order system, but whenever someone orders material, they include a job name and direct cost category. A load of framing lumber, for example, will be designated for the "Smith Job" in the "carpentry" category. If more than one category is required, they break it down right on the invoice. We try to keep invoices separate for different jobs. My field employees keep their hours daily and break them down the same way. All invoices that don't have anything to do with a particular job are recorded as indirect costs.

Stay Up-To-Date

At a minimum, you should update your record totals at the end of each fiscal year. But because the numbers in each category serve as budget figures for the coming year's expenditures, a monthly tally is better. Knowing what you spent last year on health insurance, for example, will give you something to compare this year's premium to. If the cost goes up, you can adjust your overhead to cover it.

I like to keep current with my books two or three times each week, so I always know where my business stands. It helps me to foresee problems, and it gives me time to react to them. Builders who don't keep track like this don't find out they're broke until it's too late.

Keeping good records also lets me pick and choose my jobs without the fear of not making enough money. I know that I have to make my overhead on every single job. And because I know what I need to make to cover my overhead, I can choose jobs that consistently provide that amount.

Good records also tell me which kinds of jobs my company is best equipped to do. We look for jobs where we can produce a superb product and get a good profit. If we can't be competitive on a job, we'll pass it up.

Overhead vs. Markup

Once you've made sure that all of your income and expenses are accounted for and distributed to the proper cost categories, you have all the numbers you need to find your gross income and direct and indirect costs. Now you can calculate overhead and markup.

Let's take overhead first. At the end of the year, when you total all of your indirect costs, you get a dollar amount that represents your overhead for the year. This is the amount you need to earn over and above your direct costs to cover the cost of doing business.

But the dollar amount for overhead isn't very useful when you're preparing an estimate. If you divide it into an average monthly amount and apply it to every job you estimate that month, you won't be competitive on small jobs and you'll be underpriced on larger jobs.

Because builders need to include indirect costs in every estimate, no matter how big or small, overhead is usually expressed as a percentage. *Overhead is the ratio of indirect costs to gross income.* Using the totals from Figures 1 and 2, the calculation looks like this:

$$\frac{\$91{,}200\ (\textit{Indirect Costs})}{\$285{,}000\ (\textit{Gross Income})} = 32\%\ (\textit{Overhead})$$

The most common mistake people make is to confuse overhead with markup. When they estimate a job, they figure the total cost for materials, labor, and subcontractors — the direct costs. Then, if they know that their overhead is 32%, they multiply their direct costs by 32% to get their markup. For example, if they estimate the direct costs for a job at $10,000, they add 32% of this or $3,200 and quote the price to the owner as $13,200.

That's just plain wrong. You lose money very quickly that way, and here's why. In the example, if your estimate is correct, you'll spend $10,000 for materials, labor, and subcontractors (direct costs), and have $3,200 left over to cover your indirect costs. Divide this by your gross income for the job ($13,200) and you get an overhead percentage for the job of only about 24%. That leaves you 8% short of the 32% you need.

Instead, after you've estimated your direct costs, you need to multiply them by your markup to find the amount to add for overhead. *Markup is always expressed as a percentage of direct costs.* The mathematical problem you need to solve is this: What *percentage* of direct costs, when *added* to *direct costs*, will

Figuring Your Markup

To cover this overhead	Mark up direct costs this much	Markup for every $10,000 of direct costs
26%	35%	$3,500
27%	37%	$3,700
28%	39%	$3,900
29%	41%	$4,100
30%	43%	$4,300
31%	45%	$4,500
32%	47%	$4,700
33%	50%	$5,000

Note: For other numbers, use the formula:

$$\left(\frac{1}{1 - \text{Overhead}}\right) - 1 = \text{Markup}$$

For example: $\left(\frac{1}{1 - .34}\right) - 1 = .52$

Figure 3. *Overhead and markup are often confused. To recover a given percentage of overhead, you must charge a larger percentage of markup.*

cover your overhead? In the example used a moment ago, the markup should have been 47% ($4,700), which makes the selling price $14,700. The $4,700 is 32% of the selling price (the gross income for this job), and will just cover indirect costs:

$$\frac{\$4{,}700\ (\textit{Indirect Costs})}{\$14{,}700\ (\textit{Gross Income})} = 32\%\ (\textit{Overhead})$$

Figure 3 lists the percentage by which direct costs must be marked up to break even at various overhead percentages.

Profit

These calculations establish your *break-even price — the selling price at which all of your costs (direct and indirect) are covered, with nothing left over.* But every business needs to accumulate some wealth to cover down times and finance capital expenditures, like new equipment. So you should include net profit in your markup. *Net profit is what's left after direct and indirect costs are subtracted from gross income.* My net profit is usually about 4% of gross receipts, which adds a little over 6% to my markup. In the example, a 47% markup covers overhead, and an additional 6% markup ($600) takes care of profit:

$10,000 (*Direct Costs*)
x .53 (*Markup*)
==================
$ 5,300 (*Overhead & Profit*)

The selling price for this job, therefore, should be $15,300.

Common Mistakes

Most builders don't take a salary; they just draw from their checking account when they need to pay their bills. But not paying yourself is the biggest mistake you can make. All of the things you do in the office or after hours are necessary to the success of your company and should be accounted for as an indirect cost. Just because you're not banging nails doesn't mean your hours are less productive or less costly, in terms of overhead, than those of people in the field.

I bid everything on a fixed price because it means I never have to explain my overhead to clients. The minute you do that, you've lost the job because they will never believe you. They just can't fathom it, and no amount of explanation will change that. The best thing to do is bid your price and stick with it. ■

Robert Criner is president of Criner Construction Company Inc. of Yorktown, Va.

Top Cat on the Crew: The Bookkeeper

by Judson Motsenbocker

Behind every successful contractor is a sharp-eyed bookkeeper

It's a well-known fact that most remodelers and small builders are hands-on people who came up through the ranks. They did the physical work first, and became business people later. That's why it is vitally necessary for contractors to surround themselves with good people to help run the business. It doesn't matter whether the business does $50,000, $500,000 or $1 million a year: The need for help is the same.

It is necessary to divide any company — and again it doesn't matter what the size is — into three parts: administration, sales and estimating, and production. These three jobs cannot be done by one person. Any two of the jobs can be done by an owner or a hands-on person, but one of the tasks requires outside assistance. It might be in sales and estimating or it might be in production, but the one we find most easy to give up, and the one that makes the biggest difference, is the administrative area of the business. The rest of this article focuses on how important it is to properly tend to bookkeeping and/or the administration of the business. Poor administration is the most frequent downfall of contractors. It's not that they can't get work done, and it's not that they don't know material or labor costs. What they don't know is their

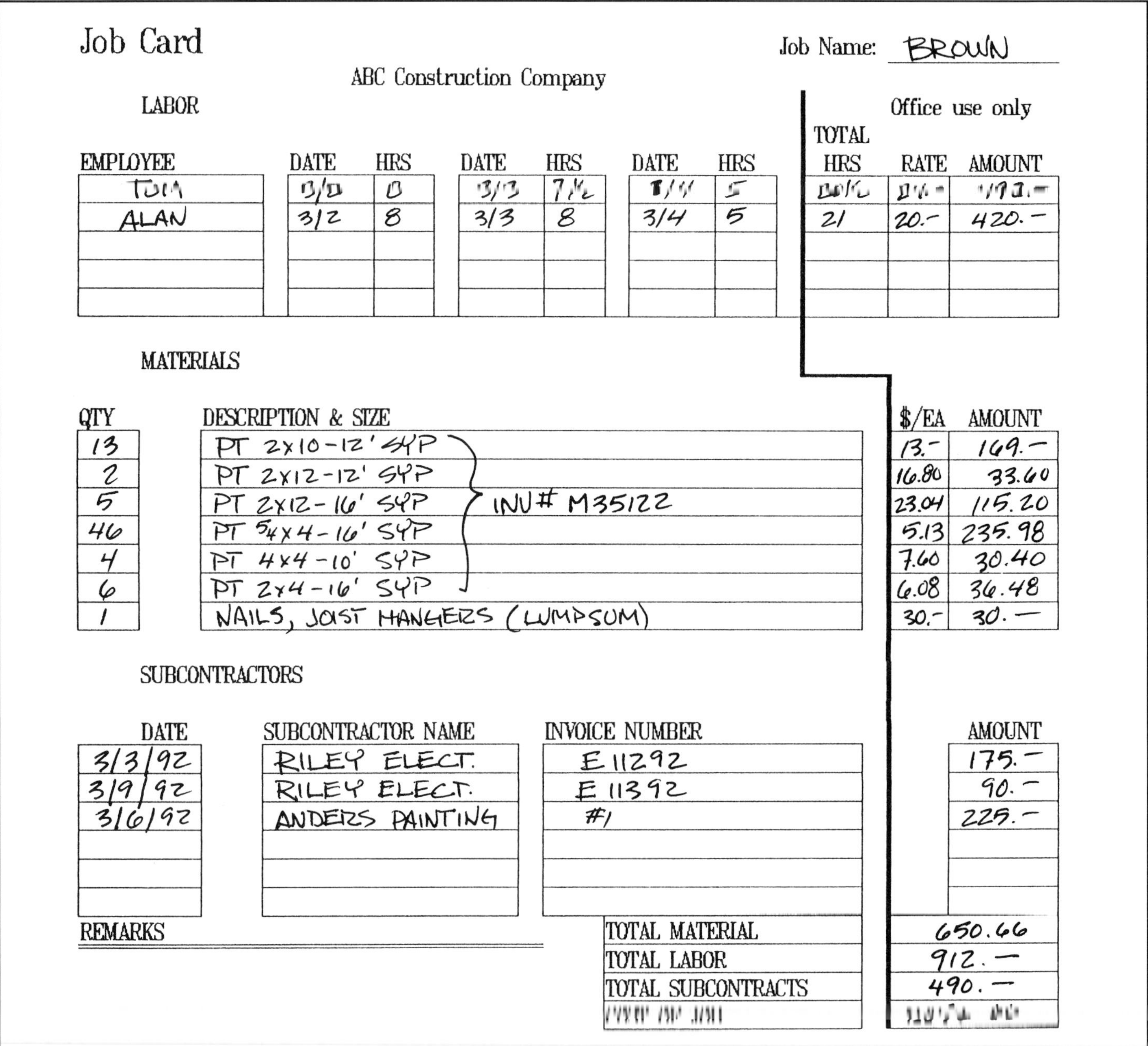

Job Card

Job Name: BROWN

ABC Construction Company

LABOR

Office use only

EMPLOYEE	DATE	HRS	DATE	HRS	DATE	HRS	TOTAL HRS	RATE	AMOUNT
TOM	[illegible]	[illegible]	[illegible]	7½	[illegible]	5	[illegible]	[illegible]	[illegible]
ALAN	3/2	8	3/3	8	3/4	5	21	20.–	420.–

MATERIALS

QTY	DESCRIPTION & SIZE	$/EA	AMOUNT
13	PT 2x10-12' SYP (INV# M35122)	13.–	169.–
2	PT 2x12-12' SYP (INV# M35122)	16.80	33.60
5	PT 2x12-16' SYP (INV# M35122)	23.04	115.20
46	PT 5/4x4-16' SYP (INV# M35122)	5.13	235.98
4	PT 4x4-10' SYP (INV# M35122)	7.60	30.40
6	PT 2x4-16' SYP (INV# M35122)	6.08	36.48
1	NAILS, JOIST HANGERS (LUMPSUM)	30.–	30.–

SUBCONTRACTORS

DATE	SUBCONTRACTOR NAME	INVOICE NUMBER	AMOUNT
3/3/92	RILEY ELECT.	E11292	175.–
3/9/92	RILEY ELECT.	E11392	90.–
3/6/92	ANDERS PAINTING	#1	225.–

REMARKS

TOTAL MATERIAL	650.66
TOTAL LABOR	912.–
TOTAL SUBCONTRACTS	490.–
[illegible]	[illegible]

A job card keeps track of all the costs for a particular job. It helps with billing and payroll, and can be used to compare actual and estimated costs.

overhead cost.

I am often asked who the most important person in my business is. I would like to think it is me, one of the carpenters, or my foreman. However, I have to admit that the most important person is the bookkeeper. I can sell the jobs, hire the people to do the jobs, and buy the material. But, if I do not know what my overhead is, everything I have done is in vain.

The double-entry bookkeeping system is the most accepted system throughout the country, but some people still use the cash-basis system. There are several different formats and various suppliers for each. The National Association of Home Builders has a system called Account and Financial Management, but many other companies have good formats as well. A good bookkeeping system must work by hand before it is put on a computer. Computerizing a bookkeeping system should be done in parts, and the parts should work together in an overall program.

An accountant who has dealt with the building and remodeling business will be familiar with the system that is most economical and the one that will give you the best results. As with many things, the least expensive route may not be satisfactory. It is important to have a good accountant to advise and assist you, not only in setting the system up, but in making sure it is working properly. Of course, you also need a bookkeeper or a secretary/bookkeeper combination who has the ability to take care of some of the general paperwork that we will discuss.

Track The Job

After a job has been sold, the proposal has been signed, and the detailed estimate attached to the proposal in the office area, it is necessary to prepare a job-cost or job-card system to keep track of each project. It does not matter how large or how small each project is. This can be done on an 8 1/2 x 11 card. Include the description of the job and its location, and the list of materials, subcontractors, and labor used on that job. Materials and subcontractors should be listed by invoice, date, and amount. Labor should be listed by date and number of hours. Then when the project is complete, these numbers can be totaled and you can determine exactly how much cost you have in that project. This job-card or costing system will not show what your overhead is. Overhead includes costs such as insurance, heat, lights, rent, automobile expense, taxes, and so forth. Therefore, this job-costing system is only a portion of the total costing system that needs to be developed.

Materials. One of the jobs that the administrative person needs to do is bill these jobs after they are completed. In order to be sure that all material for that job is billed, you should use a purchase-order system. Once a job has been assigned a number from the job card, a purchase order needs to be filled out for each vendor and/or subcontractor each time an item (or items) is purchased for that job. From this the bookkeeper knows what material has been purchased or what subcontractor has been hired to do part of a job. If it is a small project, it is very important that a purchase-order system is used so that the job can be billed before the 10th of the following month. This will provide cash to pay invoices due. It's very helpful to have the supplier send you invoices by the week. But some small suppliers and people you don't buy much from do not like to send them by the week. Therefore, it is necessary to use the purchase-order system so that the bookkeeper can call and get the amount of money that is involved on a project. Sometimes it is very easy to say "Well, I'll just remember where we got the material." That's probably true for a small operation, but when you have more than one person buying material, or more than one job going at a time, it becomes necessary to use a purchase-order system so that you can determine exactly what materials have been used on each job and so that all chargeable items will be billed.

Also, as invoices come in, it is very important that the bookkeeper check the board footage, the lineal footage, and the extensions of that footage with the price that has been quoted for the material. The invoice totals should be checked against the statements received each month. My bookkeeper earns half of her wages by finding errors in invoices. Of course, it is also important that all of the invoices get posted to the proper accounts within your accounting system.

Another vital job of the administrative department is to keep a price book up-to-date so that the estimator can keep track of the cost of materials. This book needs to be properly checked when invoices come in so that prices can be updated and kept current. Any good secretary or bookkeeper can do this simple comparative process after the procedure has been explained.

Labor. It is also necessary for the bookkeeper to keep track of all the time that goes into each job, whether you have one employee, three employees, or fifty. Their time needs to be kept by job number and posted to particular job cards. If some labor is not chargeable to a specific job, because workers were cleaning the warehouse, working on trucks, or whatever, keep a record of it so you can control these overhead items.

The bookkeeper needs to total the time for each employee after it has been posted to the various job cards so that payroll checks can be written. The payroll itself can be complex, in that all the wages and taxes need to be figured and kept separate for each employee. If your bookkeeper is qualified, these records, including the quarterly and year-end payroll taxes, can be done in your office. You may, however, wish to make your accountant responsible for this.

Other overhead. Consider other overhead and nonchargeable items. If you have more than one vehicle, it is beneficial to number each one so that any gas, repairs, or employee labor can be charged directly to that vehicle. You then have an accurate account of what that vehicle is costing per year to operate. Properly recording other overhead costs such as utilities and construction supplies (saw blades, rope, chain, and other things not charged directly to one job) will help you control the cost of operating your business.

Insurance. Another vital link in administration is insurance. It is absolutely essential that someone work with your insurance agent to determine exactly what is covered and what is not covered in your business. In today's rapidly changing times, our bookkeeper works with the agent about every six months to determine whether there have been any changes in our coverage. We check to see if we need to update anything, if we are paying too much for our insurance, or if something important is not covered.

Your Right Hand

I am often told that the owner of the company does the bookkeeping him- or herself. I find it extremely difficult to believe that a good bookkeeping job can be done by someone who is trying to run a crew, sell the jobs, and do the estimates. It is simply more than one human being can do.

I think it is easy to see that the basic administrative job that I have outlined is at least a part-time and more often a full-time job. When you add answering the telephone, operating the two-way radios, checking materials for the production department, talking to customers, and other miscellaneous jobs, it's not hard to see that a full-time person can be readily used and his wages well covered. It is often said that people can't afford a bookkeeper/secretary combination. My theory has always been that you can't afford to be without one. He or she needs to be your right-hand person to keep track of the overhead and your general expenses.

Profit

A profit-and-loss statement should be produced at least every three months, and I strongly suggest that one be prepared every month. It will show what overhead costs are, and you will readily know if your markup is adequate. The selling price of a job must be the cost of the material, the cost of labor, the general cost of doing business (overhead), and any profit that you want to make. Remember when you mark up cost, the percentage of markup must relate to the expenses indicated on the profit-and-loss statement. The percentages on the profit-and-loss statement are based on 100% of the gross amount of sales. When you are costing out a job, you are working with only direct costs (labor, materials, and subcontractors). Therefore, when you are marking up a cost at 50% you are really only marking up the total sale by 33%. If your overhead happens to be 25%, you are only making 8% profit where you thought you were making 25%.

Summing Up

To briefly review: After a written proposal has been signed, a job-costing system must be developed to keep track of labor, materials, and subcontractors on each job. The system can use job cards or a three-ring notebook.

You need a purchase-order system to keep track of the materials and subcontractors. You need time sheets filled out by your employees to document labor costs for each job and for payroll. And you need a system that makes sure the customers are billed in a timely fashion and that invoices are filed in well-organized accounts-receivable system.

A payroll system is needed to keep track of the individual wages paid to, and taxes withheld from, each employee. An accounts payable person is needed so that you know how much you owe each individual vendor, and when to pay what you owe. The general-overhead and operating-expense items need to be recorded in separate accounts so that a profit-and-loss statement can be compiled and a tax return filed.

Even if your office happens to be in your home, you still pay heat, lights, telephone, insurance, etc., to operate the house. Remember to document these items so that you can determine the portion that is attributed to your business. Successful remodelers and builders are backed by a good set of books — an accounting system which is acceptable to the profession. Remember, the job isn't completed until the paperwork is done! ■

Judson Motsenbocker is owner of Jud Construction, in Muncie, Ind., and a frequent speaker on business management for remodelers.

Why I Keep My Own Books

by Paul Eldrenkamp

Crunching his own numbers can give a remodeler insights he can't get from an outside accountant

I'm a carpenter turned remodeling contractor. I know carpentry best, but when necessary I can also do some wiring, plumbing, painting, and plastering, among other specialties. Nevertheless, I subcontract those and other trades, because my subs can do them better and more cost-effectively.

There's one job, however, that I've never subbed out during the eight years I've been in business: bookkeeping and accounting. I'm not especially skilled or talented in this area, and doing it myself doesn't save me money. But I've found that to really know what's going on in my business from week to week, I'm best off crunching the numbers myself. Keeping and analyzing my own books enables me to spot trends and react to them quickly, both within a job and within my business as a whole; to see which individual activities make money and which lose it, and why; and to gain an intuitive understanding of how my business works. This has given me much greater control, and, I believe, probably kept me from going under in the worst days of the latest recession.

My number crunching falls into two fairly distinct categories: basic bookkeeping, or recording of basic financial data; and the use of analytical accounting tools.

The Basic Books

The routine bookkeeping is mainly a matter of recording transactions and printing the simplest reports. This information lays the groundwork for the more complex, analytical tools I use. I update the basic bookkeeping data daily, weekly, monthly, and at the end of each job. I'll describe these more mundane procedures quickly. They include:

Income and expenses. Every morning I record the previous day's accrued debits and credits.

Payroll. I do this the easy way: I use a payroll service and direct deposit. The service also automatically submits my quarterly tax payments.

Billing. At the end of each week, I prepare and mail any bills payable the following week. Then I can collect at the next weekly client meeting.

Cash-flow projection. This is computed weekly. It projects how my bank balance should fare over the next 30 days, based on expected receipts and expenses.

Profit-and-loss statement. This standard bookkeeping report shows profit or loss for the month. This P&L statement includes both the actual and budgeted expenses and income, and a balance sheet showing my company's net value.

Analytical Tools

Along with this fairly routine bookkeeping, I also do some of the basic accounting reports to keep things on budget, as well as more analytical reports and ratios that reveal how my company is doing. Some of these reports are fairly standard, either in business in general or in construction. A few are things that probably aren't unique, but which I myself have developed to suit the way I do

Figure 1 — Weekly Job-Cost Accounting Report

Stimson																			
Porch Restoration																			
Week ending 9/20/91		Lead #1			Lead #2			Carp.			Mat'ls			Subs			Total		Total
		(hrs.)			(hrs.)			($)			($)			($)			($)		($)
	week	actual	budget	week	actual	budget	week	actual	budget	week	actual	budget	week	actual	budget	week	actual	budget	variance
Demolition and disposal		0.00	0.00		17.00	8.00	0	407	192		0	0		700	750	0	[illegible],107	942	-166
New footings		3.00	0.00		17.00	4.00	0	479	96		271	300		250	600	0	[illegible],000	996	-5
New framing		27.00	32.00		35.00	32.00	0	1,485	1,533		1,004	852			0	0	[illegible],489	2,385	-104
New decking		20.00	16.00		15.25	20.00	0	844	862		2,189	1,990			0	0	[illegible],033	2,852	-181
Railing, & column repair	15.00	15.00	25.00	22.50	22.50	26.00	898	898	1,221	186	225	400			0	1,084	[illegible],123	1,621	498
Primer & preservative		0.00	4.00		9.50	8.00	0	228	287		144	75			0	0	372	362	-9
Totals	15.00	65.00	77.00	22.50	116.25	98.00	898	4,341	4,191	186	3,834	3,617	0	950	1,350	1,084	[illegible],125	9,158	33

Figure 2 — Final Job-Cost Accounting Report

Stimson								
Porch Restoration								
Final unit costing update								
	Labor	Mat'ls	Subs	Total	Unit	Qty	Unit cost total	Unit cost labor
Demolition and disposal	407	0	700	1,107	sq.ft	720	1.54	0.57
New footings	479	271	250	1,000	ea.	18	55.58	26.61
New framing	1,485	1,004		2,489	sq.ft.	720	3.46	2.06
New decking	844	2,189		3,033	sq.ft.	720	4.21	1.17
Railing & column repair	898	225		1,123	ln.ft.	110	10.21	8.16
Primer & preservative	228	144		372	sq.ft.	720	0.52	0.32
Totals	4,341	3,834	950	9,125		720	12.67	6.03

Figure 3 — Profit Analysis Report

Lead #	Type of job	Sales hours invested in project	Total production hours	Final gross profit	Final gross profit per production hour	Final gross profit per sales hour	Final gross profit margin
90 033	Porch	13.00	189.25	$9,995	$52.81	$768.85	69.92%
90 036	Energy upgrade	4.00	10.00	$512	$51.20	$128.00	14.97%
90 037	Bathroom, laundry	14.00	156.25	$4,698	$30.07	$335.57	47.29%
90 038	Porch	8.00	166.00	$7,156	$43.11	$894.50	65.33%
90 040	Bathroom	7.25	108.75	$4,661	$42.86	$642.90	46.31%
90 041	Window replacement	5.25	47.25	$2,526	$53.46	$481.14	27.16%
90 043	Cabinets or trim	9.25	136.00	$6,593	$48.48	$712.76	41.24%
90 051	Bathroom	17.25	138.25	$5,363	$38.79	$310.90	33.75%
90 051	Deck	11.75	165.25	$4,735	$28.65	$402.98	49.50%
90 052	Bathroom	3.00	45.75	$1,055	$23.06	$351.67	50.24%
90 058	General cosmetic upgrade	21.75	105.75	$6,277	$59.36	$288.60	41.27%
90 063	Porch	4.50	5.00	$500	$100.00	$111.11	20.00%
91 009	Repair	4.00	11.00	$1,539	$139.91	$384.75	40.31%

Figure 1. *The weekly job-cost report uses information from time cards and invoices to compute weekly and cumulative costs by category. Comparing budgeted amounts to actual costs provides an easy check against cost overruns.*

Figure 2. *The final unit-cost update for a job summarizes the information compiled in the weekly job-cost reports. The resulting unit costs for labor, material, and subs help the author update his unit-cost price book.*

Figure 3. *The profit analysis report shows each job's profit margin and gross profit per production hour. The latter number, the author has found, is the key to meeting his profit goals. These two profit indicators don't necessarily move together. For instance, the window replacement job (Lead 90.041) produced a gross profit of only 27%, but a profit per production hour of $53.46 — well above the author's target that year of around $40 per production hour — making the job a winner. One of the bathrooms, however, Lead 90.052, brought a 50% gross profit margin but only $23.06 gross profit per production hour — a relative loser.*

business. These tools have proved most valuable, but I don't think I would have developed them if I didn't keep my own books.

Job-Cost Accounting

I have found that job-cost accounting provides a crucial check for keeping jobs within budget and estimates accurate. Every week on big jobs, and at job's end on small jobs, I record the material costs (from invoices) and labor costs (from time cards showing how much time my workers spent on each construction category) of each job by estimating category. I also compare actual to budgeted expenses for each category (see Figure 1). I plug these figures into a spreadsheet template (a software program) that also includes the original estimate. This lets me crank out quick compare-to-budget reports at any stage of the job, so I can spot cost overruns early.

The final job-cost accounting report for each job includes unit costs for the project: what it cost per lineal foot of foundation, per square foot of wall framing, per window, and so on (Figure 2). This information allows me to update our unit-cost estimating database frequently and quickly.

I also do monthly compare-to-budget reports for overhead expenses, to keep them in line.

Labor Budget Analysis

I've found that budgeting and accounting for our time, particularly our production time, has been the key to increasing the profitability of our jobs.

I have only three employees, all carpenters. Obviously, production is their main job, and most of their time is budgeted to it. But because the two lead carpenters inevitably do some warranty work, project management, and administrative tasks, we budget a few of their hours to those categories as well.

My own time is budgeted more widely: 30% sales, 25% production, 6% warranty and marketing, and so on. I spend just over 8% of my time on bookkeeping and accounting functions.

So that we can hit these targets, everyone gives me weekly time cards showing not only the hours they spent on each production category, but also any time spent doing administrative or warranty work. Once I plug this information into my computer, I can do weekly, monthly, quarterly, and annual reports showing the time everyone spent on the various categories.

More importantly, I can use this labor category information to figure some crucial ratios: profit per production hour and profit per sales hour (see Figure 3).

Profit per production hour. This is the single most important ratio I calculate, and has led to some of the most profitable changes I've made in my company.

At the beginning of the year, I project our expected number of production hours (roughly 1,750 production hours per full-time production worker) and our expected gross income. From those I figure the gross profit per production hour needed to generate a gross profit margin of 33%. For instance, with the three employees I had this past year, and myself putting in 25% of my time in production, we figured to log about 5,700 production hours. I was projecting around $700,000 in business. To make sure that 33% of the $700,000 ($231,000) was gross profit, we needed to make a gross profit per production hour of about $40 ($231,000 ÷ 5,700). Pricing to generate that gross profit puts us toward the top of what our market will bear; but pricing lower robs me of the safety provided by the 33% gross profit margin.

To see if a given job is worth doing, then, I can compare its expected gross profit per production hour to our goal. For example: If I mark up a $1,000, 40-hour job 50% to $1,500 (a job that's all labor), the gross profit is $500 ÷ 40, or $12.50 per hour. A larger job, in which we put in 500 of our own hours but gross $25,000, would give us $50 of gross profit per production hour ($25,000 ÷ 500).

Knowing this, I've learned to price jobs more strategically, and to use our time carefully. For instance, I've learned to raise our markup considerably on small jobs that are labor-intensive, because we found we were making little money per hour on those.

The lesson here is that a straight percentage markup for all jobs doesn't work. We've found that on jobs that demanded a lot of our own time, particularly in areas other than framing or finish carpentry, we tended to make little per production hour, and needed higher markups. On jobs where we subbed out a lot, we could use smaller markups and make the same profit. That's why we've gone from doing just about everything ourselves to doing mainly framing and finish work — the carpentry tasks we're really good at. Subs do almost everything else.

This ratio also provides a clear basis for comparing different types and sizes of jobs. It tells me in clear numbers what I knew only instinctively before: that we made little profit on smaller jobs. Now, when I price those with a much higher profit margin, I know I'm doing it for solid business reasons, not just to avoid what seems like a hassle.

Gross profit per sales hour. I also figure gross profit per sales hour: a job's gross profit divided by the amount of time I spend pitching it, from first call to signed contract. This tells me where my sales hours are most profitably spent, which is (not surprisingly) larger jobs.

This has also made me fairly gun-shy about design/build jobs where I have to sell the design to get the job. I've found that to hit my profit target on design work, I need to charge close to what architects do. But I don't feel comfortable doing this, because I'm not a trained designer. So I've started to pass the design on to a designer or architect to come up with the specs.

I rarely use profit-per-sales-hour figures to actually help me set prices, however. That would be attaching too much importance to these hours, which are relatively few. It's mostly a qualifying tool. It gives me the strength to say no.

Warranty hours per project. This tells me which types of projects produce the most warranty work. If one type stands out, I can look closer to see if it's a matter of our abilities there, or an inherent part of the job. Depending on what I find, I might fix something we're doing wrong, raise prices for that type of job, or avoid it when possible.

Getting Crunched By Your Own Numbers

You run risks crunching your own numbers, and some of them are deadly.

The cash-basis trap. When you don't adequately account for upcoming expenses, it's easy to spend money that you'll desperately need later. This is a common mistake when you use, as I did until recently, a single-entry, cash-basis system of accounting. In that system, the general ledger, like your checkbook, tallies only monies actually received or paid out. Accounts receivable and payable are kept in separate ledgers, divorced from the rest of the books.

Unfortunately, having your accrued receipts and payments separated from your main books makes it easy to think that you're doing well just because you have money in the bank, or that you're doing poorly because your balance is low. (The former is not necessarily true, whereas the latter usually is.) In reality, a business's health is tied as much, if not more, to accrued sales and expenditures as to those already realized.

To avoid this cash-basis trap, I recently switched to double-entry, accrual accounting. This not only provides an automatic check on your entries (the "double-entry" aspect), but integrates accrued incomes and expenses into the general ledger. This lets me see what's down the road, rather than just what's in the rear-view mirror.

The learning curve. I have had to learn by doing. This is a slow, inefficient process, which sometimes produces mistakes.

The discipline thing. Because I'm slow, I have on occasion let the bookkeeping slide.

Getting Started on Your Numbers

The realistic strategy to developing your own bookkeeping system is to keep it simple enough that you actually use it. If you don't job cost or track overhead at all now, you might start by tracking just three job-cost categories (labor, materials, and subcontractors) for each job and see how your actual expenses compare to budgeted expenses in those areas. At the same time, budget and track all overhead as one category, doing budget comparisons once a month.

If even this pared-down approach proves too much, then just track production expenses and overhead. Once you get comfortable with this, you can refine the system, breaking production and overhead into their various components for estimating, job costing, and planning. The key is to set goals you can actually meet, or you'll keep slipping up. In other words, as Skeeter Coleman, songwriter turned remodeler, once told me, eat the elephant bite by bite and eventually it's gone. — *P.E.*

This means I may not spot problems, such as an unusual but avoidable cost overrun, soon enough to mitigate them. Such lack of foresight and basic financial data has sunk many a small contractor.

Be Nimble, Be Quick

Despite these drawbacks, I feel that doing my own books has greatly improved my ability to control my business. Being right on top of the numbers means I can spot financial trouble early in a job, and react to it quickly enough to save the profit. And I have gained important insights into the financial dynamics of my business — from big things like realizing we needed to sub more things out, to little things like discovering that one way of hanging joists was faster than another.

It may be possible to get this sort of perspective and quick-response time with the help of a skilled bookkeeper and a sharp accountant who truly understand a contractor's unique needs. But I find doing it myself more rewarding. ■

Paul Eldrenkamp owns Byggmeister Inc., a remodeling company in Boston, Mass.

Time Card Basics

by Sal Alfano

A well-designed time sheet will provide a solid foundation for payroll, billing, job-costing, and more

If you're like most builders, you started small — just you and maybe one person to hold the other end of the board. You were friends, you'd worked together for a while, and you were both on the job site every day, all day. It was easy to keep track of things.

When payday rolled around, you'd tally up the hours (identical except for the day you had a dentist appointment) and submit a labor bill to the owner. A "big" job took maybe a month and a half, and you did only one job at a time.

My, how things change.

You got more work, hired more people, and started running more than one job at a time. You put all your workers on the books. In short, you became an employer. And now you find you can't be on every job every day, all day, and when payday rolls around, it's a major event.

Why You Need Time Sheets

A well-designed, thoroughly completed time sheet won't solve your payroll problems — it may, in fact, increase the paperwork slightly — but it can be a great help. It's a part of the business every employer and employee ought to pay close attention to.

The weekly or biweekly record of hours worked provides a crystal-clear record of every employee's performance. It gives you full documentation for federal and state employment records and cost-plus-labor billing, and makes writing payroll a bit easier.

Consider how many times you have come to the end of even a small job and asked your employees how many hours they spent on it, only to find they can't remember *what* they did last Monday, how long it took, or even *where* they were working. Guesswork simply won't do, especially on cost-plus work. There are plenty of owners who get a kick out of keeping track of your workers' hours, and taking you to task if your tallies don't match.

The Well-Designed Time Sheet

A well-designed time sheet can provide a wealth of information to help you with scheduling, estimating, change orders, allowance pricing, and personnel-evaluation. The time sheet should be designed to give you the *kinds* of information you want it to provide. But don't overdo it. Keep in mind the time it will take you to read it, and the time and effort it takes your employees to fill it in.

The time sheet I use (see sample) may not exactly fit your needs, but will illustrate the principles involved. The size should be convenient. Crew members should be able to carry them in their shirt pockets. (This may seem silly, but your employees won't use a time sheet consistently if they don't carry it.) Mine is letter-size, folds nicely to about 2x4 inches, and files away neatly in a manila folder.

Color makes a difference, too. Something other than white is best to easily distinguish labor records from the other

ABC Construction Company

TIMESHEET

Employee: ALAN

Pay Period: 4 MARCH – 17 MARCH | 92

JOB---->	THOMAS	SMITH				
MON	3/P	5/64				8
TUES		2/64 6/71				8
WED		8/71				8
THU		8/81				8
FRI		71/2/73 1/2/CL				8
/ S/S						
Hrs. by JOB---->	THOMAS 3	SMITH 37				40

JOB---->		SMITH	GREEN			
MON		3/73	4/110X ALLOW.			7
TUES		8/73				8
WED		71/2/73				71/2
THU		8/73				8
FRI		6/73 11/2/172 1/2/CL				8
/ S/S						
Hrs. by JOB---->		SMITH 341/2	GREEN 4			381/2

21 Excav/Drainage 22 Demolition 23 Perim. Treat.	71 Roof X-Trim 72 Felt/Roofg	91 Strapping 92 Drywall 93 Painting	155 Elect. 166 Plumbing 167 Heating	THOMAS	3
33 Concrete 44 Masonry 55 Steel	73 X-Trim/Sidg 74 FG/VB 75 Rigid foam	94 Underlay 95 Flooring 96 I-Trim	171 Misc. 172 Scaffold	SMITH	71/2
61 Floor system 62 X-Wall system 63 Roof system 64 I-Walls/Ceiling	76 Ridge vent 81 Wndws/X-Drs 82 I-Doors 83 Locksets	110 Cabinets 111 Shelves 112 Built-ins 113 Stair/Rail	G=Gopher CB=Call Back CL=Cleanup P=Punchlist	GREEN	4
S=Supervisc SV=Site Visit	ES=Estimate D=Design	BK-Books MTG=Meeting	X=EXTRA M=MILEAGE		

Add 'X' to number if work is an EXTRA. Explain EXTRA work & MILEAGE on back of sheet.

When job categories are numbered on a time card, your employees can easily record the time spent on particular tasks. The entry of "5/64" on Monday, for example, indicates that five hours were spent on interior partitions at the Smith job.

paperwork. I use green, and on those all-too-frequent occasions when someone misplaces a time sheet, it's a lot easier to find.

A space for the employee's name and the dates of the pay period are at the top. I suggest you fill in this part yourself; even the most meticulous employee will forget to, as often as not. The rest of the top half has a list of the days of the week, their dates, and columns for job names and hours worked. This is all you need in order to know who worked where and for how long.

Alphabet Soup

But what if you want to know more? The important thing is to keep it simple. An employee who is skilled at working with wood isn't necessarily good at working with its more refined form — paper. How, then, do you solve the problem of creating a time sheet that is detailed but doesn't require a college education and fine penmanship to complete?

The time sheets I use tell me not only who worked where and for how long, but *what* they were doing. Before I came up with this, I used to ask the crew to summarize the day's activities. What I got was alphabet soup.

Aside from not being able to read the writing, I found that not everyone categorized the work in the same way. I got things like "Smith job, 8 hrs., framing and vents." Framing *what*? The roof? Walls? Floors? And *what* vents? Soffit? Ridge? Gable? *Dryer* vent?

A good solution is to list job categories and phases. Have them numbered and listed at the bottom of the sheet. A simple number designates a particular activity, and the same number is used by everybody. The list is arranged according to the way a job usually progresses. There are still some areas that overlap, but the bases are covered.

A typical entry — "5/64" — tells me that five hours were spent on interior partitions. The "X" for extras helps me keep track of work that either was priced on allowance or is being done in addition to what was called for in the contract. For instance, "4/110X" means four hours were spent installing cabinets that were priced on allowance.

The last twelve categories, designated by letter, are mainly for me. (I keep a time sheet, too.) They help me keep track of how much time I spend off site, and help me evaluate efficiency and plan for future jobs. How many of us can pinpoint within eight or ten hours how much time is spent cleaning up? And who's *doing* the cleanup — the lower-wage laborers or the highly paid carpenters?

"Gopher" is another telling item. I think we are all surprised at how much time is spent running to the supplier for materials. With this time sheet, I am alerted to an excess of errand time, and can discuss with my foremen ways to manage the job more efficiently.

Of course, even a well-designed time sheet won't work for you if it isn't properly used. Make sure your workers fill in their time sheets daily. Memory fades quickly when you're engaged in similar activities every day, and all the data in the world won't help a bit if it's incorrect.

Finally, keep a time sheet yourself. For one thing, it sets an example. Also, you might be surprised to discover how you spend your time, and that you might want to use it differently. ■

Sal Alfano, formerly a general contractor in East Calais, Vt., is an associate editor with The Journal of Light Construction.

Chapter 9 • The Art of Delegation

Hanging Up Your Tool Belt

by Linda Case

How to survive the two toughest stages in the growth of your remodeling business

Remodeling companies are like children. They have their own personalities, idiosyncrasies, growing pains, and stages. Just as a toddler's or teenager's behavior is predictable, so too are the stages of your company's growth. Knowing that you're going through a difficult transition — and that it's perfectly normal — helps a little, but getting some advice on how others have gotten through it is even better.

Two very common stages of business growth are (1) when it's time to hang up your tool belt, and (2) when you need to hire a salesperson or production manager. Be aware that these transitions are dangerous; many remodelers end up losing their businesses at just these points.

Stage One: Time To Hang Up the Tool Belt

Realizing it's time to give up swinging a hammer is one of the earliest and most difficult transitions. Let's say you're small; you do about $100,000 to $200,000 of volume a year. You still work a significant number of hours in the field with a tool belt on. But you want out. Your body is beginning to send you the message that it's a limited, not a limitless, resource. You've hit the upper limits of how much work you can sell as well as how much money you can take home, because there are only so many hours in the day. You want to expand and hire others to do the physical part while you supervise and sell. But it's not that easy when you look at the numbers. The biggest barrier is finding the money for your salary once you're no longer working as a carpenter and crew boss.

First you need to assess your strengths and weaknesses in light of the four major focuses of any remodeling company: marketing, sales, production, and management.

Marketing. If the "It's time to hang up my tool belt" scenario fits you, your company has probably needed about $150,000 of work a year to run smoothly. Those jobs haven't been brought in by any organized marketing effort, but rather by referral from one client to the next. Satisfied customers do your marketing for you.

Selling. Since you work diligently (after all, it's your own business and reputation that are at stake), your services are a very good buy for the consumer. You're personable, capable, conscientious. You work from home, which keeps your overhead very low. If you underestimate the labor costs on a job, you simply work longer hours to make up for it. Thus, you haven't needed real selling skills. Between your low price and high praise from former customers, you are very likely to sell 75% of the jobs you price.

Production. Production is your strong point. You are both production supervisor and lead carpenter on your jobs. Since you also bid projects and sell them, you are in on discussions with clients from the beginning. This keeps communications straight.

Business management. This may be your Achilles' heel. You may not know all the local, state, and federal regulations that affect you. Your contracts are simple one-page proposals purchased ready-made, and you carry change orders around in your head. However, success is stalking you. Your good reputation is beginning to pull in more inquiries than you can handle in either sales or production.

Replace Yourself in the Field

You have two options. The first is relatively uncomplicated: you can refuse to grow. The good news is that you can raise your prices, hand pick the very best of the leads, and let the others fall to the wayside. The bad news is that your fortunes depend on your good health and ability to continue doing physically demanding work.

The second option is to ride the growth wave. This will eventually help you to hang up your tool belt, but it raises the stakes and much of what you'll be doing will be risky and unfamiliar. Here are some of the things you'll want to consider, along with their pitfalls:

Step 1. First you'll have to replace yourself in the field. This is probably your first big hiring decision, so take your time. What you're looking for is a skilled lead carpenter. You'll phase yourself out of the field over six months to a year, which will give you a chance to thoroughly train your new hire.

Pitfalls. There are two major errors to avoid here. First, don't hire a lead carpenter who has technical knowledge, but no managerial skills. If you do, you'll be creating a monster. You'll get a carpenter who, instead of being independent, is in constant need of your help. Second, be aware of how hard it can be to turn over the reins to someone else. The art of delegation is a fine one, but you can't be successful in business without learning it. And you might as well learn it with your first major hire.

This means carefully setting goals with your employee and giving him some leeway in achieving them. You have to give your new hire permission to succeed *and* to fail as well. Since you're doing the training and monitoring the progress, things can't go too awry.

Step 2. As you move out of the field, your salary becomes overhead or part of the cost of sales; when you were still working with

INTERVIEW

Staffing For Growth

To get a first-hand account of what kinds of systems and procedures keep a growing company functioning smoothly, we talked to Kelley Hale, of El Cerrito, Calif. Hale runs a high-end residential remodeling company, which successfully doubled its business volume in 1991 from $600,000 to $1.2 million, and went from five to eleven employees.

JLC: *How fast has your company grown?*

Hale: In January of 1991, we had four major remodeling jobs, worth a total of about $600,000, plus one smaller job worth about $40,000, come in all at once. That meant that we would have to complete our normal year's volume in half a year's time. As I signed the contracts for all these jobs, I realized I was signing commitments to make my company grow.

JLC: *How did you staff all these jobs?*

Hale: I was able to spread the starting dates for the jobs over a five-week period. This gave me time for hiring. We were fortunate to find some really good people, partly, because the economy was slow and there were some good people out of work. We hired two new foremen and made a foreman out of our lead carpenter, who had been working with us for nine years. So, including my original foreman (my first employee, Mike), this gave us four foremen — one for each major job.

JLC: *Hiring the right kind of person to work on a remodeling job isn't easy. What kinds of qualities do you look for?*

Hale: The hard part is finding people who have a good attitude — workers who can convey respect for the owner's belongings and sympathy for the difficulty the owners face when their house is being torn up. I look first for that. If the construction skills are not fully developed, they can come later.

JLC: *That must be hard to interview for.*

Hale: I talk to people for a long time and usually have them back for a second interview to discuss some of the things we miss the first time through. I don't conduct the typical interviewer/interviewee conversation, since people will respond with what they think you want to hear. Instead we walk around our shop and talk about a new tool or some different types of moldings, or whatever they seem interested in. You can tell if somebody knows what they're talking about — OSB, tools, what have you — and you get them to open up.

I also let people know up front that I'm looking for a long-term commitment. Once people realize you are sincere about that, their attitudes start to change. They see that you are serious.

your tools it was a direct job cost. In order for the company to carry you in overhead, your volume will have to increase, and this work needs to be done at an adequate markup.

Here's a "for instance." Let's say you hope to do $300,000 in volume. You want to pay yourself $30,000 and end up with a 10% ($30,000) net profit. Your other overhead is very low at 10% ($30,000). Therefore, you need 30% ($90,000) gross profit, and that converts to a 43% markup on your hard costs (in this case, $210,000).

Pitfalls. Again, there are two major pitfalls. First, you may need to sharpen your labor estimating skills. In the past, it was easy enough to work extra hours without pay if your estimate was low. Now you must pay an employee, and that can throw a huge wrench into your financial plans. As a precaution, consider taking a crash course in estimating, buying some manuals, and adding some extra contingency money to your bids until your range of accuracy on labor is within 1% to 3%.

Second, selling jobs at a 43% markup is certainly not as easy as selling jobs where you subsidized labor overruns and marked up perhaps 15% to 20%. You now need some real selling skills. You have to be able to convince the potential buyer that buying from your company is a good decision, that you are reliable, pleasant to work with, and deliver excellent craftsmanship.

Many remodelers have trouble convincing *themselves* that they are worth the increased charges. But this is the first step to convincing others. In fact, your overhead will continue to go up as a percentage of your sales volume as you get larger in remodeling, so you might as well get used to it now.

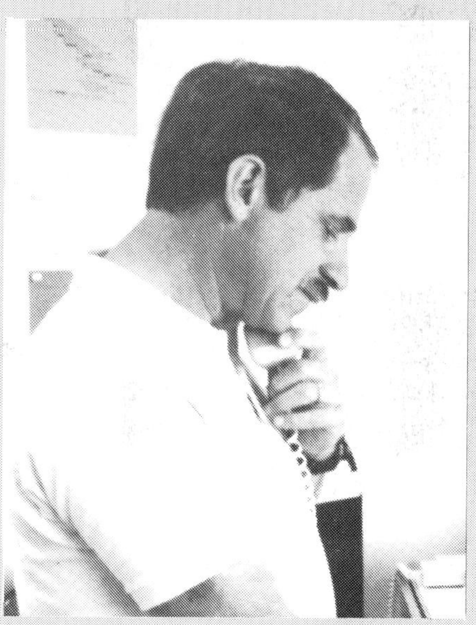

JLC: *With four or more jobs going, you can't be at any one site for much time. How do you delegate?*

Hale: I'm learning to give the people I hire the space to do the best work they can. Because we've grown so quickly, it's forced me to let go—with good results.

But even the best employees can perform well only in a framework they understand. There is nothing worse than being the foreman on a job and not knowing, for instance, where exactly you are expected to get materials or additional help. So I'm trying to give people some clear procedures and schedules to work with.

JLC: *How do you develop those procedures?*

Hale: Scheduling is an important one, of course, so everyone will know what's expected. We use the Master Builder software package, which allows us to derive our schedule directly from our initial estimate of the job. One copy goes to the foreman and one goes to the owner. We update it as often as we feel is necessary or as the owner requests.

If something starts to fall behind, the foremen and I will figure out why — was this a realistic schedule from the start, or is there a real problem? — and try to fix it.

To order materials, we use purchase orders. To avoid delays, we now have a full-time driver just delivering material and making sure no one needs to run to the store. It's deadly to have to pay a carpenter to run to the lumberyard. So the driver really helps keep the materials flowing efficiently.

The driver also helps serve as a conduit — he's the everyday physical link between the company's different parts. He comes to my office in the morning to pick up messages and a list of purchase orders, such as materials that were not delivered with the main load or special-order items.

The driver has a pager and everybody on the job has that pager number. The driver calls every job at just after 7:00 in the morning to see if that job needs any tools or materials, and he routes himself accordingly. He also shuttles time cards and runs debris to the recycling center and the dump.

Continued on next page

Step 3. Because you need to increase your volume, you should start doing some simple marketing like using site signs, sending "thank you" notes for every referral, staying in touch with former clients, printing up fliers and brochures, and doing some home shows.

Pitfalls. Effective marketing — the kind that produces quality leads inexpensively — is a year-round task. If you wait until you need your next job to do this marketing, you'll be pushed into advertising. Advertising produces lower quality leads at a higher cost.

Step 4. As your company grows, office systems, bookkeeping, contracts, and job costing reports all become more crucial. The major difference between a $150,000 company, a $500,000 company, and a $1 million company comes down to increased systemization. This helps others carry out your vision, and allows you to better predict your ability to deliver and achieve your goals.

Pitfalls. Production people like you hate paperwork; real work is done on the job site. Moving from the field to the office can be a difficult adjustment, but the sooner you understand that you make your money in remodeling with a pencil rather than a hammer, the better off you'll be.

Step 5. Communication becomes increasingly important. Where once you sold the job and then delivered it, you now will sell and someone else will produce it.

Pitfalls. Immediately, this requires more complete plans and clearer specifications. Without them, you'll end up spending much of your time straightening out errors and answering questions.

This important transition of leaving field work is usually gradual for the remodeler. Although you can keep your business life fairly simple at this stage, it's crucial that you plan carefully and stay on top of it since each future stage will be more complex and the financial ante will be higher.

This is another fateful moment in the growth of a remodeling business. Let's say that your remodeling company is selling and producing about $500,000 in projects. You wear all three major hats: You're the top (and only) salesperson, the production supervisor, and the administrator. You

INTERVIEW *(continued)*

If he gets done early, there is always a job that needs a laborer for the remainder of the day. There's never a shortage of that somehow.

The driver is really a very important part of the whole job. It's not a high-paying position, but it is a job with a lot of responsibility.

JLC: *Do you have daily meetings with your foremen?*

Hale: Years ago we did that, but it cost too much to pay the people to drive from the office to their jobs after the meetings. So now I meet just once a week with each foreman to go over costs, scheduling, and other details.

We also talk about the customers and how they're coping. It's easy to get wrapped up in the nuts and bolts of construction and ignore your customers. So I encourage my foremen to stay focused on keeping people happy and keeping them informed. I also tell them to keep an eye out for any bad reactions to the progress of the job. That's where the sensitivity to the client comes in. We need to watch for those small shifts in attitude that indicate a customer is growing dessatisfied with the job—that he didn't like what somebody said or the way something looks.

We set up a job phone and a job mailbox at every job so we can always reach each other. This helps move information between me and the site—invoicing, change orders, and that sort of thing. I also developed an index card file to track which tools and which people are on which jobs. Each job has a tabbed section that has cards for each employee and each tool on that job, so that I can see exactly where everything and everybody is.

JLC: *What do you enjoy most about seeing your business grow?*

Hale: Trying to keep four or five jobs going and making sure every client we have is pleased with our work.

I try to talk to every client every day. When I honestly feel that we have pleased those people as much as if we were doing their job only, that makes me feel good and like we really are doing something right. Because without our customers we have no work. They are where it all comes from. I try to pass that on to our employees. I don't pay them, our customers pay them. That is where their livelihood comes from. That is the bottom-line recognition. ■

probably have a part-time bookkeeper or spouse who pitches in to help. But you're still beat. Your week is 65 to 75 hours long. You're late doing everything. Quality is slipping, but not because you don't care. It's because you can't be everywhere and do everything the company needs.

You are ready to hire your first management employee who will take over either production (the actual building) or your selling duties. By now, the experience that you've had hiring and delegating to production employees will be of immense help in learning to shift a major portion of your workload to someone else.

Stage Two: Making a Good Hire

Your new salesperson or production manager should be able to fully take over all sales or production in a $500,000 company. In general, a good, full-line, remodeling salesperson should be able to sell $600,000 of volume, and a good production manager should be able to handle $600,000 to $700,000 of work as a traditional superintendent. If the company uses well-trained lead carpenters who earn bonuses based on job profitability, a sharp production manager should be able to oversee as much as $1.2 million.

Here are some of the steps and precautions that will help get you there:

Step 1. Make sure your new salesperson or production manager is an A+ employee. No remodeling company can survive for very long if either sales or production is crippled by a bad hire.

If possible, look for a remodeler (with strengths in production or sales) who is leaving his own business. Former remodelers have the advantage of understanding the entire business.

Pitfalls. Even if you make the perfect hire, your new employee will still require a lot of training in the way your company does things. This means that for awhile you will be busier than before you hired him. For the first month this training should be very intense. For three months consider that employee probationary; keep up the training and give him lots of feedback. But remember: you probably won't think he's doing the job as well as you did it.

Step 2. Although you're turning over this phase of your job to the new employee, you're not turning over all the control of it. You'll want to monitor his work, and that means much more documentation, which may be a change from the seat-of-the-pants methods you're used to. You didn't have to answer to anyone. But the new employee has to answer to you.

Pitfalls. Taking over part of the boss's job is an unenviable task. Write out a clear job description for your new employee. That will help you clarify just what you are turning over, and provide the new salesperson or production manager with a road map to follow.

Step 3. Do some simple budgeting to assure yourself that you can pay this second management person (you are the first). Will your current markup at your target volume cover your overhead and provide a net profit for your company? If not, you will either have to raise the volume or the markup or both. Either that or you'll have to find a way to reduce costs.

Pitfalls. At this stage it's crucial that you abandon the "I'm only providing a job for myself" business mentality. You owe it to your business and to your employees to make a net profit that you can invest in the future and use to bridge cash shortfalls. Quite simply, you have to replace seat-of-the-pants management with good future planning.

Step 4. You will also need a strategy to develop additional volume. Budget for simple marketing that will generate enough leads to produce the extra work. That marketing needs to continue throughout the year even if you reach your target number of leads. You'll also need to set up simple ways of tracking the marketing. By keeping data on number of leads, source of leads, lead costs, and sales costs you'll be able to budget better and market smarter.

Pitfalls. As your marketing increases the number of leads coming in, it's easy to treat these leads with equal attention. Learn to qualify leads so that your time is spent on the ones that are likely to bring in work.

Step 5. You may remain extremely busy because you are not using your time as efficiently as you could. For two weeks keep a time card of all you do; break it down into 15-minute increments. Then analyze that list and find someone else to do anything that can be done by a person who makes $15 an hour or less. Thus drafting, bookkeeping, running to the lumberyard, typing a contract, etc. should all be farmed out to employees or freelancers.

Pitfalls. This not only requires good delegation skills but may also mean that you have to give up some of the things you've always reserved for yourself because you really like doing them.

Step 6. At $500,000 you'll probably have a number of field employees and a full time bookkeeper/ secretary/ assistant. Set a goal of no more than one field employee per $100,000 in volume and work to make that output higher every year.

Pitfalls. Production will ultimately be the make-it-or-break-it part of your business. Assuring good client relations and a well-crafted product is ultimately the hardest part of remodeling and should be looked on as the part of the business that all other parts serve. Consider giving your lead carpenters a performance bonus based on how their labor costs match those in the estimate. Create a full loop where everyone, including carpenters, helps to feed data into estimating, and then track those estimates against actual costs.

Step 7. Also plan to invite a computer into your business life. If you are running your business properly, you are producing more and more data on lead production, lead cost, job costs, production output, sales goals, etc. The earlier you computerize, the easier the transition from paper to computer will be.

Pitfalls. Net profit shouldn't go into your pocket (although it's tempting). It should be used as seed money to help your company grow. Switching to computer is an excellent example of spending money and time now — without immediate return — in order to reap later benefits.

When you are inside your company riding the growth wave and trying to cope, the changes can come quickly and seem almost random. But from outside they are predictable and the relationship between them is easily seen.

"Growing" your company is never easy. But if you plan carefully, you'll gain the necessary management experience as the company increases its volume. Each new stage in the life of your company is a challenge that can prepare you for even greater ones down the road. ■

Linda W. Case is president of Remodeling Consulting Services, in Silver Spring, Md., and is a nationally known writer, speaker, and consultant for the remodeling industry.

You Can't Do It All

Delegating administrative tasks to a capable employee can prevent a host of problems — and free you to do what you do best

by Judson Motsenbocker

When I started my remodeling business 22 years ago, I knew there was more to it than just rounding up some jobs and banging nails. But I have been surprised by the variety of skills I have needed to be successful.

Good people skills, for instance, are as basic to this line of work as the ability to work with tools and wood. You can't sell your business without them, and they're essential to maintaining good relationships with your clients.

But perhaps most important, and most foreign to many contractors, are the administrative abilities necessary to keep the business side running smoothly. People who start businesses are notoriously negligent about tracking the day-to-day administrative tasks such as bookkeeping, correspondence, and filing. Somehow these activities don't seem as productive as making bids or doing the actual remodeling—they distract us and keep us from focusing on the big picture. However, these activities *are* important and need attention.

Don't Neglect Administration

Every construction business may be divided into three distinct areas: sales and estimating, production, and administration. Even in small companies, no one can do all three. But trying to do it all is probably the most common mistake small contractors make. When one person tries to do all three of these things, all three areas suffer. But it's usually the administrative end that suffers most.

It's easy to underestimate the importance of these administrative tasks. Consider what they actually include: answering the phone, handling mail, typing letters and proposals, filing, verifying and paying invoices, posting verified prices to job cards, placing and verifying purchase orders, billing, handling payroll and taxes, keeping insurance up to date, tracking accounts receivable and accounts payable, and working with the company's attorney, accountant, insurance agent, and banker.

Every construction business must perform these tasks, and in all but the tiniest companies they are a full-time job for at least one person. Neglecting them can cause all sorts of problems: lost sales leads because of unreturned phone calls; credit problems because of late payments to suppliers; overpayments to suppliers because you don't have time to check invoices; inaccurate estimates because of poor cost records; and IRS audits or late payment penalties because of late or sloppy tax work.

Make the Right Hire

Even if you're convinced you need one, hiring an administrator is easy to put off. It's expensive, it's a hassle to train someone, and it's hard to find the right person, someone who understands the construction business. It's also difficult to delegate when you're used to doing it all yourself. But I learned early that neglecting administration is a mistake. You'd never dream of handling production that way.

When I started out, even though I had only four people in the field, I hired a full-time administrative person. That person's salary was added in with the other costs of doing business, such as other salaries, insurance, gas, transportation, and tools. I knew I'd have to sell more to cover that salary, but having an administrative assistant allowed me to concentrate on sales, estimating, and production.

You may have to look hard to find the right person, and when you do, that person should be well paid. You want to hire someone capable and willing to take responsibility for this whole area of your business. You'll probably need to pay more than if you were hiring clerical help. If you aren't sure of the going rate, check with other companies or ask around at a construction trade association meeting.

Define The Job

Before hiring an administrator, you should define that person's duties. The following items should be included:

Phone calls and correspondence. These are your links to the outside world. The administrator should answer all calls promptly and pleasantly and take good, detailed messages. It's easy to underestimate the advantage of a live voice versus an answering machine, which many people won't talk to. According to some estimates, answering machines scare off as many as 50% of all calls. The administrator should also keep up on correspondence. Prompt, well-written correspondence on attractive letterhead does wonders for your business's image.

Leads. Teach your administrator as much as possible about the business. The more an administrator knows about construction and your company's capabilities and expertise, the better he or she can answer phone inquiries and help you qualify and pursue your leads.

Verifying invoices and purchase orders. Suppliers do make mistakes on invoices and purchase orders. All invoices should be compared against receipts, and everything should be filed promptly and neatly.

Tracking costs. This goes along with verifying invoices. I have my administrator keep my materials price book current. This saves me many phone calls and helps with job costing and estimates down the line.

Bookkeeping. Your accountant may generate your year-end numbers and quarterly taxes, but the administrator should handle the day-to-day bookkeeping duties, such as payables, receivables, payroll, and monthly profit and loss statements. He or she can also stay on top of late receivables.

Scheduling. Since the administrator is in the office all day, he or she makes and verifies appointments with clients.

Working with other professionals. Every business needs its "advisers" — an accountant, an attorney, an insurance agent, and a banker. The administrator can maintain contact with these people and make sure all relevant paperwork, accounts, and insurance coverage are up to date.

Office procedures. This is the rest — the filing, bill paying, processing of time cards, and general paperwork that will otherwise pile up on your desk.

What About Family?

Many contractors hire a family member as the administrative person. This works well for some. But for many, working with a family member causes all kinds of problems, from fights at home to resentment among other employees. If you hire a family member, it's especially important to define their duties and boundaries of responsibility clearly, and make sure you both stick with them.

Do What You Do Best

Hiring an administrator may help your operation run more smoothly and professionally. But its real value is that it allows you to do what you do best: sales, estimating, and production. It also frees you to concentrate on the big picture so that your business can keep growing, and so that you can anticipate and adjust to changes in the business climate. The way I see it, there isn't any way you can afford not to have this kind of help. ■

Judson Motsenbocker is owner of Jud Construction in Muncie, Ind., and a frequent speaker on business management for remodelers.

"Lead Men" Can Solve Management Problems

Using experienced carpenters to supervise job sites helps this contractor run multiple jobs efficiently and profitably

by Stephen King

When I started my design and build remodeling company, Common Vision, I had two enthusiastic partners who could each shoulder one-third of the management burden. We often had three or four jobs going simultaneously, and as long as one of us was on the job site, we didn't need to hire crew leaders. But after a period of time one partner dropped out, and then another. This forced me into a different management role, and meant I could no longer supervise at the job.

To ensure that operations continued to run smoothly, I had to change the company's management approach. The key to the new system has been the use of experienced carpenters as "lead men" to supervise job sites.

With this management strategy, a good computer system, and a few tricks to keep jobs organized, I've been able to do all the sales and estimating and keep six year-round employees busy. Our company has done about $860,000 worth of business each of the past three years, and even now, in the lean times of the New England building recession, we're getting jobs from repeat customers and referrals.

The Lead Man Concept

Contractors who start small generally work on the job along with another carpenter or two. After a full day at the job site, they do the paperwork on their kitchen tables at night. But there comes a point when, to maintain the company's business volume, you need more than employees — you need job leaders who feel they have a real stake in the success of the company. You need "lead men."

A real transition occurs at this point, when you turn over the job to a lead man or woman. You are still the sole owner of the company, but you now have responsible employees. This new era brings with it a need for a real management "strategy," and what follows are some of the elements of a strategy that works for me:

Bonuses. On top of my lead carpenters' salaries, I add year-end bonuses based on how the company does. This gives them an extra incentive to keep the operation running smoothly and efficiently.

Training and education. I also invest considerable money in education. Employees who know more about their jobs contribute to company productivity. I budget roughly $300 to $500 per employee per year; employees can use this money for anything they enjoy learning — it doesn't even have to be directly related to construction, although it usually is.

This money pays for magazine subscriptions, books, day or weekend courses, seminars or conferences, or semester evening courses in anything from framing to management to nonconstruction topics. If a lead man can make an argument that a course would be especially valuable, we'll go beyond the $500. This money adds up, but it encourages our lead men to expand their professional and personal horizons.

Minicompanies. Perhaps even more important, job leaders are given a certain degree of autonomy. They handle every job from start to finish, essentially acting as the contractor. I'm involved mainly in the preliminaries of sales and estimating, leaving the building and customer relations to the lead men. They make the work schedules, deal with the subs and owners, coordinate the different stages of construction among themselves, and generally take responsibility for the actual construction. In a way, they run their jobs as minicompanies.

Sometimes our lead men pass the baton of job leadership from one to another within a given project. This lets each of them bring his strongest skills to bear on every job. On a large job, for instance, our framing lead man will handle the early stages, then pass the job on to the cabinet and finish leads. A third lead man might come in to do the siding and exterior work. This can mean overlaps of a day or two as transitions take place, but it ensures that the work is always of the highest quality. And it keeps everyone working on what interests them most.

Of course, on smaller jobs, such as a kitchen renovation that's light on framing, one person can usually take it all the way through.

When we developed this system, my big worry was that the lack of continuity would bother the client. But we haven't had any complaints. The day or two of overlap between lead people helps. We make it a point at that time to have both lead people talk to the clients and answer any questions. In addition, I drop by all jobs at least once a week to make sure everyone is happy.

Finally, to keep communication sharp among the seven of us, we have monthly meetings where we share anything new, fine-tune procedures, and resolve mistakes, gripes, or other problems.

Defining my role. To give the lead men the independence they are promised, it's important that my role is well-defined. I do all the sales — making sure the six lead men stay busy. I also do design.

In addition, I do the estimating, and I'll usually place the first large materials order on a big job. I track costs, checking the expenditures on larger jobs every two weeks. But I don't pressure anyone to work faster just to keep up with my estimate. There is generally nothing I can do to make a job move faster anyway, and haranguing people usually just discourages them and forces mistakes. So I don't make a big deal about it when we're behind schedule. Of course, I want to find out why it took longer, so I can take it into account in the next estimate.

Getting the Right People

The crucial element of this system is good people. When I hire, I look for people who know how to schedule jobs and read specs and plans — that is one reason I've hired three former independent contractors. But I also consider personal qualities heavily. I look for experienced people who can act independently.

The author, at bottom left, and his lead men: All are capable and able to act independently. Several are former contractors.

Of course, company policy has to foster this independent thinking, or it won't happen. No one is punished for taking a risk or trying something. We chalk up mistakes as learning experiences. Knowing they can make occasional mistakes encourages people to solve problems on their own.

Staying Organized

Another key to making this system work is organization. We keep things organized by keeping them as simple as possible, and by documenting every job well.

Job folders and diaries. We keep all the job information on each job in a single folder, which gets passed from lead man to lead man at any transition. The job folders have material suppliers' phone numbers, the cost of materials, change orders, and any other information the job leaders will need on the job site.

In addition, we keep a sort of "job diary" either at the job site or with the job folder.

Each day's work is recorded in the diary, along with anything of note that comes up: problems that resulted in changes of plans, questions from the owner, and anything else out of the ordinary. Having this type of information on record helps later when the next lead man is asked, "What about that bay window we talked about adding on?"

The diaries also keep me in touch with what is going on, since I collect and review them at the end of each week.

Standard construction techniques. We also stay organized by using standard construction procedures whenever possible. We've put about half of this information in book form — I'm working toward having it all on paper.

One example of this is our standard details for foundations, framing, and interior finish work. We don't have ten ways of doing things. If a person learns a better way of doing a job, he or she passes it on, and it becomes standard procedure. This not only makes for consistent work, but helps prevent conflicts when two project managers are working on the same job.

The job bucket. Trips to the hardware store or the lumberyard for miscellaneous supplies are standard fare in most remodeling companies. To reduce these runs, we have several six-gallon plastic trash cans with snap-on lids that contain miscellaneous supplies such as a first-aid kit, light bulbs, staples, and caulk. A list is taped on top, and one of these trash cans goes to every job. Anyone taking stuff out has to replace it on the next lumber run.

I look on the list of activities on the time card to see how much time people spend on material pickup. If I see people spending two hours on miscellaneous runs to the hardware store, I find out why. Sometimes you run short on shingles, which is a legitimate reason. But the job bucket eliminates most of the miscellaneous runs and makes the work day more productive.

Low overhead. Finally, we reduce overhead by foregoing an office. We use a 14x14-foot space in my home, and it's packed with three work stations and all the filing cabinets we need. We now have two computers in this space, and a lot of our job information is on disk. A separate phone line and good stationery help provide a professional image.

Running our "lean and mean" company has brought us plenty of work, satisfaction, and satisfied clients. I'm happy with the efficiency with which we run an $800,000 a year company. But most of all I'm impressed by how much the lead men in our company care about their jobs. It shows in their work, and the clients see it — often a client or referral will ask for a particular lead man. ■

Stephen King's company, Common Vision, remodels residences in the New Haven, Conn. area.

Chapter 10 • Scheduling

Planning With Your Schedule

by Sal Alfano

With good scheduling practices, you can manage increased volume and earn greater profits

Scheduling is the key to keeping a building business in order. Good scheduling is especially crucial when you're lucky enough to have more work than you can comfortably handle. Scheduling involves more than planning individual jobs. It also means coordinating multiple jobs of varying size and scope, and with overlapping start and completion times. Obviously, the two activities are closely related and add up to a monumental task. To get started, you must first gather preliminary information.

Boom Or Bust

Preliminary scheduling requires a little crystal-ball gazing to determine what work lies ahead. But predicting the coming building climate is not all guesswork. The trick is to use every resource at your disposal to anticipate the future demand for your services. Start with architects and designers for whom you regularly bid work. Find out what projects they have on the boards and what's ahead. Excavators are another source. Often, owners will talk to an excavator about their upcoming project long before they've decided on a builder. The same is true of real-estate agents and bankers.

Also gauge the demand for repeat business. The owner for whom you built a house last year may want to add a deck or covered porch this year. The key is to find this out early enough to make room in the schedule. Often the best approach is to simply come out and ask your former clients directly. They will likely appreciate your professional attitude.

Small projects can be scheduled in advance for slow times. Or you can split a job into two parts. For example, you can schedule the footings and piers for a deck to be constructed anytime before frost, and save the carpentry for winter when nothing's doing.

In a bust year, you may have to take on a greater number of small projects. In a boom year, you can pick and choose the work more carefully, concentrating on projects that will make the most money with the fewest headaches.

When, How Long, and Who

To schedule successfully, you must answer three questions: What are the start and completion dates? How long will it take to perform the work? Who will be available to do the work?

Start dates. Most customers are only concerned about when you'll be finished. They want the addition done by their daughter's graduation day or some other milestone. But don't fall into the trap of promising a finish date until you can nail down a starting date.

Before you sign a contract with the owner, notify your key subcontractors — excavators, concrete crews, well drillers, roofers — of your best guess for when the job will start. Be up front with them, and let them know that it might be revised. They'll appreciate the help it gives them in their own planning. And they'll be able to tell you if they can meet your schedule.

Owners and architects almost always underestimate the amount of time it will take to draw up the plans and specs, estimate costs, make revisions, and write the contract. Since all of this must be done before you break ground, be realistic when you commit to a start date. If you need two weeks to estimate costs or finish up work in progress, say so, and adjust the starting time accordingly.

Completion dates. When it comes to figuring out how long a specific project is going to take, don't let anyone else's opinions affect your own tried-and-true judgment. Don't consent to the owners' unrealistic timetables; educate them as early as you can. You're better off losing the job than agreeing to an impossible deadline.

When considering calendar-day completion schedules, be sure to allow for delays. Weather will interfere with the work of several key operations — excavation, concrete, and roofing, for example. Two weeks of rain just before your job starts will likely put you just as far behind as two weeks of rain after you start, because the subs you're planning to use will fall behind on the projects they're doing prior to yours.

The same is true for changes in the work. Almost all change orders are written for additional work, which takes additional time. Make sure you allow for this time in your schedule.

Finally, some materials, like windows and doors, require long lead times. If you can't install the siding because you haven't received the windows yet, you're in trouble. Certain other items — light fixtures, plumbing fixtures, and locksets are good examples — are often not chosen until well after the job has begun. If you see a dangerous delay developing because of late orders or slow delivery of key elements, make the appropriate adjustment as early as possible. For instance, you can provide for extensions of time like this in your contracts. But whatever you do, keep your clients informed. They'd much rather hear about a delay sooner than later. Also, if they need to make choices to keep things moving, a little gentle prodding can work wonders.

Labor requirements. You must also determine who is going to do the work. In boom times, you'll need to hire extra help as required, preferably from a pool of people you have hired before. Known degrees of skill and rates of production are important since it is almost impossible to determine how much time an unknown crew will take to complete a given job.

Make sure your long-term employees will be available to supervise the extra help. They will save you time by explaining your normal methods for you, and they'll insure a higher quality, more consistent product. And be sure to tap the extra help as early in the season as possible, before the pool dries up. This will depend, of course, on the kind of season you anticipate.

If it looks like a really busy season ahead, consider subcontracting more parts of the job than usual. If you sub the roofing and drywall, you free up your crew for exterior trim and siding. If you sub the finish work or cabinet installation, you will have more people available to get the next job going.

	Task	Date	Personnel
1.	Design	9/1	Office
2.	Select new entry door	9/1	Office
3.	Select new exterior lights	9/1	Office
4.	Select paint & stain	9/1	Office
5.	Schedule excavator	9/2	Office
6.	Schedule electrician	9/2	Office
7.	Schedule painter	9/2	Office
8.	Order new entry door	9/2	Office
9.	Order new exterior lights	9/2	Office
10.	Order material delivery	9/2	Office
11.	Excavate for piers	9/8	Excavator
12.	Place footings and piers	9/8	Crew #1
13.	Framing	9/9	Crew #1
14.	Lay deck boards	9/11	Crew #1
15.	Build bench and railing	9/14	Crew #1
16.	Build stair and railing	9/14	Crew #1
17.	Break through new doorway	9/16	Crew #2
18.	Install new door	9/16	Crew #2
19.	Install new exterior lights	9/16	Electrician
20.	Trim and punchout	9/17	Crew #2
21.	Paint and stain	9/18	Painter
22.	Cleanup	9/18	Crew #2

Figure 1. *This simple schedule breaks down the job into specific tasks, puts them in order by start date, and shows who will perform the work.*

Use A System

Since every client wants to know when their project will be finished, you need a reliable system for predicting progress. Ideally, your system will be easy to establish and maintain, and will incorporate features that allow you to adjust to the inevitable delays.

A good scheduling system accounts for two basic kinds of tasks. *Administrative* tasks include ordering materials and equipment, selecting products (this often involves helping the owner make choices), and scheduling other tasks, including subcontractor work. *Field* tasks include all of the work performed in the actual construction of the building: framing, sheathing, siding, trim, and so on.

Each of these broad categories can be broken down into smaller subdivisions, and each builder must decide on the level of detail required. As the detail increases, so does the accuracy of the schedule, but it also becomes increasingly difficult and more time consuming to maintain.

Task List

Let's use a simple example to illustrate how all the parts of the schedule come together: a small exterior deck with a new door from the existing house. The tasks might break down as follows:

Administrative
- Design
- Schedule excavator
- Schedule material delivery
- Select new entry door
- Order new entry door
- Select new exterior lights
- Order new exterior lights
- Schedule electrician
- Select paint and stain
- Schedule painter

Field
- Excavate for piers
- Place footings and piers
- Framing
- Lay deck boards
- Build bench and railing
- Build stair and railing
- Break through new doorway
- Install new door
- Trim and punchout
- Cleanup
- Install new exterior lights
- Paint and stain

The simplest way to develop this schedule on paper is to first put each task in the proper order, and attach a date to each one. This means reshuffling the original list. You should also note who will perform each task (Figure 1, facing page).

This simple listing provides a basic schedule. It shows every task related to the job, when it starts, and who will perform the work. But it leaves a little to be desired. For one thing, it doesn't explain the gaps between dates, or indicate when certain portions of the work will be complete. It's also difficult to see how individual tasks relate to each other.

Task September	T 1	W 2	T 3	F 4	S 5	S 6	M 7	T 8	W 9	T 10	F 11	S 12	S 13	M 14	T 15	W 16	T 17	F 18	**Personnel**
1. Design	X																		Office
2. Select new Entry pool	X																		Office
3. Select new exterior lights	X																		Office
4. Select paint and stain	X																		Office
5. Schedule excavator		X																	Office
6. Schedule electrician		X																	Office
7. Schedule painter		X																	Office
8. Order new entry door		X																	Office
9. Order new exterior lights		X																	Office
10. Order material delivery		X																	Office
11. Dig holes for piers								X											Excavator
12. Place footings and piers								X											Crew #1
13. Framing									X	X									Crew #1
14. Lay deck boards											X			X					Crew #1
15. Build bench and railing														X	X				Crew #1
16. Build stair and railing														X	X				Crew #1
17. Break through new doorway																X			Crew #2
18. Install new door																X			Crew #2
19. Install new exterior lights																X			Electrician
20. Trim and punchout																	X		Sam
21. Paint and stain																		X	Painter
22. Cleanup																		X	Sam

Notes:
11. Send transit to site.
17. Remind Sam to check for buried wires.

Figure 2. *Graphing a schedule makes it easy to read and understand at a glance. This graph shows both start and finish dates, and allows you to track progress by shading each task bar as the work is completed (tasks 1 through 12 are completed).*

Seeing Is Believing

A more visual presentation often solves these problems. Using a piece of 1/4-inch graph paper or similarly ruled ledger paper, you can display the same information more graphically and more clearly (Figure 2, previous page).

This kind of schedule can be read and understood at a glance. It's now obvious that the gap between 9/2 and 9/8 is the lead time required to schedule the excavator. And the reason "Task 14, Lay deck board" takes four days is because a weekend intervenes. Start and finish dates are shown as well, and you can track progress by shading or coloring each task as it is completed. There's also plenty of room to revise the schedule. If it rains on Wednesday 9/9 and the framing has to be postponed, it's easy to move everything down a notch. Using different colors for revised start and finish dates will help keep things straight. You'll also end up with a clear record of when the job was delayed and how it affected the completion date. You can append notes, footnote style, to the bottom, or write them in alongside the appropriate block.

Many people find it helpful to set up this type of chart on a wall, using movable markers to make changes in the schedule. I use colored 3x5 index cards and push pins, using one color for subs, one for my own crews, and a third color for key dates, such as the completion of framing, plumbing, or rough-in.

Mix And Match

You can combine or modify these methods depending on what you're comfortable with, the size of the project, and the number of jobs you have to schedule in a given period. The basics are a list of tasks, some way to assign them to days of the week, and a simple method of reassigning them as the need arises. A graphically laid out schedule will greatly improve your ability to visualize how the job falls together and to make adjustments before it starts to fall apart. ■

Sal Alfano, formerly a general contractor in East Calais, Vt., is an associate editor with The Journal of Light Construction.

Simple Scheduling for Remodelers

Two simple charts will help keep employees, subs, and suppliers informed and on track

by Larry Bussey

For years our company has struggled with scheduling our remodeling projects. Not so surprisingly, all jobs seem to take on identities of their own. What works for a bathroom remodel doesn't necessarily work for a bedroom addition. And even "standard" kitchen remodels can be as different as day and night.

Besides, as most of us have discovered the hard way, scheduling a remodel is a little like charming a snake — as soon as you think you've got it made and relax a little, the snake gets restless and bites you. So why spend your valuable time scheduling for a business dominated by Murphy's Law?

Well, as long as your company is doing only a few jobs at a time and you can personally supervise all of them, you can get by without too much structure. But as soon as things get a little more complex, you can no longer supervise every job yourself and you have to begin to delegate management responsibilities. When that happens, a good working schedule that builds out each project on paper becomes a necessity.

This is true in my own business — with a slight wrinkle. I grew up learning the "kick 'em in the butt" system of motivational management. While I was perfecting this technique, however, the management industry switched tracks and began to employ the "pat 'em on the back" method of motivation.

So there I was, completely competent in an outdated motivational skill, and totally frustrated in my vain attempts to master the new technique. Out of necessity, I have developed a scheduling system that allows others (more skilled than me in the new motivational approach but less skilled in remodeling) to run our projects smoothly, cover our overhead, and still make a profit.

Getting Started

The best-kept secret in creating a schedule is to keep it simple enough that your office staff will be able to use it to keep your jobs on track.

The second best-kept secret in scheduling is how to use the schedules themselves — they should be right up there on the wall.

They also should be in the hands of your subcontractors, and can even be shown to your customers. But I'll get to that in a minute.

The best time to schedule is when it's quiet and there will be few or no interruptions. Write down any questions that arise as you create the schedule, and get the answers later, if possible. Of course, if you have several people capable of scheduling, you may want to employ the think tank method and plan together. The key here is to experiment until you find out what works best. For our company I usually do the rough draft and then call in one or two superintendents to work through it behind me.

Over the years our company has tried many different types of scheduling systems. What we've settled on is a hybrid of PERT (Project Evaluation and Review Technique) and CPM (Critical Path Method), with a healthy dose of SWAG (Sophisticated Wild Guessing) thrown in for good measure. Sounds impressive, but it couldn't be simpler.

First, we make up a master schedule for the entire company. This helps to coordinate all employees and subcontractors, and includes all materials deliveries. Then we create an individual schedule for each job that shows which employees, subs, and suppliers will be on the site each day. Here's how to duplicate what we use.

Cut And Paste

Start with a large desk calendar (pad type), cut it into horizontal strips, and then tape it back together in a single, horizontal line. You'll need as many days as your longest wall will hold for this master calendar, and other different-length calendars for the individual job schedules.

We have a blank page made with several rows of seven squares labeled Monday through Sunday. From this master we can run off as many copies as we need to make up our own calendars. It's also great when we have to add a few days to an individual job schedule.

For both master and individual calendars you'll need a good selection of colored highlighting markers (if you use a computer, this system works well with color graphics) to color-code the different subs, suppliers, and in-house trades on your schedules.

Master schedule. This schedule gives you "the big picture." Once your blank schedule is all taped together, select a color for holidays and weekends, then fill them in.

Now choose a shade for each major trade and each supplier. You'll also need different colors for each of the different types of work to be done by your company employees (painting, carpentry, etc.) Then, choose one last shade to represent all minor subcontractors.

Now add the date to each square in the appropriate day of the week. The remainder of each square will contain the color-highlighted names of all the subs, suppliers, and employee trades scheduled to work any of the jobs that day. I list my subs and deliveries at the top of each square, and the employee tasks near the bottom (see Figure 1, next page).

Individual job schedules. These are similar to the master schedule, except that their lengths will vary according to the job.

But before you cut your job calendars to precise lengths, let me share a trick with you that works for our company. To allow catch-up time, don't schedule anything for Mondays and only plan for half a day on Fridays. Also, add 15 percent of the total working days scheduled (including Mondays and Fridays) to predict a realistic completion date. However, this new completion date is only for you and your client. If your employees and subcontractors know they have extra time, they'll use it. Work always expands to fill the time allotted for it.

Now that you've got your horizontal calendar ready with weekends and holidays color-coded, you can start filling it in. In the first square, list the job name, job address, job telephone number, and start date (never the completion date) in bold print as shown in Figure 2 (next page). In the next square, list the names and telephone numbers of the subcontractors you'll be using on this job. In the third square, do the same with all suppliers, and in the fourth and fifth squares list all employees for that job. Each sub, supplier, and employee trade should be color-coded with the same highlighting marker used on the master calendar. At this point, it is not necessary to fill in the calendar dates; in fact, you shouldn't do so until the construction permits are in hand.

Beginning with the sixth square, lay out the job from permit through punch list and final cleanup. Remember to consult your master schedule to prevent conflicts. Also,

Master Schedule

Ralph's Plumbing Boyd Electric ABC HVAC Valley Drywall Ace Roofing Misc. Subs Harvey Lumber Davidson Concrete Carpenters Painters Laborers	Tuesday 5/9 Plumbing rough-in Roofing installed Electrical rough-in Carpentry	Wednesday 5/10 Plumbing rough-in Roofing installed Electrical rough-in Painters	Thursday 5/11 Roofing Installed Electrical rough-in Drywall del.

Figure 1. *On the Master Schedule, each sub, supplier, and trade is highlighted with a different color. This lets you quickly see who is busy on a given day. If their name and color appear on this calendar, you can consult the Individual Job Schedules for the particulars.*

Individual Job Schedule

J.R. Smith 1102 5th Street 555-4011 May 9	Ralph's Plumbing 555-4341 Boyd Electric 555-4815 ABC HVAC 555-6611 Ace Roofing 555-3650 Black Diamond 555-4611	Harvey Lumber 555-2658 Valley Drywall Davidson Concrete 555-4161	John Jay-Supt. Jesse-Carpenter Shane-Carpenter Greg-Laborer Les-Laborer Will-Laborer	Bill M.-Painter Ron-Painter	Demolition DAY 1
start Date	Subs	Suppliers	Employees	Employees	

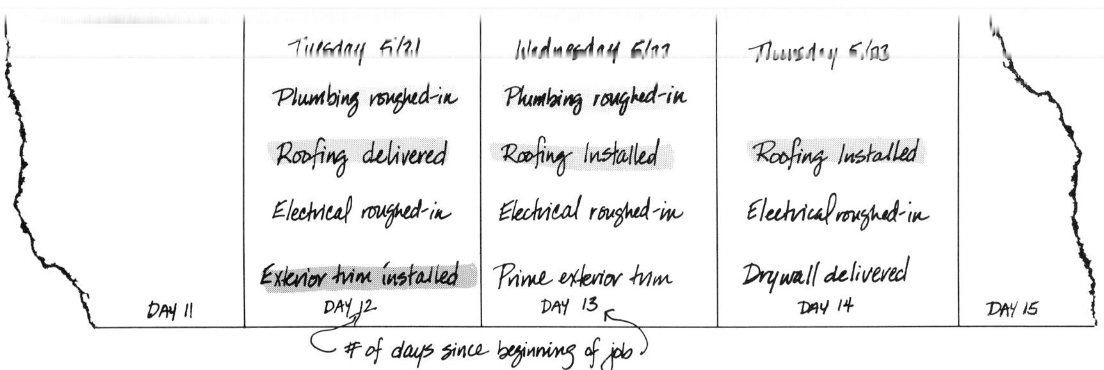

Figure 2. *The first five squares of an Individual Job Schedule contain all names, phone numbers, and employee assignments for that job. Starting with the sixth square, the job is laid out from demolition through cleanup using the same color-coding as the Master Schedule. At the bottom of each square, write the number of days since the beginning of the job. Days 11 through 15 are shown at bottom.*

account for holidays and vacation days that your clients, key employees, suppliers, and subs will be taking.

Once you've finished with this individual schedule, transfer all sub, supplier, and employee references with their color-coding to the corresponding days on the master schedule (if they are not already listed for another job) to make sure it reflects the sum total activity for each day.

Using Your New Tool

Knowing when to start a job is critical. This is where your master schedule again comes into play. By consulting it, you can give realistic starting dates to your prospective customers. I've found over the years that even the most impatient clients are willing to wait, provided you can show them why. And a schedule can be a very persuasive tool. Why? Very simple. How many remodeling contractors or subcontractors do you personally know who use a scheduling system? My guess is none.

But this kind of scheduling really shines with subs too, particularly the ones that I've encountered. Although very competent, their own scheduling revolves around the age-old principle that the "squeaky wheel gets the grease." That's why we have developed such a "squeaky" system.

First, one of my job superintendents or I call the subcontractors that will work on a specific job (after a client's contract is signed) and fill them in on our schedule. Often we invite them into our office to go over it and make any needed adjustments. (There shouldn't be too many surprises at this stage though, since we already have asked the subcontractor or supplier about their current and upcoming workload at the time they submitted their bids.) They are then given a copy of the new job schedule along with an updated master schedule.

But just in case the subs forget, our office manager/secretary calls one week before each sub's scheduled arrival or supplier's delivery and then again the morning before. This is a precaution against somebody else's "squeaky wheel" getting their attention.

Simply put, a good scheduling system will shape up even the most disorganized company because it forces you to think ahead instead of reacting to the disasters of the moment. And like using a computer, it becomes so much a part of your routine you can't imagine having done business without it. ■

Larry Bussey has been a remodeling contractor for nearly 20 years. His company, The Larry Bussey Group, now focuses on the purchase and renovation of older properties.

The Critical Path Method of Scheduling

by David Carns

With CPM, you'll know when each task must be started and finished to complete the job on schedule

Simple construction projects rarely have formalized schedules. But as projects increase in size and complexity, involving more labor, materials, and equipment, you'll want a scheduling technique that gives both you and the owner a better ongoing picture of a job's progress.

A good scheduling system brings several benefits. First, it provides an accurate financial picture of the project. Both your costs and the owner's are closely related to a project's duration. The sooner a project is completed, the sooner the owner can use the building, and the sooner you, as builder, can direct your resources to other jobs.

In addition, a good scheduling system:

- Establishes critical "must-do" dates and other milestones, such as the start and finish dates for the job, and the latest dates to obtain permits, order materials, and call subs;
- Helps subs and suppliers see how they fit into the overall picture;
- Helps prevent resource conflicts, such as scheduling two jobs for the same time period, each of which requires three carpenters when you have only four carpenters available;
- Forecasts cost and payment information to help with cash flow on a job;

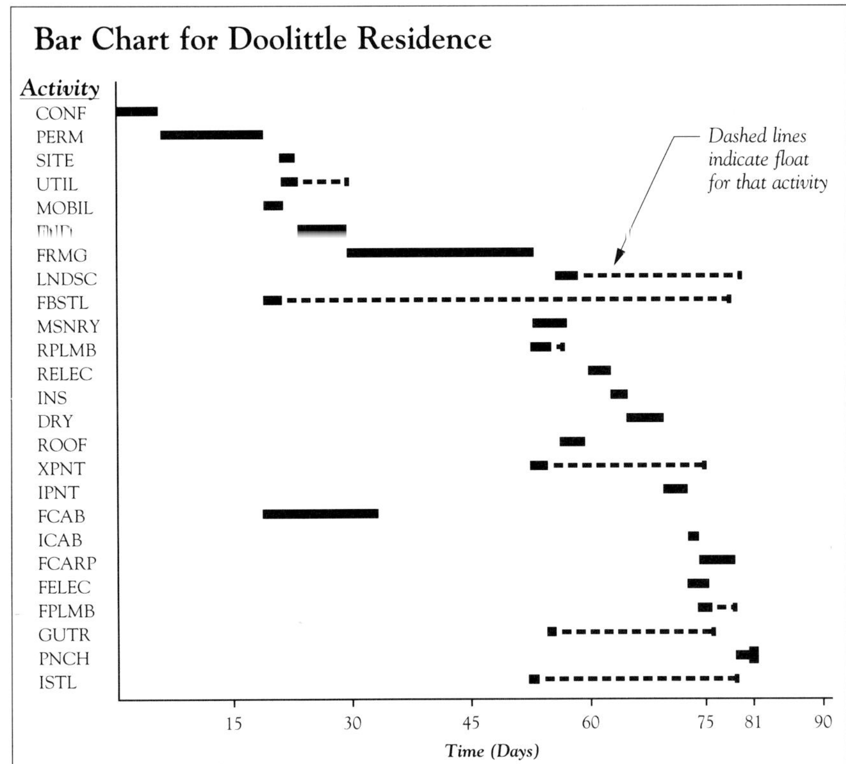

Figure 1. *The bar chart is a common scheduling tool useful for showing the start and finish time of each activity. It is limited in that it doesn't show which tasks must be completed before another can begin.*

- Serves as an excellent sales tool, convincing both owners and financiers of a potential project's value. If you've taken the time and made the effort to analyze a job's progress before it even starts, you are probably one step ahead of the competition.

Finally, the more sophisticated scheduling systems can show how the various activities in a job depend on each other, as well as their expected durations. This shows you at a glance which items you must start and complete at any given time in order to finish the job on schedule.

Bar Charts vs. Networks

There are several common scheduling methods used by contractors. One of the most basic, the bar chart (see Figure 1), uses horizontal bars that run above a time line to show the start and finish time of each activity. The bar chart is a good basic tool, but it does not show which activities depend on which others. This can lead to trouble, since it's easy to lose track of what steps in the process must be completed before others can begin.

A better scheduling system shows the interrelation between different stages of construction. Rather than having each activity shown as a separate bar, they're portrayed as boxes linked in a *network* representing their relationships and interdependencies.

For small and medium-sized builders, the most relevant network technique is called the Critical Path Method, or CPM. This method was designed for large commercial projects. But small builders working on complete houses or moderate to large remodels can also benefit from the extra organization — and therefore flexibility — that CPM provides.

Linking the essentials. The main goal of CPM scheduling is to clearly identify and link the stages of construction that form the project's "critical path" — the sequence of interrelated activities that determine the project's minimum duration. A delay in any one of these activities will mean a delay in the project completion date.

A CPM flow chart (see Figure 2) links these critical activities along a central critical path that forms the backbone of the construction process. Activities that are not critical (in the sense that there is some flexibility regarding their completion time) are represented on work paths that branch off the critical path.

CPM can look intimidating at first. But once you know how to create and read a flow chart and have worked through a job or two with the system, you should be able to use it easily. I'll walk through a sample project here to show the steps involved in constructing and reading a CPM schedule.

Creating A Schedule

Let's say you need a schedule for a custom home you are planning to build for the Doolittle family. Here's how to begin:

Step 1: Define activities and their durations. This is something you probably already do for every job, at least in your head. Break the project into individual work activities, and give each stage a three-to-five-letter abbreviation, as shown in Table 1 (page 140). The level of detail you use for these categories depends on the project's size and complexity, as well as the intended use of the schedule. You must also establish time durations for each activity. These are usually given in work days and will depend on the available

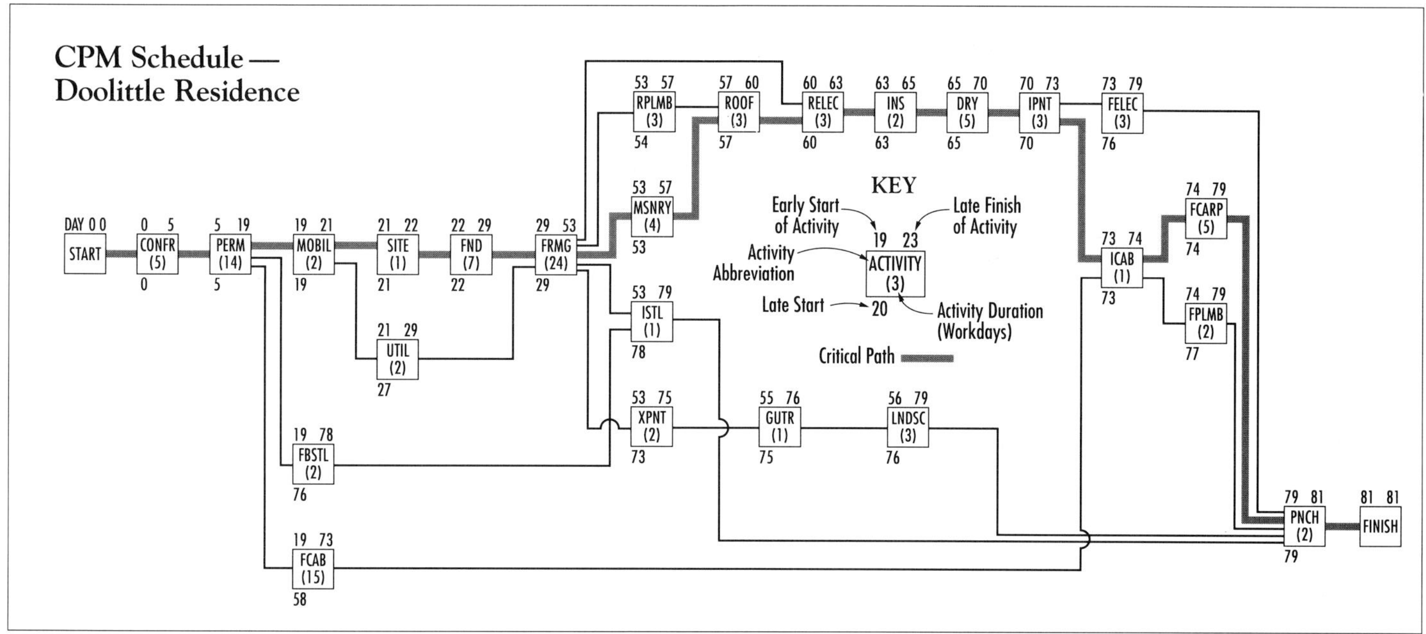

Figure 2. *A CPM flow chart links activities along a central "critical path," which forms the backbone of the construction process. Activities that are not critical (in the sense that there is flexibility regarding their completion time) are represented on work paths that branch off the critical path.*

work crews, the equipment needed, the assemblies to be fabricated off-site, and the subcontractors required.

The order of the list doesn't matter at this point, though it's usually easiest to list items in the rough order in which they will occur. Table 1 shows the list of activities and their durations for the Doolittle home, beginning with the conference with the Doolittles and their architect and engineer.

Step 2: Establish relationships. Next, establish the relationships between the activities you listed in Step 1. To do this, make a list of prerequisites — the activities that must immediately precede each activity on the list (see Table 2, page 141). For example, you don't want to obtain your permits until you've finalized things with the owner, so the conference activity (CONFR) becomes a prerequisite for the permit stage (PERM). And you can't begin framing (FRMG) until you have completed the foundation and brought utilities to the building, so foundation (FND) and utilities (UTIL) are listed as prerequisites for framing. Don't list everything that comes before a step, only its immediate prerequisites.

Establishing these relationships requires visualizing how the project will be built, including the interaction between subs, work crews, and suppliers. Take some time with it. It requires a lot of thought and forms the basis of the schedule.

Step 3: Build the network. The next step is to create a flow chart or network that schematically shows how each step in the building process relates to all the others, based on the prerequisite relationships established in Step 2. Each box represents a construction activity, as indicated by its abbreviated symbol. The different "paths" for the network are shown by the lines connecting the boxes.

Each box should contain the abbreviated name of a construction activity and, in parentheses, the number of days the activity will take to complete. The boxes go in the order in which their activities occur, from left to right.

As you work from left to right, you'll find that parallel paths branch off at certain points, where one activity's completion allows the simultaneous start of several others. The completion of framing, for instance, makes possible the start of several other activities, each of them beginning a new path. Some of these paths will rejoin the critical path later.

For now, your goal should be to get the boxes lined up in their proper order. It may take two or three tries but you should be able to construct an easy-to-read network with as few parallel paths as practical. Always move left to right.

Table 1. Activities and Durations

Activity	*Symbol*	*Duration(days)*
Confer with A/E owner	CONFR	5
Obtain permits	PERM	14
Sitework and excavation	SITE	1
Utility work	UTIL	2
Mobilize workforce and equipment	MOBIL	2
Foundation work	FND	7
Framing	FRMG	24
Install steel railing	ISTL	1
Landscaping	LNDSC	3
Fabricate steel railing off site	FBSTL	2
Masonry chimney	MSNRY	4
Rough plumbing	RPLM	3
Rough electrical	RELEC	3
Insulation	INS	2
Drywall	DRY	5
Roof	ROOF	3
Exterior paint	XPNT	2
Interior paint	IPNT	3
Fabricate cabinets off-site (includes order time)	FCAB	15
Install cabinets	ICAB	1
Finish carpentry	FCARP	5
Finish electrical	FELEC	3
Finish plumbing	FPLMB	2
Install gutters	GUTR	1
Punch list	PNCH	2

Analyzing The Network

With the network properly constructed, you now have a flow chart showing how the activities follow one another, their interdependencies, and the duration of each. From that information you can glean several other crucial facts about each stage of the job so that you'll have not just a flow chart, but a true critical path schedule.

Establishing critical path. Using what is called a "forward pass," you can work through the network to establish the project's critical path — the path in which any delay will cause delay in project completion. To get the critical path, you'll need to estimate the early start (ES) time for each activity. The early start is simply that — the earliest that an activity can be started, based on the durations and interrelationships of the preceding activities. The forward pass is simply a matter of working your way left to right through the network to find the *longest path through the network.* This will identify not only the early start time of each activity, but the network's critical path. This will tell you the soonest the project can be expected to be completed, assuming your estimates of the duration of each task are correct.

Start with the first task, which in this example is the architect and owner conference (CONFR). We assign it an ES of 0 (zero). The ES of any other activity is found by choosing the *longest* possible path in days to that activity. This is done by moving left to right through the network, adding the ES of each activity to its duration to find the ES of the following activity.

For example, consider the following simple network:

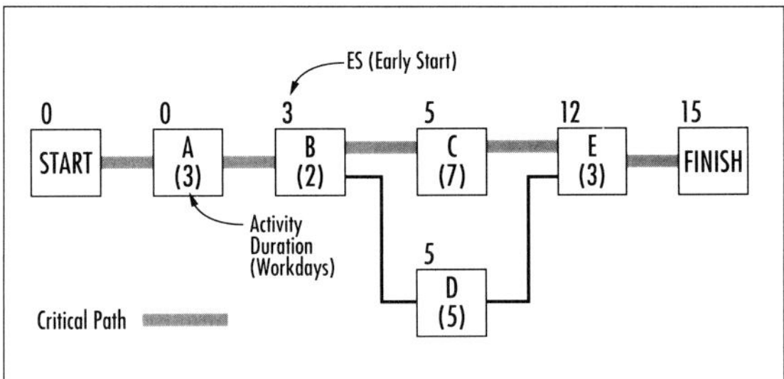

Since there is only one path to activity B, the early start of B is found by adding the ES of A plus its duration (0 + 3 = 3). The early start of C is then found by adding the early start of B (3) to B's duration (2), yielding 5 days (3 + 2 = 5). D's early start is found in a similar manner.

Activity E is tricky, because it has two paths leading to it, one from C and one from D. To find its early start date, you take the longest path: Hence E's earliest start is found through C, which has an ES of 5 and lasts 7 days (5 + 7 = 12 days).

In this example, the soonest the project will be finished is 15 days. All the activities except activity D are on the critical path, which is the longest path through the network. The activities on the critical path are known as "critical activities," and their delay will delay the project. Activity D is noncritical and hence has some "float" or "slack" time, which will be discussed later.

Completing a forward pass for the Doolittle residence shows that the house can be completed in 81 work days, since the last activity, the punch list (PNCH), can be started after 79 days and takes two days to complete.

The Backward Pass

A "backward pass" through the network yields the late finish (LF) of each activity. That is the latest an activity may be completed without delaying the overall completion date of the project.

In the backward pass, start at the end of the job, by assigning a late finish date to the very last activity in the network equal to its ES — in the Doolittle's case, 81 days for the punch list. (By definition, this is the last day the job can be completed without delaying it beyond its 81-day schedule.)

Next, follow all the paths backward, one activity at a time, assigning late finish dates (see Figure 2). Do this by taking the late finish of the following activity (the activity to the right) and *subtracting* that activity's duration. Then write the result above the upper right hand corner of the box for each activity. If there is more than one path back to an activity, perform this same subtraction for each of the paths, then choose the lower value as the late finish.

For example, in the Doolittle job, Install Cabinets (ICAB) is followed by boxes with durations of 5 days (FCARP) and 2 days (FPLMB), both of which have late finish days of 79. Subtracting these durations from their common finish day of 79, we get figures of 77 and 74; we use 74 as the late finish day for ICAB, for that is the last day we can finish that activity without delaying the project's completion. If we went with 77, we'd delay everything an extra three days.

Table 2. Activities and Prerequisites

Activity site	*Symbol*	*Prerequisite*
Confer with owner /AE job	CONFR	Start of
Obtain permits	PERM	CONFR
Sitework and excavation	SITE	MOBIL
Utility work	UTIL	MOBIL
Mobilize workforce and equipment	MOBIL	PERM
Foundation work	FND	SITE
Framing	FRMG	FND, UTIL
Install steel railing	ISTL	FRMG, FBSTL
Landscaping	LNDSC	GUTR
Fabricate steel railing off site	FBSTL	PERM
Masonry chimney	MSNRY	FRMG
Rough plumbing	RPLMB	FRMG
Rough electrical	RELEC	FRMG, ROOF
Insulation	INS	RELEC
Drywall	DRY	INS
Roof	ROOF	RPL
MSNRY		
Exterior paint	XPNT	FRMG
Interior paint	IPNT	DRY
Fabricate cabinets	FCAB	PERM
Install cabinets	ICAB	IPNT, FCAB
Finish carpentry	FCARP	ICAB
Finish electrical	FELEC	IPNT
Finish plumbing	FPLMB	ICAB
Install gutters	GUTR	XPNT
Punch list	PNCH	FCARP, FELEC FPLMB, ISTL LNDSC, UTIL
Finish job		PNCH

Late Start

From the late finish, we can easily derive one of the most important pieces of information about each activity, its late start. This is the latest an activity may be started without delaying the project. To get it, we simply subtract its duration from its late finish day. For instance, for installing cabinetry — ICAB — the late start is 73 (which is the late finish day minus its duration of one day). That's the last day we can start that job without delaying the project.

Write the late start beneath the lower left hand corner of the activity box (Figure 2).

Late start dates are quite useful, since they let you know exactly when each activity *must* begin. They also tell subcontractors and suppliers how they fit into the overall schedule.

Float or Slack Times

The float or slack time of an activity is the length of time its actual start may be delayed beyond its early start without delaying the project — in other words, it's the slack between its early and late start times. You get it by subtracting an activity's duration plus early start from its late finish: LF -(ES + Duration).

By definition, all activities on the critical path have zero float time, because their cumulative early starts and durations provide the length of the critical path itself.

However, many activities outside the critical path have float times, and it's

CPM On Computer

You can comfortably create a CPM schedule by hand for very small construction projects. But when it comes to projects the size of the Doolittle house or larger, you'll want to do it on computer.

The Power of Software

No computer program can divide your project into its separate activities or assign durations to, and recognize relationships between, the activities. These are decisions that come from your own experience and judgment about the project.

CPM software can, however, do the following:

- Perform all calculations, including the forward and backward passes, to determine the early and late starts, float times, and so on, almost instantly.
- Translate "schedule days" into actual work days on your calendar. Most CPM software considers work and nonwork days (weekends, national holidays, etc.) when calculating the schedule. So instead of "Day 62," for example, you would see "Tuesday, Aug 12."
- Allow easy updating. What if durations change? What if the precedence relationships change? What if activities are added (or deleted) through change orders? All of this can be easily handled with most CPM software, whereas it could be a nightmare to revise by hand.
- Print both bar charts and CPM networks.

In short, computer software makes CPM scheduling much more usable — a real tool, and maybe even a pleasure, rather than a chore.

Features to Look For

Choose software that is designed with you in mind, not software designed for huge commercial projects or industrial processes. Define your needs and compare them to what is available. Consider the following:

- *Ease of use*. Will this software become useful without being overly complex?
- *Hardware requirements*. Does it require greater hardware than you have (such as more memory or a hard drive), and if not, are you willing to spend for the difference?
- *Graphics capability*. Will it print both bar charts and networks? Networks are most useful for visualizing a project, but bar charts can be useful for sizing up progress at a quick glance. With some programs, these graphic capabilities are sold separately.
- *Output*. Can the network be printed on your printer? Some may require expensive "plotters," or large printers that can print on wide paper.
- *Flexibility*. Other than the basic schedule, what other reports can you derive and print from the program? Are these useful to your business?
- *Documentation and support*. Is the manual easy to use? Is technical support readily available?

You might also check whether a program has some of the more advanced scheduling features, if you think they'd be useful. These include:

- *A resources feature*. This allows you to add labor, equipment, and material costs to the activities. It can be useful as an estimating tool and as a means of forecasting and tracking cash flow throughout a project.
- *A resource leveling feature*. This helps level the personnel and equipment requirements on the job, so that you don't have peaks and valleys of demand for labor or equipment.
- *Integration*. This refers to a program that's designed to interact with other related software packages, such as estimating programs.

Software Suggestions

Much of the CPM scheduling software on the market is geared to large-scale commercial and industrial projects. Programs that may prove the most valuable to the smaller volume contractor include:

- *Advanced Pro-Path 6* (SoftCorp Inc., 2340 State Rd. 580, Suite 244, Clearwater, FL 34623; 813/799-3984; $199 for the program, $59 for network diagram option). For IBM-compatibles. This is an affordable, easy-to-use program with calendar, resource, and reporting capabilities.
- *Microsoft Project Version 4.0* (Microsoft Corporation, 1 Microsoft Way, Redmond, WA 98052-6399; 800/227-4679; $495). This is the program I've had the most experience with. It's well-documented with an easy-to-follow manual and a self-paced tutorial and good support from Microsoft. The program is complete to the point that it should be able to handle the needs of most mid-sized commercial contractors. But it's simple enough to prove useful on smaller jobs as well.
- *Claris' MacProject II* (Claris Corp., 5201 Patrick Henry Dr., Santa Clara, CA 95052; 408/727-8227; $499). For Macintosh users, construction computer consultant Eric Freed of Carefree, Ariz., recommends MacProject II, which can generate flow charts and do backward and forward passes and other computations. —*D.C.*

worth knowing those, for they represent the areas where you have some leeway in scheduling. Knowing that you have 39 days of slack on the Doolittle's FCAB (fabricate cabinets), for instance, means that if you want to, you can delay the order if some design change is made or if the Doolittles aren't quite sure yet which cabinets they want.

Using the CPM Network

Besides the obvious advantage of helping the builder to visualize a project, a CPM network can provide answers to some very important questions.

For example, say the finish carpenter wants to know when he must be on the job and if he has flexibility as to how long he can take to complete his work. The schedule tells us that the finish carpentry (FCARP) is scheduled to begin at the end of day 74 (or the beginning of day 75) and to take 5 days. This activity lies on the critical path, so any delay will delay the project.

Another example: You realize at the end of day 62 that the painters haven't begun painting the exterior, and you seem to remember that they were supposed to start around day 50. Is this a problem?

A look at the schedule shows it is not. The early start day for XPNT is day 53, but because the only thing waiting on it is the punch list, XPNT has a float of 20 days, with a late start day of 73. The nine-day delay so far is not a problem. The painters still have 11 float or "free" days left to begin their work before the completion of the project will be delayed. ■

David Carns is a licensed civil engineer and teaches construction management at Central Washington University in Ellensburg, Wash.

Calendars On Computer

by Morris D. Carey, Jr.

A realistic schedule, easily created on computer, can yield great rewards when shared with everyone involved on the job

A general contractor friend of mine has harped on me for years about the importance of job scheduling. "Schedule is everything," he'd say. He insisted that an otherwise profitable subdivision could prove to be a disaster if the schedule was lost. I had trouble with his philosophy. I wasn't building subdivisions. I was doing room additions and remodels, the kind of work that's over not long after it begins.

The Importance of Sharing

Time has proved my friend right. Until about a year ago we employed a superintendent whose job was to schedule our in-house crew and our subcontractors. He was required to maintain a written calendar of events and organize work for each job by that schedule. Jobs weren't coming in on time, customers were angry, profits continued to drop, the crew was confused, and our subs were pulling their hair out trying to keep up their end of the bargain. I began to feel that our super didn't know how to create a schedule.

But I was wrong. He *did* know how to schedule. The problem was he didn't know how to share his schedule with others — important others like the crew and the subs who were ultimately responsible for performing the work. No one in the field knew any more than what the superintendent laid out for a given day or week. No one knew the total scope of the job, or who was responsible for what.

Today, no job begins in our company without a schedule. And everyone involved gets a copy: the plumber, electrician, sheet-metal contractor, crew, salesperson, customer, even the accounting department. Now that everyone knows what's planned, it happens with less hassle, and it happens on time.

The moral of this story is simple: A schedule works if it is realistically created, but most importantly, if everyone involved with it gets a copy. The results are very positive. Our referral business is up as a result of a happier clientele, our crew and our subs are able to plan earlier, and our jobs have become easier to manage — and that makes them more profitable.

Like other business functions, scheduling can be done manually or on a computer. With the right software, the computer can make it a lot more efficient. Computerized scheduling programs abound — here are just a few that I'm familiar with: Project Scheduler 2000 (Scitor, Sunnyvale, Calif.), Time Line (Symantec Corp., Cupertino, Calif.), and Quick Schedule Plus (Power Up Software, San Mateo, Calif.). Each of these programs will print a Gantt (horizontal line) presentation for a single job or combination of jobs (see Figure 1).

A Simple Calendar

The program I use to keep my jobs on track isn't classified as a scheduler at all, but really does the trick for me. It's called Calendar Creator Plus (CCP), and is offered by Power Up Software. It's available for an easy-to-swallow $39.95.

CCP is nothing more than a simple calendar maker. The program comes with a schedule of standard holidays, which is simple to alter, and which helps prevent scheduling conflicts.

First, I use the program to print a blank calendar (with holiday schedule) covering the general period of time the job is expected to take (see Figure 2, facing page). This is a real plus, because the calendar I create can be made to begin on the same week the job starts and can include a weekdays-only (five-day) presentation, or a seven-day presentation with weekends included. I select the seven-day presentation to flag work that our customers elect to do. This helps our subs and crew to be aware of customer participation, and forces our customers to plan for what they are responsible for.

Next, I handwrite my schedule onto the blank form, making changes as I go. And finally, I enter the handwritten data into the computer. It's quick and easy.

CCP works on about a dozen popular brands of printers and allows the user to create calendars covering different periods of time, such as one year, six months, or three months. It also prints calendars in six-week, one-month, three-week, two-week, one-week, and either one- or two-day formats. Each calendar can be given a title (for example, Jones Production Schedule), or the same calendar can be printed with different titles. This is nice if your type of construction includes a lot of repeat work with only minor modifications. Events can be entered into the program by a given date or as floating events. Floating events are ones that repeat, such as every Friday, or the second Tuesday of every month, or the fifth of December every year.

Data is easy to enter or change, but, unlike true scheduling programs, items of work cannot be moved forward or backward in time when you need to change the schedule due to other events.

My big attraction to the program is simple: When it's finished it looks like a calendar. And everyone can read a calendar. It doesn't offend anyone or insult their intelligence. Just a plain old calendar. I use this system to juggle up to 15 jobs at the same time, using a separate calendar for each job.

There is also an important role for the more sophisticated programs that produce Gantt and CPM presentations, although I feel that their best place is on

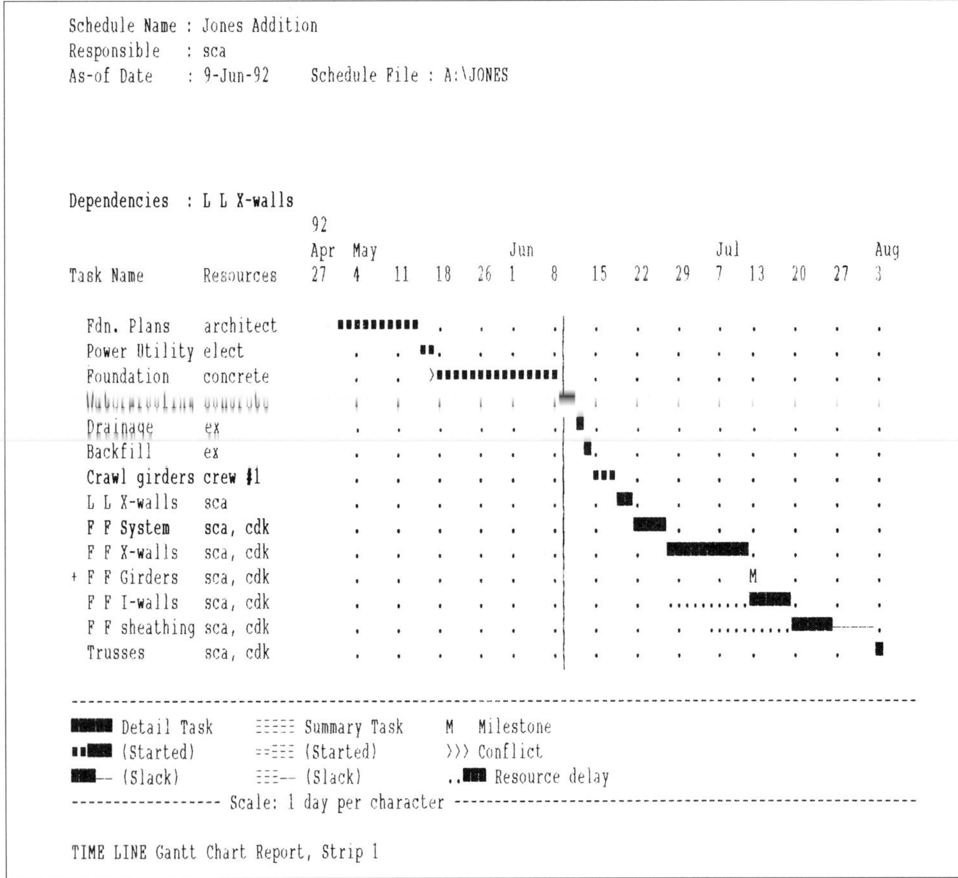

Figure 1. *Most computerized scheduling programs print out a time line, or "Gantt chart," like the one above, for a single job or several concurrent jobs.*

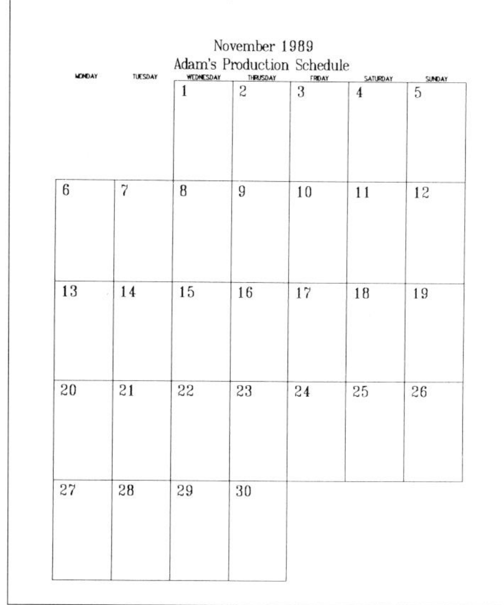

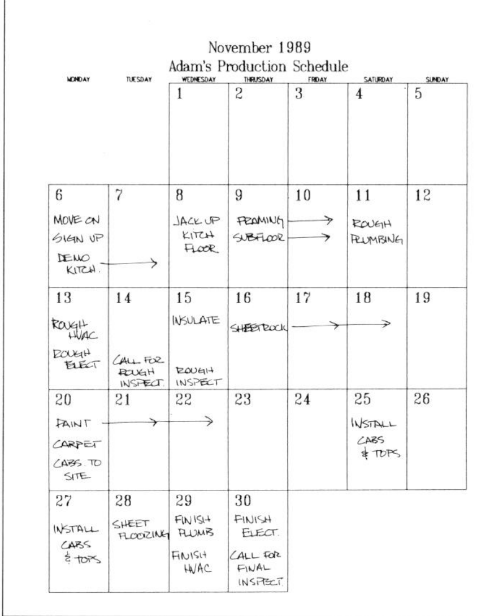

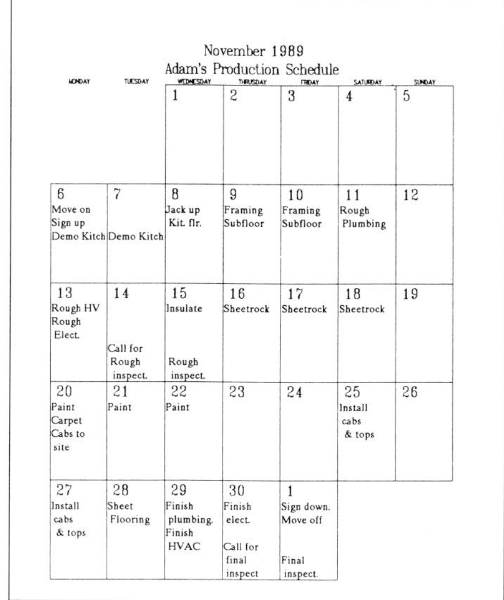

Figure 2. *The first step with Calendar Creator Plus is to print out a blank calendar, with holidays, for the general period of the job (left). Next, pencil in your schedule on the blank form (middle). Finally, key in the data and hand out copies (right) to everyone involved with the job.*

the wall in the production office and in the job shack of larger-scale operations. Improved features will probably bring these into the custom-home and home-improvement markets someday as well. Who knows? Maybe one of them will have the good sense to offer a calendar presentation option for their schedules? ■

Morris D. Carey is a partner with Carey Bros. Construction based in Pittsburg, Calif. He is also an author and lecturer, and co-authors the monthly computer column for The Journal of Light Construction.

No-Fault Scheduling

by Doug Immel

A master list detailing every step of the job can reduce surprises and boost profits

I love some surprises — unexpected snow on Christmas Eve or old college mates calling out of the blue. But surprises on a construction site? No thanks. Construction "unknowns" consume time, energy, and profit. And nothing is more humbling than bringing five subcontractors to a halt after spending a week coercing them to hurry up.

After years of designing and building, my partner and I realized that carpenters are only as good as their tools and general contractors are only as good as their systems. So we got serious about planning.

As my partner Marianne put it, what we needed was "a case of boredom": a scheduling system designed to make our work brainless. We figured the more tasks we could make routine, the less energy we would need to spend remembering the 25¢ decisions that, if forgotten, can stop a $250,000 job.

The Idiot List

To achieve this no-miss scheduling system, we looked to the fundamental teaching technique of breaking a task down into its tiniest and most discrete parts.

We created what we called the idiot list, which included every step in the process: every design decision, every specification needed, every material order, every sub to schedule, every detail to tell our subs, and so on. Distilling the entire contracting process in this way—starting with the first client contact and ending with the champagne toast at the open house—yielded a list of "prompts" for keeping track of the niggling details.

The items, or prompts, are grouped under the appropriate headings, such as design, framing, or electrical. The prompts begin with such flashy verbs as *order*, *call*, *write*, *talk to* (clients, subs, inspectors), and *go to*. The list covers nine pages and is

Date	Num	Description	Memo	Category	Clr Amount
9/2	E 870	NOTIFY ELECTRICIAN	verify contract signed		0
			spec smoke detector locat.		0
			spec cable outlet locat.		0
			supply airtight box		0
9/13	E 880	ROUGH ELECTRIC DONE	sched rough inspection	SUB/21	-1,200
			notify insul 3 days		0
			notify drywall 1 week		0
			call for 99: doors/wind/siding		0
			call for 99: rough electric, etc.		0
9/13	890	ORDER FIREPLACE MANTEL	time delay ***		0
9/13	I 900	CALL TOWN - ROUGH INSPECT	electric		0
			plumb		0
9/16	910	NOTIFY AIR SEALER			0
9/16	920	PAY HVAC ROUGH		SUB/20	-1,800
9/16	930	PAY ELECTRIC		SUB/21	-400
9/16	X 940	CLAP 3 WALLS COMPLETE	notify painter to start		0
9/17	N 950	INSULATION SUB STARTS	foam board & acoust seal		0
			foam airtight boxes		0
			foam penetrations		0
			explain winter warm side		0
			sub indoctrination		0
			notify drywaller		0
			order gypsonite		0
			notify int. painter to prime		0
9/17	I 99	INSPECTION - BANK	1 day grace for payment		0
9/17	970	CHECK EXT TRIM			0
9/17	I 99	INSPECTION - BANK ROUGH	1 day grace for payment		0
9/17	P 980	EXT PAINTER STARTS	spot test color for client	MAT/24	-250
9/17	990	CLAP LAST WALL			0
9/17	P 1000	CALL PAINTER TO PRIME	prime uninstalled millwork		0
9/18	I 1010	ECH INSPECTION	sched after insul installed		0
9/18	I 1020	TOWN INSPECTION: INSUL	call town foundation inspec		0
			call 99 driveway/water/sept		0
9/19	I 1030	VERIFY ROUGH INSPECTION	verify cards signed off		0
9/19	1040	PAY DIG-FORM		SUB/03	-2,100
9/19	1050	CHECK SIDING			0
9/19	1060	INCOME: DOOR/WIND/ROOF		SALES	4,500
9/19	1070	INCOME: ROUGH PLUMB/HEAT	chimney/insul	SALES	2,500
9/19	1080	PAY FORMS		SUB/04	-2,200
9/19	1090	DRIVEWAY-PAVE		SUB/04	-1,950
9/19	1100	GET FRONT STEPS		SUB/04	-300
9/20	F 1100	ORDER FINISH MATERIALS	closet shelf? cup or pine		0
			mantels		0
			shelving		0
			special mouldings		0
			door swings		0
			door handle types		0
			casing types		0
			notify int. painter to prime		0
		Total [illegible]			
		Total Inflows			7,000
		Total Outflows			-10,200
		Net Total			-3,200

The author's 600-item "idiot list" (excerpted above) is written on Quicken and is designed to eliminate surprises. The construction process is broken up into its most basic elements and chronologically organized on the list. Cash flow is also tracked for subs, materials, and progress payments.

about 600 items long.

We've compiled this list into a computerized bookkeeping program called *Quicken* (Intuit, P.O. Box 3014, Menlo Park, CA 94026; 800/624-8742), a $69.95 program we use for check writing, recordkeeping, budgeting, and financial reporting. We run the program on a Macintosh, but it is also available for IBM-compatible computers.

Although Quicken is not really designed for this type of scheduling function, it's easy to use and gets the job done. Any computerized spreadsheet program, however, would work as well.

You don't have to have a computer to make the idiot list work — it works with pen and pencil too, though you won't be able to sort items easily by categories. But you can put together the crucial checklist and revise it occasionally.

Coding Nitty-Gritty

We had to adapt our bookkeeping program a bit to make it work as a scheduler. Quicken categorizes its "records" (each corresponding to a particular check, deposit, or withdrawal) by several pieces of information (called "fields") such as date, amount, payee, or memo—the usual things you can enter on a check. These fields can also be used to sort or organize the various records. For instance, the program will list all the checks written to a certain supplier or on a certain date.

To use the program for our scheduling, we made each record contain in its "memo" section a single task, such as "Notify drywaller" (see sample list). Along with the task, the record contains a one-letter code indicating what phase of design or construction it is related to (such as X for exterior or N for insulation). Where appropriate, we also include information about what sub to contact or hire, and what larger component or process the task is part of. The "Notify drywaller" prompt, for instance, is one of several items included in the prompt "Insulation Sub Starts."

Finally, items that involve spending or receiving money have a dollar figure attached to them, so that the list doubles as a detailed projection of expenses and billing. This lets you check your projected cash flow at any given point in the construction process.

Making It Work

Compiling this list hasn't been easy, and it's never finished. On the first try, we spent hours visualizing the process from start to finish, hoping to catch as many tasks as we could. It became a game of who could reduce a task to its most basic, critical elements.

Inevitably, we forgot some tasks and had to add them later. For example, one day the electrician asked where we wanted the cable television outlets to be. We didn't know and had to go back to the client, resulting in a time loss. To avoid this in the future, we made an entry in our steno notebook (which we have on hand at all times for this reason) to add the prompt "Ask client for cable TV outlet specs."

In fleshing out these items, it is critical to note and specify the relationships between tasks, trades, and subs. For instance, the cable outlet specs, once obtained, need to be included in the "Notify Electrician" section. This then requires an additional prompt in the design section to enter cable television outlet specs in the "Notify Electrician" list, right after the prompt to ask the client about cable outlet locations.

Vigilance Buys Control

As the list has been refined, it has made the contracting process much smoother for us and for our subs.

Using the list, we have eliminated more than half of our lost days, and have more than doubled our profit margin. Our typical projects are custom homes that cost $100,000 and up to build. Before we developed the list, we often had jobs fall a month or two behind. On a house costing about $150,000, that's nearly $1,000 a week. We were breaking even on most jobs, regardless of size. On some jobs, we made a few bucks, and on some, we lost money. Now we're on time and make sure profits. On our last job, we lost only a few days in a 12-week schedule, and cleared a substantial profit—typical of the improvement we've seen since adopting this system.

Being this well organized shines your image. People in construction are used to things going so-so or even poorly — thus the frequent cries of "Murphy strikes again." But that attitude is demoralizing, and clients don't buy it. ■

Doug Immel owns Douglas Blake Immel Inc., a design/build, construction management, and energy-consultation company located in Woonsocket, R.I.

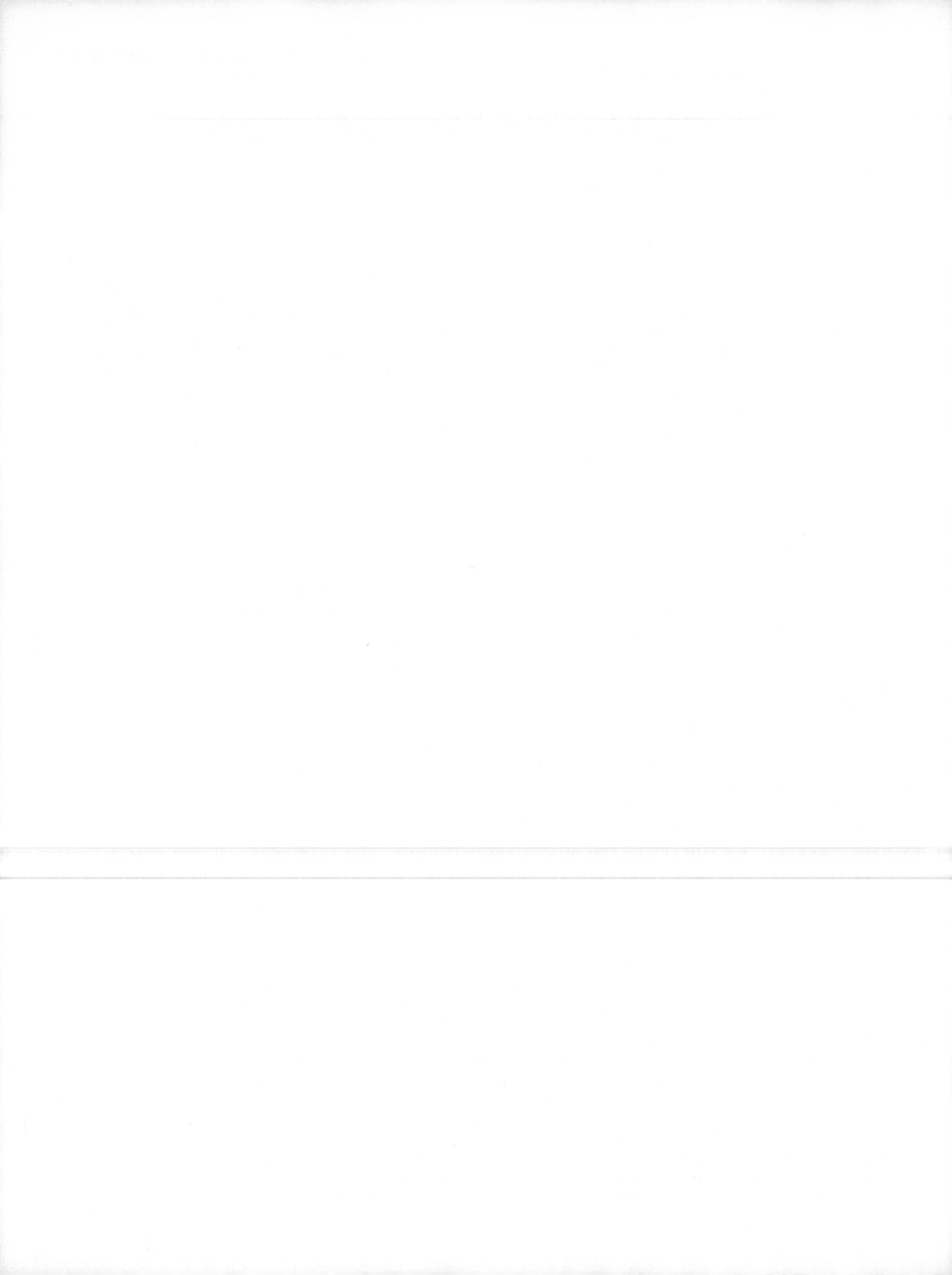

Chapter **11** • **Customer Service**

A Customer Service Program That Works

by Ross Robbins

How to keep a close eye on the last project while rushing ahead to the next

The main focus for most young companies is naturally on the next project, rather than on the last. The immediate problem of cash-flow always seems to take priority. That's the basic survival instinct. Because of that, customer-service and warranty-service programs tend to take a back seat. But they are critical to long-term success.

Starting Out

When my company started out four years ago, we had zero starts, zero inventory, zero sales — and substantial personal debt. So we were highly motivated to move ahead. The first thing we had to do was to convince bankers to lend us money, and to get homes built.

So for the first year, the whole focus was on production. Our first sale closing occurred about five months into our first year. That was the first cash we got. It was also our first customer, and our first customer-service problem.

Suddenly, we had two responsibilities instead of one. On the one hand we had the responsibility to the future of the business: designing homes, meeting with clients, negotiating contracts, estimating costs and budgets — tasks I find exciting and challenging. On the other hand, we had a customer saying, "You know I've called three times about that scratch in my floor, and my paint touch up, and I'm not getting the attention I deserve. All you care about is your next sale."

It didn't give me a warm, cozy feeling when I got those phone calls, so I tended to avoid them. I think that's human nature, especially when I had been insulated from these kinds of calls previously.

Not Taking Phone Calls

Early in my career, I had the opportunity to work for a fairly good-sized company with a marvelous warranty-service department. I operated the company's custom-home division and took care of service problems by paying a nominal fee to the warranty-service department. I thought all they did up there was fill out forms, send workers out, get the forms back, and people went away happy.

When I went into business for myself, my perspective changed. Customer calls became a burden. By my second year of business, it reached a point where I didn't take phone calls. Even if I was in, my secretary asked to take a message.

Not that my intentions weren't wonderful. I really wanted all my customers to be happy. I wanted them to be good friends. I wanted them to tell all the world about what a good job we did. But I just didn't have the time. I was going to get to the customer-service problem "tomorrow," if I could just get through the crisis today.

Yet we needed our customers' good will. Like most small builders, we had no organized model-home program. We needed our customers to let us come into their homes with prospective clients to show what kind of work we do. But I wasn't taking care of my new homeowners, so I was on a collision course.

Satisfied Customers Needed

Things got worse until I hit bottom and realized that I had to take time away from day-to-day operations, get organized, and set up a system.

The first step was to send out a letter to our subs and suppliers. It was a statement of our commitment to service and our intention to be Denver's top-quality home builder. This letter also stated that we would require our subs and suppliers to meet certain standards of quality, timeliness, and call-back service. And if they didn't, the letter made clear that we were serious: We would hold checks, and look for new suppliers who wanted to make that kind of commitment to the program.

The letter was not earth-shaking, but it helped me to crystallize my thoughts, and get my subcontractors and suppliers on board with us. The interesting thing I found about taking this step is that when I put my expectations in writing and told everybody what I was going to do, I realized that I had better follow through and do what I said I was going to.

The first six months were pretty rocky. We had a lot of squabbles about whose responsibility a job was, about why the construction manager didn't give a sub the right instructions, and so on. Was it a warranty problem at all? Or was it a walk-through problem that never got fixed and was the construction manager's responsibility?

We had finger pointing. In many cases, subs claimed that we had specified materials that they could not control. Therefore, we should have known better. For example, we gave them a window that was difficult to install square, or a brand of stain or tile grout that didn't perform appropriately.

Administrator Needed

Once we had set up forms, assigned responsibilities, and worked out the kinks in the system, I thought I was home free. Now I could get back to the business of building houses, which is what I enjoyed.

Wrong! The suppliers and subs under-

The Expectations Game

Bainbridge, Inc., seeks to promote reasonable expectations on the part of their customers, and then to meet them. A key tool in this effort is the *Expectations Book*, which lists what constitutes warrantable claims, what does not, and how problems will be treated. Below is a brief sampling from the extensive list that is given customers upon signing a sales contract.

CATEGORY: SITEWORK

■ ***ITEM:** Excavating and Backfilling*
- **Possible Deficiency:** Settling of ground around foundation, utility trenches, or other filled areas.
- **Construction Standard:** Settling of ground around foundation walls, utility trenches, or other filled areas shall not interfere with water drainage away from the house.
- **Builder Responsibility:** Upon request by the buyer, the builder shall fill settled areas, one time only, during the first year warranty period. The owner shall be responsible for any grass, shrubs, or other landscaping affected by placement of such fill.

CATEGORY: CONCRETE

■ ***ITEM:** Expansion and Contraction Joints*
- **Possible Deficiency:** Separation or movement of concrete slabs within the structure at expansion and contraction joints.
- **Construction Standard:** Concrete slabs within the structure are designed to move at expansion and contraction joints.
- **Builder Responsibility:** None.

CATEGORY: THERMAL AND MOISTURE PROTECTION

■ ***ITEM:** Roofing and Siding*
- **Possible Deficiency:** Ice buildup on roof.
- **Construction Standard:** During prolonged cold spells, ice buildup is likely to occur at the eaves of a roof. This condition occurs when snow and ice accumulate and gutters and downspouts freeze up.
- **Builder Responsibility:** The buildup of ice on the roof is a homeowner maintenance item.

CATEGORY: MECHANICAL

■ ***ITEM:** Heating*
- **Possible Deficiency:** Inadequate heating.
- **Construction Standard:** Heating system should

stood the philosophy, and I had their commitment, but I still had no way to administer the program. I thought the secretary in my office could just send out the service requests to suppliers and subs when we received them from homeowners.

But it didn't work that way. Nobody was there to evaluate the nature of the problem and see which sub it belonged to. Nobody decided whether it was a builder problem or a homeowner-maintenance problem, or not a problem at all.

My problem was partly a financial one. How was I going to find someone in a four-person company to administer the program at that level? I had to find somebody 10 to 20 hours per week, I figured. Through a reference from my interior-trim sub, I located a kind of handyman — an older guy who called himself the House Doctor, and was looking for part-time work. Boy, I thought, that guy's perfect. He could go in, assess the problem, take care of it, clean up any loose ends, and then just send the form back to me. And he could deal with these things before they became big problems.

The House Doctor took care of the little things fine. But he ran into problems when he found something that he couldn't repair himself. For example, a tiny leak in the dishwasher (a manufacturer's defect) had dripped through the bottom of the appliance over a period of months, and

be capable of producing an inside temperature of 70°F, as measured in the center of each room at a height of 5 feet above the floor, under local outdoor winter design conditions as specified in ASHRAE handbook. Federal, state, or local energy codes shall supersede this standard where such codes have been locally adopted.
- **Builder Responsibility:** Builder shall correct the heating system, as required, to provide the required temperatures. However, the owner shall be responsible for balancing dampers, registers, and other minor adjustments.

■ **ITEM:** ***Plumbing***
- **Possible Deficiency:** Noisy water pipes.
- **Construction Standard:** There will be noise emitting from the water pipe system, due to the flow of water.
- **Builder Responsibility:** Builder can not remove all noises due to water flow and pipe expansion. However, any "water hammer" is the builder's responsibility.

CATEGORY: DOORS AND WINDOWS

■ **ITEM:** ***Wood and Metal Windows***
- **Possible Deficiency:** Malfunction of windows.
- **Construction Standard:** Windows should operate with reasonable ease as designed.
- **Builder Responsibility:** Builder to correct or repair, to provide reasonable ease of operation.

CATEGORY: WOOD AND PLASTIC

■ **ITEM:** ***Finish Carpentry (Interior)***
- **Possible Deficiency:** Quality of interior trim workmanship.
- **Construction Standard:** Joints in moldings or joints between moldings and adjacent surfaces should not exceed 1/8 inch in width.
- **Builder Responsibility:** Repair defective joints, as defined. Caulking is acceptable.

CATEGORY: FINISHES

■ **ITEM:** ***Gypsum Wallboard***
- **Possible Deficiency:** Defects which appear during first year of warranty such as nail pops, blisters in tape, or other blemishes.
- **Construction Standard:** Slight "imperfections," such as nail pops, seam lines, and cracks not exceeding 1/8 inch in width are common in gypsum wallboard installations, and are considered acceptable.
- **Builder Responsibility:** The builder will repair only cracks exceeding 1/8 inch in width, one time only, during the first year warranty period. Builder is not responsible for color variations in the paint.

down into the oak floor, cupping it all the way across. So he had to get the hardwood floor sub back in. But that required spending time on the phone rounding up people and coordinating things. When he ran into trouble with that, he came back to me with the problem. I was back at square one.

The Flypaper Dilemma

This problem seemed to stick to me like flypaper. I'd take one hand off, and it would stick to the other. So I looked around and found a well-recommended warranty-service guy who had recently been laid off (the Denver market was hurting at the time). I couldn't afford to put him on my payroll, but came up with a solution: We put together a group of four small builders to share the guy.

We set it up on a contract basis on a per-house enrollment fee. Between the four of us, we figured, he could keep busy and make a good income.

Here's how it works: The warranty-service manager takes the report of the problem, calls the sub that's responsible, and asks if he's seen it or if he is familiar with the problem. If not, he goes out personally, and makes a site visit with the homeowner.

After seeing the problem, he decides whether it is in fact a warrantable item, or whether it's a homeowner-maintenance item. If the repair is questionable, but he thinks it will make the customer happy, he is instructed to agree to the work if it will cost less than $100. If it's questionable, and will cost over $100, he says, "I'll have to check with the office on this and get back to you." Within the limits of this one guideline, he's got the field authority to make all necessary decisions. I think that's a key element in the system.

If the service manager finds that there's a subcontractor mistake or something that needs repair, he calls the sub, sets up a time for him to meet with the customer, and sends a confirmation letter to the sub and the homeowner telling them when the appointment will be. Included in the confirmation letter to the sub is a sign-off sheet for work completed. The sub takes that with him to the home, does the work, and gets it released by the homeowner. Then it goes back to the customer-service man, and finally into my files.

Reasonable Expectations

After ironing out a few wrinkles, everything was working very well except for the grey areas. Our new customer-service manager kept coming back to me with questions: Is this covered? Is this alright? What are we supposed to do about this?

It was a question of expectations. He needed to find out what type of product I intended to deliver to my homeowners, and to convey to them what was a warrantable item. So we put together a full warranty book, which we call the "Expectations Book," and which the customer acknowledges receiving at the time that he or she signs the purchase contract. It describes the warranty coverage itself, and lists each item with a description, a possible deficiency, the standard we expect to meet, and the builder's responsibility to correct it to that standard. It covers many of the components of the home. The package also contains all the warranty-request forms.

In addition, we send out a 30-day letter, a 6-month letter, and an 11-month letter. The 11-month letter tells the owner he has one month left on the one-year warranty. Is there anything he wants us to pay attention to before that expires?

The Best Salesperson

With this program, and personal contact with my customers, I now have almost carte-blanche to go into any home we've ever built. In fact, I have a salesperson on staff who has discovered that the best thing to do is simply walk into a home, introduce the prospective client, and then just shut up. Our customers do the selling. For one thing, the customer's credibility with that potential buyer is a lot better than ours.

So we've found this program to be a very solid benefit. I've got control of a problem that was driving me nuts. My customer-service program generates future business, rather than detracts from it. And finally, I don't have any skeletons in my closet anymore. I can answer any phone call without worrying. And that's a very comforting feeling. ■

Ross Robbins is the vice-president of Bainbridge, Inc., of Englewood, Colo., which builds about 30 new homes a year.

Service Manuals Make a Good Impression

by Kathy and Gary Wheatley

A service manual that combines maintenance tips with warranty procedures makes good business and good marketing

Like most remodelers, we're always looking for ways to improve service for our customers, especially after our work is completed. We want our customers to feel free to call us if they have any questions or problems with our work. We also want to maintain our reputation with them since they are a good source of referrals.

With annual sales of about $600,000 and seven employees (including ourselves), our company is not large enough to have a customer service crew. Although we would like to personally call all our past clients to find out how our work is holding up, we don't have the time. Instead, about two years ago, we created a warranty and service manual that encourages customers to get back to us if they are having a problem.

We give out the manual for jobs totaling $5,000 or more. We limit it to this size job since it coincides with the five-year Home Owners Warranty Corp. (HOW) remodeler warranty that we provide.

The manual consists of 14 pages bound in a standard folder that we purchase from a business supply catalog. We use the folder pockets to supply brochures and an application for the HOW program, as well as preaddressed envelopes for customers to return the forms.

The manual is divided into four parts: home maintenance tips, emergency procedures, 30-day, 6-month, and 12-month warranty lists, and a record of service calls.

Home maintenance tips. Few people bother to read through the owner manuals for all the products used on a particular job. Doing so would take hours. Yet without proper care, many of the items we have built or installed may not last as long as they should. Also, if problems develop, some customers might blame us. So our manual includes eight pages of maintenance tips to help our clients take better care of their homes, as well as the work we've done.

Our tips cover exterior work, roofing, plumbing, wells and septic systems, hvac, electrical, flooring, windows and doors, bathrooms, and basements. For example, we discourage wallpapering until a year after the job is completed to minimize damage from normal settlement cracks and nail pops, and warn customers that abrasive cleaners will scratch and dull fiberglass and acrylic tubs and showers. We also cover general maintenance items, such as recommending that tree limbs growing on or over a roof be trimmed, or that the water heater be set to the temperature recommended by the plumber or utility company.

Emergency procedures. It seems emergencies always happen after business hours. So we've provided a list of emergency phone numbers where we can be reached, as well as numbers for the electrician, plumber, hvac contractor, and anyone else that was involved in the project. If the listed contractors are unavailable, our clients are instructed to call elsewhere for service. They should pay the bill and then submit it to our office. If we decide the problem is under our warranty, the client will be reimbursed.

Warranty lists. No matter how thorough we are, problems can still develop after we've left the job site. Bathroom fixtures may not function properly, for instance. Normal settlement and the shrinkage caused by

Mr. John Brown
111 Lakeview St.
Anytown, MD 11111

December 26, 1991

Dear Mr. Brown:

Thank you for your business! We hope you are pleased with the alterations, our workmanship, and our company in general. We enjoyed working with you.

Attached please see our Service/Warranty Manual. We have developed this package to more efficiently service your warranty needs. You will notice the booklet contains a 30-day, a 6-month, and an 11-month warranty list. Please list items requiring attention on these lists and return them to our office when appropriate. While it is unreasonable to expect us to make a trip for each item individually, you may certainly feel free to contact us before the scheduled report is due for any items that are of a "semi-emergency" or emergency nature.

Should there be any "open" service items, they are noted near the back of the booklet on the RECORD OF SERVICE CALLS page. These items will be attended to as soon as possible.

Additionally, please find an application for the HOW 5-year insured warranty. Please fill in any items checked, sign where indicated, and return to us in the enclosed envelope. Your policy will arrive 6-8 weeks after processing in Texas.

Again, thank you for entrusting your home to us.

Very truly yours,
WHEATLEY ASSOCIATES, INC.

Kathy L. Wheatley
President

WHEATLEY ASSOCIATES, INC.
6-MONTH WARRANTY LIST

NAME: Ms. Susan Brown
111 Lakeview St.
Anytown, MD 11111

TELEPHONE: 555-5555

COMPLETION DATE: 03/16/92
HOW COVERAGE: Yes

1. ______________________________

2. ______________________________

3. ______________________________

4. ______________________________

5. ______________________________

6. ______________________________

7. ______________________________

Please include all items still outstanding, whether reported previously in writing or by telephone. Please hold any caulk, drywall repairs until the 11-month punchlist, to allow for normal settlement and heating/cooling shrinkage. This is a complete list of all items outstanding, both inside and outside as of 6 months after completion. Please return this list at your earliest convenience, at approximately 6 months after completion.

Wheatley Associates, Inc. Home Care and Maintenance Information*

Your home represents shelter and financial security. It is important to protect your home by careful maintenance. Please read carefully for there are many important items pertinent to the care of your newly completed project. If you have any questions, regarding any item, please do not hesitate to call.

Exterior:

1. Fill ruts in stone driveways annually to avoid larger ruts. On blacktop drives do not drive close to the edges. Driveway should be resealed annually. SPRING

2. Rock salts should not be used on brick, stone or concrete walkways, as this substance will deteriorate these materials. Sand, sawdust or kitty litter are good alternatives. WINTER

Roof:

1. Clean gutters at a minimum, annually. Some will need to be done monthly so that they can flow freely. It is imperative that the gutters and downspouts be free of leaves before winter sets in. FALL/SPRING

2. High winds, snow and ice can damage a roof. Check after heavy storms and every spring for water stains in attic and on the overhang. PERIODICAL

Plumbing:

1. Before freezing temperatures arrive, you must turn off the exterior hosebibs. The cut-off valve is usually inside the basement directly opposite the hosebib. First turn off the cut-off valve. Then, open the hosebib outside to allow the water to drain. Then, close the hosebib. *You must remember to close the hosebib or the cold temperature can run through the pipe and freeze the lines inside.* FALL

2. Fiberglass and acrylic tub, shower, and whirlpool units should *never* be cleaned with abrasive cleaners. They will scratch and dull if these cleaners are applied. HELPFUL HINT

Heating/Cooling:

1. Clean or replace heating and air conditioning filters. SPRING/FALL

2. Heat pump thermostats should be set at a predetermined comfortable setting and left alone. Frequent readjustment of the thermostat is a very inefficient use of the heat pump principle. HELPFUL HINT

Flooring:

1. Even stain resistant grout, used in your ceramic tile and marble areas, will darken with wear. To minimize this occurrence, always change cleaning solution frequently and rinse with clear water. PERIODICAL

2. *Never* use abrasives or steel wool pads on ceramic tile. HELPFUL HINT

Windows and Doors:

1. Vinyl or rubber weather stripping's life can be prolonged by applying petroleum jelly. PERIODICAL

Based on a guide by the Home Owners Warranty Council of Maryland

Wheatley Associate's service manual contains eight pages of maintenance tips (excerpts shown above). The tips help the new owners take care of their investment and help reduce callbacks.

lumber drying out are inevitable after the first year and may require caulk and drywall repairs. Screws and knobs may come loose.

As a project nears completion, our superintendent and the clients develop a final quality-control punch list. It is stipulated in the contract that after the items on the list are completed, the client will make the final payment. Any other items found after the list is completed are included on the thirty-day, six-month, and twelve-month warranty lists in the manual. The cost for repairing any items included on these lists is covered by our company's one-year warranty.

The warranty lists not only reassure our customers that problems will be solved, they also keep us from trotting back to past jobs every time a nail pops. We are able to take care of many problems in a single visit. Also, we encourage customers to hold all caulk and drywall repairs until the 12-month list.

Record of service calls. This sheet is intended to help clients track any work that remains. For instance, if something is broken during construction and must be reordered, the list serves as a record. It reaffirms to the client that we have not forgotten the item and will attend to it as soon as possible.

Other Benefits

Producing the manual is neither expensive nor time consuming. We use our word processor to customize each page with the homeowner's address and the date the project concluded. We also customize the emergency repair list since we may use different subcontractors depending on the job. We also write a cover letter thanking the customer for their business and encouraging them to contact us if they need assistance.

Ideally, every one of our customers will read the manual and follow all the tips offered. But, we've found that only about 10% of the warranty lists are returned on time. We'd like to believe this is because our work is so good that one year later there are no problems. In reality, it seems that most customers handle the little repairs, such as tightening screws or fixing the caulk, themselves.

The warranty and record of service calls provide another advantage: They create a paper trail that comes in handy in case there are any major problems down the line. We've never had this happen, but there are plenty of remodelers who have wound up in court after clients alleged that some of their work was never completed or that repeated complaints were never answered.

The manual also serves as a selling point. We keep a copy in our portfolio to show potential customers that if they work with us, we will take care of them.

We have learned that differentiating ourselves from the competition is one of the best ways to get more jobs. Our manual tells customers that we are both professional and committed. Many people don't expect remodelers to be either. ■

Kathy and Gary Wheatley are the owners of Wheatley Associates, a Monkton, Md., remodeling firm specializing in additions, window and door replacements, and light-commercial alterations.

Customer Questionnaire

by Terry Brocious

A simple after-the-job questionnaire can help your company improve its performance and its image

Our company has a motto: Building the Future, Reshaping the Past. The motto, of course, refers to the business of remodeling houses, but it also says something about how this business learns from its customers. When we sent performance questionnaires to our customers, we learned that the best way to build our company's future was to reshape the way we had done things in the past.

Most of our $2 million in business comes from multi-family remodeling, but we continue to remodel many single-family homes, which is how our business got its start. With a simple 11-question form (see questionnaire, next page) we ask our customers what we do that pleases them and where we can improve. In as long as it took for a brainstorming session, we came up with a form that we can use for all our customers.

The survey form costs no more than paper, ink, and postage. To make it look professional, we print it with our logo. Each survey arrives with a self-addressed, stamped envelope, which is probably why we get

Big River Construction and Remodeling Co., Inc.

A Valuable Evaluation for Big River Construction

1. How did you find out about Big River Construction and Remodeling?

2. If by personal reference, what was it about our company that prompted you to call us?

   Who referred you to us?

3. What are your observations about telephone communications before, during, and after your work?

4. From our first visit to your last construction detail, were our company's representatives courteous, helpful, and knowledgeable?

5. When there were others working on the job (i.e., subcontractors), how do you feel they respected you and your property and how well did they uphold our company's standards?

6. A. How would you rate our company's overall construction performance?
   poor 1 2 3 4 5 6 7 8 9 10 excellent
   B. How would you rate our administrative performance?
   poor 1 2 3 4 5 6 7 8 9 10 excellent
   C. How would you rate our professionalism?
   poor 1 2 3 4 5 6 7 8 9 10 excellent

7. During construction, how did you feel we handled the day-to-day phases of your job?

8. What could our company have done to make your project run more smoothly?

9. Are you happy with your completed project? Yes No

   What, if anything, would you have done differently?

10. How effectively and timely did we handle any problems?

11. May we use you as a telephone reference? Yes No

Please feel free to include any additional comments.

Send Free Gift to: ______________________________

Company: ______________________________

Address: ______________________________

City/State/Zip ______________________________

Thank you very much for your valuable input.

about an 85% return. We also offer respondents a company hat or T-shirt, which we promptly send after receiving the completed questionnaire.

Half of our customers respond without any prodding. The other half we may prompt with a phone call, but no more than that—we don't want to annoy anyone, and some people just don't fill out questionnaires.

The Payback

What we get in return is valuable information which we use in two ways:

• *In running our business.* If our customers tell us we're doing something right, we continue to do it. If they tell us we need improvement, we evaluate their concerns and make the necessary changes.

• *In advertising.* We are proud to show that most of our customers rate us highly, something the survey lets us measure. We put the figures in our brochures and newspaper ads. We also ask permission, after the survey is completed, to use favorable quotes in our ads. Most people love to see their names in print.

One benefit of a questionnaire is that customers are more comfortable writing their feelings down on paper than expressing them verbally, face to face. That's especially true with homeowners, but even commercial customers tell us more in writing than they do in person. It's this kind of information that helps us make managerial decisions.

The most important change we've made as a result of the surveys is in the way we supervise our subs. We found out that our subs were approaching the owners in our absence to ask questions, and some were not being careful enough about keeping a neat job site. Now we make it very clear before hiring subs that all inquiries go through us, and that we have certain expectations about job-site conduct such as smoking and parking.

Creating Your Own

Creating a questionnaire is relatively simple. To get the necessary detail, however, you should ask questions that encourage your customers to open up. Ask them how they feel about your crews, subs, administrators, and workmanship. Ask for specific examples. Allow enough space between questions for brief answers. Assure them that they are welcome to use the back of the form or other paper if they need more space.

Don't require written answers for every question. A couple of simple yes-or-no questions will quickly gauge the attitude of your customers and help to keep the survey short. Try to keep it to one page.

Make it easy for your customers to evaluate your performance in ways you can really measure. We use a scale of 1 to 10, with 1 rated "poor" and 10 rated "excellent." Our consistently high grades from customers go into our ads. We can say — and prove — that 90% of our customers rate us a perfect 10.

One thing we don't do is ask for job leads. This is not the appropriate place to do that. Customers would perceive it as the real motive of the survey. The questionnaire is the customer's opportunity to speak their piece. Make it just that and you'll leave your customers with your real message—that you are interested in what they have to say. You'll generate lots of goodwill and good referrals. ■

Terry Brocious was operations director of Big River Construction Co., Inc., in Baton Rouge, La.

Chapter 12 • Marketing and Sales

Letting Them Know You're the Best

by Len McAdams

You do good work. Now take these practical steps to make sure your customers know it.

Doing good work just isn't enough if your goal is to increase the size and range of the business you do. Why not? Because we live in a world of impressions, and how you are perceived has everything to do with the number of customers that will come your way.

Can you imagine a Fortune 500 company —no matter how superior their product — without a marketing department? These giants got where they are because they know their markets and aggressively promote themselves with both old customers and new prospects.

Yet good marketing is even more critical to remodelers. As a service industry, we rely heavily on how customers feel about us. And most of us aren't big enough to grow based solely on reputation, nor do we have the size to sweat out the lean times if we are passively waiting for our next client to call.

But marketing your company doesn't have to mean hiring high-priced consultants or changing the work you do. It does mean becoming more conscious of how others perceive you and making sure that your quality and reliability create a lasting impression.

I spend a lot of time in my nine-year-old design/build remodeling business making sure we do the best work possible, but I put just as much energy into marketing and promotion to make sure that an increasing base of homeowners in our area know about that good work.

The "M" Word

So what is marketing anyway? It's hard to pin down, but here are four things that come to mind:

- Research on what business opportunities exist for your company;
- Knowledge of what consumers want in goods or services and how to approach them for their business;
- Improving your company and its product so that it can deliver what the consumer wants at a price he or she is willing to pay;
- Promotion of your company and its product and services through a range of media.

It's no accident that only one of the four items involves promotion. Promotion is crucial if growth is your goal, but until you've done your market research, most promotion will miss the customers you want most and may even clog your system with unwanted inquiries.

So where do you begin? Try answering the following questions completely and accurately:

- Who are you?
- What is the product or service you sell?
- Who are your customers?
- How do you differ from your competition?

First impressions are vital to how you're perceived by clients and prospects. That means ties and jackets for the boss and sales staff, uniforms for the crew, and crisp paint for company vehicles.

Now boil down your answers to no more than three sentences each. Done? You've just begun your new marketing program. Actually, this little exercise isn't easy if you are being thorough, but if marketing were easy, everyone would do it well, and there'd be no advantages to gain.

Applying a Marketing Approach

I frequently meet people with a good track record in small-job remodeling and a little capital who want to broaden their customer base and build a company that will break the hand-to-mouth cycle. The question for them is how to get started. Here are some of the suggestions I give them.

Analyzing your business. Just because you haven't been promoting your work doesn't mean it isn't a fine base to build on, but first you need to know what value it has. Take a look at your last few years and pay particular attention to the jobs that have produced the best profits and the happiest customers.

Let's say that in the last four years, your 15 most successful jobs were high-end bathrooms. Looking at them more closely, you discover that the customers on these jobs were married couples in that "thirty-something" category. Okay, now it's time to start asking questions that will help you assess the strengths and weaknesses of your company.

Start with those past customers. Tell them that you are planning to expand and need candid answers to help you build a better company. Ask questions that will help reveal what they particularly valued in your relationship with them. Is is workmanship? Or your sense of design when it comes to tilework? Your ability to make them feel taken care of as a customer? Also ask questions about perceived value, what they'd like to have seen handled differently, etc.

But don't stop with your successful jobs. Take the time (and find the courage) to talk to some tough customers too; friends won't report the most important shortcomings you'll need to overcome. And then turn the same questions on your employees, your suppliers, and yourself.

Drawing some conclusions. Some pretty revealing facts will begin to appear as you gather others' impressions. You already know that your most profitable jobs have been master bathrooms for "yuppie" couples, but you may also find out that a lot of your customers mentioned your lead carpenter as being friendly and helpful. And your part-time office person has revealed that over half of the referrals calling in mentioned the tilesetter you use.

Going back over time cards, you discover that you haven't spent much time supervising these jobs, but you did work closely with the customers on the design. And finally, let's assume research shows that your geographic area has a substantial number of two-income professional families who live in homes over 15 years old. And there are enough of these potential customers to support twice your current master-bathroom projects without your taking over more than one percent of that market.

Developing a promotion plan. So you've figured out your strengths, and you know there's a market for what you do well, but how do you get that across to the people who should be dialing your number?

Start with your name. If bathrooms are your speciality, make sure they get mentioned. But to avoid restricting yourself in the future, use *"So-and-So" Remodeling & Design*, and then follow that with "Fine Bathroom Construction" in all your bathroom promotion literature.

You'll also need a logo. Consider arranging with a local technical-school marketing class to pay a prize fee to the student who comes up with the best one. Students should also make suggestions about color schemes for promotional material, but check these with printers, jacket suppliers, sign painters, etc., to make sure your choices won't cost too much or create delays in restocking these items down the road.

Now you can order stationery and forms, business cards for yourself and your employees, door hanger flyers for houses in the neighborhood where you're working, and a simple one-page brochure. Given the market you've defined, tailor all your promotional literature to your upper-middle-class clientele in language and type style. Consider using buzz works like *personal service*, *expert tile installers*, *friendly*, and *professional*. Graphics should be architectural in nature, with no cute "clip art" or amateur drawings. Above all, keep colors, type styles, language, and presentation consistent.

You might also hire a sign painter to make job-site signs. If you use a material like corrugated plastic, it will keep unit costs low and make the signs easier to discard as soon as they begin to look shabby.

The importance of appearances. Like it or not, your customers (and potential customers) will judge you not on your expertise as a builder, but on the things they are most familiar with; your appearance, their contacts with your employees, how clean your building sites are, etc. Here are a number of things you should consider to make sure you are giving your customers the impression that matches the careful work you do.

If your company truck has accumulated a few years and a few dents, have it repaired and then painted with the company colors, name, and logo. Tell your employees that the truck is now a symbol of the new company image and ask them to take some personal pride in its appearance.

Also explain to your field personnel what you are doing to bring in additional business for the company. Talk about the types of jobs you're targeting and explain the customer profile. Make sure that everyone understands that quality and referrals are the lifeblood of the company and give them authority to stop work on *any* job at *any* time if they don't feel that they'll be able to produce top-notch work with the manpower, tools, and materials on hand.

After explaining to your crew the value of appearances to the homeowners, ask your crew to help choose a uniform—something as simple as a T-shirt and cap with the company name embossed on them—for the jobsite. Be prepared to insist, even if it means firing a skilled tradesman who doesn't want to wear "a monkey suit." Appearance and attitude are that critical to your success.

Over the next six months, add personal grooming hints to your regular training program for field employees. Order company jackets to be given to employees who stick with the company six months or more — they can become a kind of status symbol. Key employees should be allowed to buy jackets for family members, or given additional jackets for outstanding performance.

And the boss isn't exempt either. Even if you began your career as a tradesman with a vow never to wear a tie, purchase some dress slacks, dress shirts, sport jackets, and neckties, and wear them. If you do have to work on site, stop at home to shower and put on a tie and coat before visiting current or future clients to discuss business. Sound extreme? It isn't. By dressing like a businessman, you'll be taken as a businessman, and you will set yourself apart from your less professional competition before a word is said.

Putting it to work. Now that you and your company are spruced up, it's time to get out there. Begin by updating your company's mailing list. Gather the names of all former customers, recent prospects, suppliers, subcontractors, and anybody you or your employees know even casually.

This handout gives the prospective client something to take home from the home show. But it's the personal contact at these shows—and follow-up phone calls—that bring your company to the customer's mind when he or she is ready to act.

Send announcements to the people on this list heralding your name change and your commitment to the master bathroom market. Somewhere in the announcement, ask for referrals by telling these friends that you depend on their help. You should continue to ask for your customers' help every time an opportunity arises, and you should make sure the opportunity arises often.

You should also design a form for taking telephone leads. It should include key questions about the source of the lead, the timing of the customer's construction plans, budget, financing, etc. Have your office help fill out this form for every incoming lead (and don't forget to do it yourself), and then tally these leads monthly.

You may also want to design a form to send to each customer after you've completed their project, asking for their help in improving company service. These report cards are very helpful in fine-tuning

Company newsletters should be interesting and entertaining. In past issues, the author's remodeling newsletter passed along useful tidbits on design and mortgage money, introduced key employees, advertised new services, announced home show appearances, and even put the word out that the company was looking for a new office manager.

project management, a good source for customer testimonials to use in advertising, and often a real boost for everybody's morale.

Some Finishing Touches

The steps I've described so far are a good beginning. You have discovered who your customers are, that they buy on a referral basis, and that the skills and personal handling your company provide are highly regarded. You've begun to build an image for your company that is carefully tailored to that set of criteria, and you've opened communications with your most valuable asset, your established client base.

To date, you've spent a very small amount of money compared to "shotgun" advertising in the newspaper and/or Yellow Pages, and you should begin to see some immediate results if you have correctly interpreted your survey and made wise promotional choices.

But you'll need some additional promotional schemes if you've set rapid growth as a goal. Here are a few resources which have been useful to me in promoting my business.

Home shows. There is no environment quite the same as a home show. We meet most of our nonreferred customers at one of two home shows annually. Since we are a personal service company, we like to meet customers on a personal basis.

Two caveats about home shows: First, these homeowners are seldom ready to do business when they come to the show. Make sure they go home from the show with your literature, but solid, long-term follow-up is critical to being there when their itch to remodel finally becomes irresistible. Also, if you hold a drawing at a home show, make sure it is for your product or service. A drawing for a microwave identifies you as an appliance salesman.

Newsletter. Our company newsletter has kept us in close contact with the people we already know. We keep it news- and people-oriented. The real key for a newsletter is to have it read, so it must entertain and be of value to the reader.

Trade associations. Participate in your local NAHB or NARI affiliate remodeling association. If there is no organized contractor referral system in your area, help start one. Learn to compete with the highest caliber contractors, and you will soon see that there is more quality work available than all of you can do.

Public relations. This is really subject matter for another entire article. Nothing as inexpensive can do as much for your company image — and ultimately your bottom line — as becoming the recognized expert in your field. When you are in print or on the radio or TV, the consumer confers immense credibility on you. If they know how lucky they are to have an important person like you working on their proposal, how many "thinking-it-over" or "waiting-for-another-bid" objections will you have to hear?

Public relations is just plain hard work, like any marketing effort. Get started by entering every contest you can — if the association does not have a contest, start one. The sooner you are an award-winning remodeling company, the better. Send letters to the editors of local newspapers and the columnists who write on home-related subjects. Always respond to a remodeling article with a compliment (this is sometimes difficult to do) followed by some added facts, then volunteer to help in the future.

Local leadership. Get to know the powerful people in your community. Start close by with church and school, then branch out into chambers of commerce or other civic groups. Connections in this power base will help to keep you informed of additional opportunities for public recognition. As you become more successful in your business life, these organizations provide opportunity to "give back" something to society. At the same time, many of those civic leaders will fit *your* customer profile.

Promotion of your company, your product, and your services is a continuous task that should be made a significant part of every work day. Decide to be the best at what you do, and then never stop telling yourself, your employees, and your customers how well you intend to serve each person who hires your firm. Prove it to yourself by making the tough decision to provide warranty service to a difficult customer. Prove it to your staff by praising and promoting those who cater to the customer's wishes, and by sacking the ill-mannered and inconsiderate, regardless of technical skills. Prove it to your customers every day — they are the future of your business.

The ambitious marketing strategies practiced by Madison Avenue seem rather removed from Mrs. Johnson's kitchen remodel, or Mr. Monroe's reroofing project, but using these techniques can help you to stand out from your competition. However intangible qualities like professionalism, reliability, competence, and permanence may seem, consumers will gladly compensate you for the value they add. ■

Len McAdams' design/build remodeling firm, The McAdams Company, is based in Kirkland, Wash. He has won numerous awards for his work and is currently serving as president of the Seattle Master Builders Association Remodelors Council.

Market Research For Small-Volume Builders

by John Schleimer

You can profit from the same marketing information big builders use if you're persistent and know where to look

Whether you're ready or not, the small-volume spec builder is going to be facing even greater marketing challenges in the 1990s. While the new buyer pool shrinks and the existing owners age, the big guys just seem to keep getting bigger ("...they have more consultants than I have employees"). How does a small-volume builder compete in a market that is more and more dependent on sophisticated strategies for targeting and pleasing the buyer?

The answer is plain, old sweat-of-the-brow homework. It doesn't take a lot of money, but it does take time and some knowledge of how to get there from here.

Asking The Questions

The first challenge is to really know your marketplace. Here are ten questions that will test your knowledge:

1. How many new homes (in the product type you build) were sold and closed in the last year in your specific market area?

2. What was the monthly sales rate for the builders and/or projects of those homes?

3. How many existing homes in each product type and price range were sold in your marketplace last year?

4. What is the projected annual demand for new housing by product type and price range in your marketplace for the next 12 months? For the next 24 months?

5. Does your marketplace have enough

qualified buyers to adequately absorb your proposed product and that of your competitors?

6. Can you accurately identify the predominant age groups and average household incomes of the new home buyers in your marketplace?

7. Do you know the current exterior elevation design and color preferences of the anticipated buyers of your new or proposed projects?

8. Do you know which current floor plan layouts and room configurations are preferred by the buyers in your projects' price range?

9. What standard features or amenities do the buyers for your product really want and expect?

10. Do you know the average retail price per square foot of your three top competitors' projects?

If you cannot correctly answer at least eight of these questions, you may end up making development decisions about your project and marketplace that lead you down the well-trodden path to failure. But don't give up hope. The answers you need can be found—without spending a lot of money — by asking the right people the right questions.

Getting The Answers

First, identify your specific market area by outlining its boundaries on a map. Note the location of each new home or development within that market area that represents direct competition. Pick normally slow traffic days to visit each project and walk each model that's available for viewing. Develop your own *competitive analysis form* where you note square footages, room configurations, elevation design, standard and optional features, base pricing per home and any lot premiums assigned.

Sales info. Spend time with sales agents or brokers discussing the floor plans that buyers favor and the standard features that generate the most favorable comments. Ask for general buyer profiles for the overall project and for each plan or house sold. Ask the sales agent what floor plan or plans he or she would like to see the builder offer if the builder already doesn't, and why. Ask how many of each type of plan they've sold. Remember that your approach and attitude will probably determine how helpful the sales agent will be and how much valuable information you will get.

If you don't have much luck in approaching the sales agent and builder, contact your local title company. They should be able to provide you with the number of closings recorded for each competitive project. The only information they'll need is an assessor's parcel number, which is readily available from local and county governments. Your title company should be able to give you a specific address, lot number and recorded sales price for each unit sold. Just take this data, drive around the project, and note which plans have been sold.

Still another source is your real estate broker. Typically he or she will be a member of the local multiple listing service (MLS). Have your broker run an MLS competitive analysis of your competitor's projects. The MLS report should give you all the essential information you need, including how many days each home was on the market.

From these three sources you should be able to determine the answers to Questions #1 and #2. The answer to Question #3, on resale activity, also lies with your real estate broker and his connection with the local real estate board/multiple listing service.

An analysis of prices and size trends can tell you what demands aren't being satisfied by new home development. Also, a specific area with strong resale home activity that brings equity to sellers often indicates a good move-up market for buyers — particularly if the area has excellent schools.

Demand and its profile. For marketplace depth, contact government agencies that forecast population growth and new job growth for your area. Take the forecasted growth for the next one- to two-year period and divide that figure by the average household size for the area to determine *housing unit demand.* Household make-up estimates can also be obtained from governmental sources or data retrieval firms.

To "qualify" that demand depth, and quantify it by age and household income level, you can again turn to retrieval firms. For $100 to $200, they can provide you with demographic printouts for specific market areas. In larger metropolitan areas, there are also private or quasi-governmental agencies such as chambers of commerce

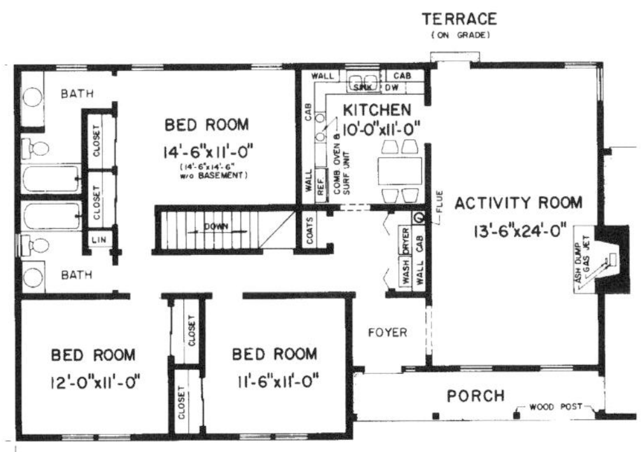

Before you build on spec, find out which elevations, floorplans, and features are selling best in your market, and also what features customers are asking for that are missing in the standard plans. Use this information to create your own mix of popular existing plans plus one or two new ones. Make sure your models are priced competitively to create a high perceived value.

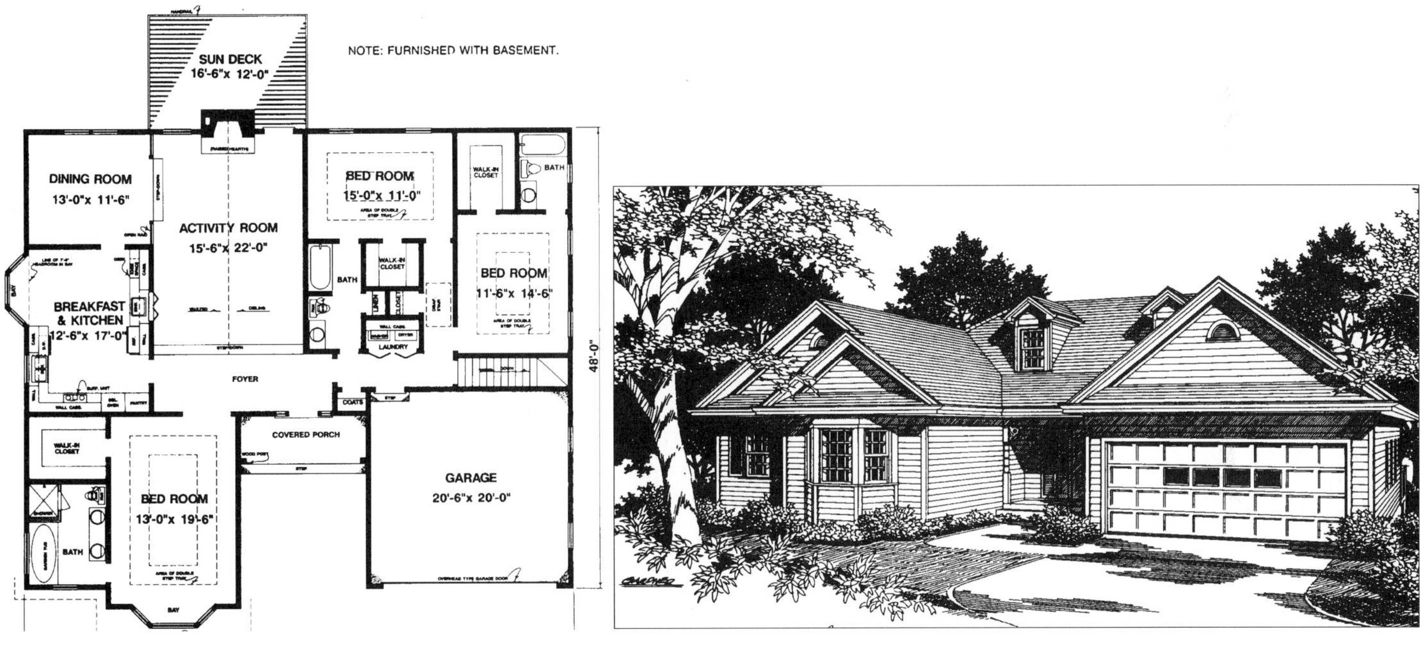

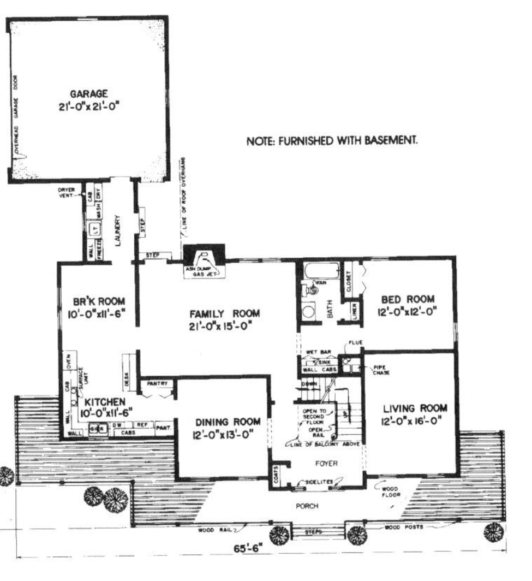

Courtesy of *W.D. Farmer's Homes For Pleasant Living*/W.D. Farmer & Associates

or trade development associations that have most of the essential demographic information you'll need. Just ask.

The right package. To determine exterior elevation design and color preferences for your anticipated buyers, again use your in-field survey of primary competitors. Unless an existing project is selling rapidly because of pricing advantage or a superior location, those projects with higher sales rates generally have the elevation designs and exterior colors that are preferred by most buyers. You can also contact local paint stores to ask what exterior colors are most in demand.

Now for what I consider the key for the small volume builder: offering a better product. The main components are offering the right floor plan designs, including the standard features expected by the buyers, and correctly pricing the product to create *perceived value.*

Go back to the information you compiled from surveying your competitors and talking to their sales representatives. Those on-site salespeople normally tell you what their buyers love in terms of plans and features, as well as what buyers have been asking for that the standard plans don't have. This is an excellent starting point for determining your product design and floor plan mix (as well as pricing structure). Next, check with your local building department and take a look at what future competing projects will be offering in their floor plan mix. Remember, once a builder has submitted his plans for review and approval, they become public documents that you have a right to view.

After looking at existing and future projects offered by your competitors, you may find one or two plans that buyers have been asking for but that aren't being offered. However, don't end up "reinventing the wheel;" if there are several existing floor plans that are selling faster than others, make sure to include them in your mix. Buyer preference trends take at least 18 months to change, so a competitively priced mix of popular existing plans with one or two new ones should be well received once built.

Other Information Sources

In-field research is not the only answer to your product determination needs. You must also *read* and *attend* anything that even remotely addresses these subjects.

Magazines and newsletters that cover buyer profiles and preferences are often excellent sources. Magazines such as *Builder* and *Professional Builder* have numerous articles every year — from market researchers, product development specialists, and architects — which discuss both existing and future housing trends. And these publications are also filled with analyses of the marketing approach used in successful projects. Look for projects from your area that have been written up, then go to school on their product lines, floor plans, and buyer profiles.

Seminars and courses that deal with housing market trends can also be very helpful. This is particularly true if these seminars and courses feature industry professionals or experts on your local market.

You will also want to attend regional and national industry conventions such as the builders' show that NAHB sponsors. From their extensive menu of seminars, choose programs dealing with buyer product and feature preferences, new product design concepts, and how to conduct your own research or analysis. If there are experts from your local market featured, so much the better.

Another excellent source of help at industry gatherings is "workshop programs." These usually allow attendees to sit down and informally discuss special situations or problems dealing with research, product design, marketing strategies, advertising, etc. They often feature prominent names in these respective fields. In a limited sense, you're getting the expensive consultants the big guys are *paying for*.

Still, there is no easy way to predict demand for any given market area or buyer profile. However, if you use some of these suggestions, you should know more about your market area and buyer profiles than most of your competition. Although it's a shopworn and hackneyed dictum, knowledge *is* power, particularly in gaining a competitive advantage in your market. ■

John Schleimer is the president of Market Directions, Inc., a Sacramento, Calif., firm specializing in market research, product development, and master marketing programs.

Using Previous Clients To Win New Ones

by Larry Hayden

Don't hesitate to ask former clients for testimonial letters. They're the most effective marketing tools you have.

How do you convince new clients that your company is the one to go with? One approach is a presentation book full of beautiful photographs detailing your handiwork. Such a book can be helpful; but more important, I've discovered, are the words of satisfied clients. Their descriptions of your company and your work carry a lot more weight with your prospective clients than any claims you can make about yourself.

Like many contractors, you probably give your prospective clients a list of names and phone numbers of people you've worked for in the past, but they don't often call more than one or two. And who knows what kind of mood your old customers will be in when they're called?

Enter the testimonial letter or, better yet, a batch of testimonial letters. As a design/build contractor, I hit on this idea while trying to develop a brochure to inform prospective clients about my company. I decided that a brochure becomes outdated quickly and has room for only a few excerpted testimonials, like the ones we've all seen in magazines. "This company did a great job for us! — J. L., Turlock, Calif." or "Couldn't have done without your fabulous advice! — L.P., Greenacre, Ontario." Convincing? Sure, about as believable as a left-handed spirit level.

Excerpts Are Limited

At first, I devised a testimonial summary — similar to excerpts, but with the full name, city, and state of each client. A client list with about 50 names and phone numbers, including those quoted, accompanied the summarized testimonials so that prospective clients could call for verification or clarification. I knew it would help to use real names of real clients to tell our story. But would the excerpts do a good enough job of communicating to prospective clients the capabilities of our company?

Even this type of testimonial excerpt, however, has a limited appeal, a sort of staged appearance. To the skeptic, it raises more questions than it answers. Prospects wonder what the *rest* of the letter said. What about the screw-ups? How were they handled?

I followed up with prospective clients to find out how reading the excerpts affected them. Let's just say I was underwhelmed by their responses. So, on impulse, I grabbed copies of the complete, original letters and gave these to the next five prospects along with the sheet of excerpts. When I checked back with them, I was pleasantly surprised to find that all five preferred the original letters, and recommended that I abandon the excerpts. I took their advice.

Using the Original Letters

Here's how we currently use these letters in our company. We copy each one and staple together a packet of about 25. We arrange them in chronological order with the most recent on top. Then we send a packet to each prospective client, along with a business card, a list of about 50 client references (names, phone numbers, city, type of project), and a note confirming their appointment to visit our showroom.

When asked if they received the material we sent them, many of our prospects volunteer that they not only received it but read every letter. It makes me wonder if there isn't a *Dear Abby* syndrome, something in all of us that makes us enjoy reading other people's mail. In any event, the response continues to be very positive.

These letters are a natural way for your clients to say thank you for a job well done. But it isn't easy or cheap to produce and deliver them. Doris, our office manager, says postage to send the letters is now approaching $1.44 per packet, and this doesn't include the cost of the manila envelope, copy paper, and her time. So much for this idea being less costly than a brochure. The results, however, are worth it.

In Their Own Words

There is a distinct value in quantity here: Up to a point, the more testimonials the better. How do you amass enough testimonial letters to create the desired impression? The first step is to do a fabulous job

Dear Larry,

Home remodeling can be one of the most painful experiences anyone can go through. The dislocation, the disruption, the dirt, the anxiety, the uncertainty, the cost, etc., can throw any family into turmoil. But Federal Building Company made it as painless as possible. We believe this was due to your professional approach to project planning. From the very beginning, you sat down with us and outlined the entire scope of the project. There were never any surprises. Everything was clearly spelled out in writing. At all times, we knew where we stood with the project costs, which is critical when dealing with a big home remodeling job. Scheduling was also carefully planned, and schedules were adhered to. This also helps reduce owner anxiety.

Of particular importance were the hard-working, courteous workmen on the job. They paid special attention to our needs and welcomed our questions and input. Perhaps most importantly, they cleaned up after themselves every day and kept the job site as neat as possible.

As the job progressed to the finish work, we were impressed by the attention to detail. If we noticed an imperfection, no matter how small, you unquestioningly went back and corrected it. This may be the most important aspect of the job and an area where client and contractor can run into difficulty. There are many horror stories where the contractor refused to fix something, feeling it was done well enough. Or cases where the item is fixed, but the client is made to feel that he/she was being too fussy. This was never the case with Federal. The outcome was a beautiful home with all the details attended to and only warm and respectful feelings for the contractor.

Mark and I would wholeheartedly recommend Federal to any of our friends who wished to do a home remodel. Again, the most important component that you bring to the job is your professionalism. Everything is clearly spelled out, there are no surprises, and the work is done to the highest standards. Friends who have used other contractors cannot claim that they have received the same high level of service.

Again, thank you for a job well done, and done on time. Hopefully, we'll be here for a long time, enjoying the space you created for us. But should we ever need another remodel, we will come straight to you.

Good luck in the future and many thanks.

Sincerely,

Mark & Betty

Federal

OUR CLIENTS TELL OUR STORY BEST

Katy & Glen Hilton / Castro Valley — "Everyone on your staff was a complete professional, a real craftsman, and an absolute pleasure to be around. I brag about how easy this remodeling job was… the way Federal Building Co. performs is… exceptional… the quality team every homeowner should have the pleasure of working with…

Willie & Angela Archie / Kensington — Federal restored our faith in contractors; thanks for a great design and execution.

Nadesan & Linda Permaul / Oakland — Your quality craftsmanship, attention to detail, and sound approach to project management all coincided to make our dream a reality… I selected you because the approach you took, managing the work from cradle to completion is the best approach… You remained in budget, on schedule and worked to keep our home as clean and comfortable as possible.

Estrellita & Raleigh Redus / Montclair — I wanted a 'WOW' kitchen and with Federal Building Co.'s help I got it. …and I love my 'new' living room, dining room and entry.

Robert F. Bole & Artenzio DiOrio / Montclair — …a job WELL DONE! …final result surpasses our expectations… completed on schedule and within contract budget… workmanship is superb… cleanliness maintained was remarkable.

Penny Cooper / Berkeley — What a pleasure… professional approach all the way from design to completion. This bathroom is fabulous!

Norgie & David Walkley / Lafayette — …most amazing of all, the project—which included a family room addition and kitchen remodeling, was finished ahead of schedule and stayed on budget!

illian R. & Arthur Weil / Piedmont — The kitchen is perfect… comparing notes with friends… I realize that I was very fortunate to have chosen Federal as our contractor. Many thanks for a well done job. Your crew are professionals… the product is a fine testimonial to your firm.

& Betty Simpson / Piedmont — Scheduling was carefully planned and adhered to… hardworking and courteous workmen… cleaned up after themselves every day… we were impressed by the attention to detail… wholeheartedly recommend Federal… the most important component that you bring to the job is your professionalism.

Jane Lenschau / Berkeley — We not only appreciate the design and quality of your work on our new kitchen and bath, but also your dedication and follow-up to see we remained satisfied.

e & Nancy Bauer / Montclair — The true measure of a contractor… Federal responded to our desires and created the perfect kitchen for our home.

e Trimble / Piedmont — Our new walk-in shower and whirlpool bath dramatically transformed our master bedroom suite… great design and execution."

Even a lengthy excerpt is no substitute for a complete copy of the original testimonial letter.

for your clients. As designers/builders, we include a lot of detail on the plans so that misunderstandings are rare. It's easier to do the most pleasing work the first time. When problems do arise and client satisfaction is at issue, we have what I call a "rubber spine." That is, we bend over backwards in our desire to please.

It's also important to tell your prospects how many testimonial letters you have received from your clients since you started collecting them as a marketing tool. Explain that you and your crew will be working very hard to please them, and will be checking with them during the course of construction to see how your company is doing.

There is nothing wrong with telling your client before, during, and after construction that their testimonial letter will be an important marketing tool, and that you are counting on their help. Most people will be delighted to assist you in this way. A contractor friend of mine even goes so far as to coach clients to use certain phrases that he wants to see in their letters. I think it's better to let people use their own words, and to encourage them to "let it all hang out."

Prospective clients want and deserve to hear the truth about you. If you make mistakes on the project, your clients might be afraid to write a letter for fear of expressing some negative information. Don't let this stop them. Encourage them to tell it like it is. Your prospects will appreciate reading the letters all the more and will really learn about your "rubber spine." It's better advertising than any words *you* could say.

Recently, we received two letters from the same client. One was nicely typewritten because she thought it would be used as a testimonial. The other was handwritten and, while it was complimentary overall, it covered the parts of the job that she thought could have been handled better. Even though we had gone over her constructive criticisms during and after the job, it was good for everyone in our company to reflect on her letters. We now include both letters in the packet we send to prospective clients. What an effect. Showing *all* the letters we get goes a lot further in establishing our credibility than the fanciest showroom, the most polished presentation, or a few hand-picked or excerpted testimonials. ■

Larry Hayden is the president of Federal Building Company, a design/build firm in Oakland, Calif. He is also the co-owner of Damage Control Mediation Service, which resolves construction disputes.

When Sales Hype Becomes A Warranty

by Reina A. Calderon

If you're not careful, your marketing statements could be construed as an enforceable guarantee of workmanship

Good builders of new homes are often proud of their work and zealous of their reputations. Frequently, when marketing their services to a new home buyer, they are tempted to treat the purchase and sales contract as a marketing tool. Statements in their purchase and sales contracts which promise that the new home will be "a first class dwelling with first class materials" or constructed "in a workmanlike fashion" may be construed by courts as creating express warranties. These statements have more than a marketing impact — they can create legally enforceable guarantees of workmanship that will survive, under the terms of most purchase and sales contracts and under state and federal law, for at least some time after the closing.

Where There's Smoke...

The case law surrounding express warranties can be confusing. One of the primary distinctions that courts typically make is whether the statement — and express warranties can be created verbally as well as in writing — was one that a purchaser could justifiably rely upon as a guarantee of the builder's performance. "Is it merely sales talk and puffery — or is it an express warranty?" courts will ask. If the statement is the former, the buyer will not be able to recover — at least not on a theory of breach of express warranty.

But the distinction is not very clear. In a case where the builder's sales brochure had stated that the builder had a "good reputation," that it "had been building homes for 35 years," and that "all homes were good homes," the court held that the purchaser was *not* entitled to recover from the builder on a breach of express warranty theory. (*Shapiro v. Kornicks*, 124 N.E.2d 175, Ohio, 1955.) Contrast that case with one in which the builder told the purchaser that an uncompleted house would be "completed right," that he "took pride in his homes," and that he "watched every detail." In that case, the court held that the purchaser could recover on the express representation of the builder.

Don't Go It Alone

Sound conflicting? The warranty question is one area where it does not pay for the builder to negotiate the purchase and sales contract without legal assistance. Without guidance it's easy to be pressured by the buyer's lawyer into giving his client "a little comfort" or assurance that the building won't fall down next year, the brickwork won't crack, the basement won't be filled with water next spring, or the roof won't leak.

Moreover, builders in many states have to worry about warranties that courts in those states — either because of state statutes or judge-made law — *imply* in construction contracts, regardless of what the builder expressly warrants. Typically, the warranty implied is that the builder warrants that the new home is habitable and/or that it is constructed in a good and workmanlike manner or using reasonable construction methods. Implied warranty cases and statutes run against the grain of the familiar "caveat emptor"— or "buyer beware" rule that courts have traditionally applied in the context of a real-estate purchase.

Because jurisdictions are divided on whether an implied warranty of habitability is recognized, it's particularly important to get competent legal advice during the contract negotiation stage. In addition, some states permit limitations, under particular circumstances, of implied and express warranties. It's important for the builder to know which warranties he can limit and how he can limit them legally under his state's laws. In some cases, the federal Magnuson-Moss Act, a consumer warranty statute with complicated requirements regarding warranty and service contract programs, may set limits on how warranties are limited or disclaimed with respect to certain personal property classified under the Act as "consumer products."

Avoid Making Problems

From a risk management point of view, knowing something about the law of express warranties — and particularly the unexpected ways in which they can be created — may prove very useful to the builder in managing his everyday sales and marketing efforts. Unlike implied warranties, which the courts or the legislatures impose, the creation of express warranties is, with proper education of the builder's sales and marketing staff, almost entirely within the builder's control.

The basic rule is a common-sense one: If there's a chance that the project as delivered won't conform to a statement or representation made in advance about it, the statement or representation shouldn't be made. It's important to keep in mind that "statements" and "representations" can be other than oral or written. Models, plans, specifications, and other graphic or visual depictions are representations creating express warranties, the breach of which will create liability. (*Wheaton Park, Inc. v. Lombardi*, 149 A.2d 422, D.D.C., 1959.) Sales and marketing personnel should be educated so that they avoid using pictorial images in sales brochures or other graphic depictions that "set the mood" but may ultimately be misleading to purchasers. For instance, if your project is located in a resort town on the ocean, but the project is situated three or four miles from the beach, it may be misleading to present photographs of swimming and sailing in a brochure without a statement that the project is located three or four miles from the beach.

Builders may create express warranties unexpectedly where, in an effort to define what it is they have contracted to build or deliver, they attach a set of plans and specifications to the purchase and sales contract and agree to build in accordance with those plans and specifications. In such circumstances, courts have held that such plans and specifications create an express warranty that the building is *properly* constructed according to them. (*Lipson v. Southgate Park Corp.*, 189 N.E.2d 191, 1963.) In Connecticut, if a builder contracts to deliver a certificate of occupancy at the closing, that promise creates an express warranty that the building will be constructed in a workmanlike fashion and according to accepted building practices. (*Maier v. Arsenault*, 100 A.2d 403, 1953.)

In summary, builders would do well to consider all possible forms of representation as capable of creating express warranties, and to think carefully about their chances of living up to the promises they make. ■

Reina A. Calderon, Esq., is the vice president and general counsel of Calderon Energy Company of Bowling Green, Inc., in Bowling Green, Ohio.

Chapter 13 • Design/Build

Design/Build Remodeling: One Company's Approach

by Matthew DeTroy

With a unified approach you can offer design creativity while keeping costs under control

Six years ago, we realized that in order to thrive in the very competitive "Gold Coast" region of Connecticut, we needed to do something different. Because we had already established a reputation for service and quality, it made sense for us to build on that. We felt that a design/build approach would expand our services to the customer, and offer us a way to provide flexibility, while controlling quality and expenses. Since then, we have grown from a small, family-run business to a $1 million-plus company. We've also learned a great deal about the benefits (and difficulties) of this approach. Among other things, it requires new billing, scheduling, and management procedures.

How We Got Started

Before we incorporated, we had a great deal of involvement with architects. They knew the design of a job inside and out — that's why we hired them. But they often seemed unable to have a job completed on time and within budget. Generally, they had little idea of the true cost of the project until they bid everything out. And often they created innovative designs that turned out to be impractical to build and in need of expensive, budget-busting modifications. As builders, we would catch all the customer's heat, while we waited for the architect to come survey the situation and find a remedy. This would often take days, much to the frustration of the homeowners and ourselves.

In addition, whether practical or not, blueprints and designs can be misinterpreted. The architect designs his plans to be read one way, while the customer may have something else in mind. Meanwhile the remodeler may interpret the plans yet another way and builds as such.

Our decision to offer design services and avoid these problems led me to take courses in architectural design, drafting, and interior design at local colleges.

Slowly, we incorporated what I was learning in many of our smaller assignments, such as porches, dormers, and great rooms. I would consult with my instructors, many of whom were licensed architects, who would review my drawings and suggest modifications. Eventually, I didn't have to depend on my instructors' critiques anymore and we were on our way to becoming design/builders.

At first, I combined design responsibilities with production. My brother, Gregory, who is a tremendous help in evaluating the practical application of a design, became the production supervisor. As a family-run business, this worked well.

But as our company matured and grew larger, so did our responsibilities. We found it necessary to delegate duties to other professionals outside the family circle. Again, because of our emphasis on service and quality, it was very important for us to recruit someone who really cared about home remodeling as an art form. Most young graduates of design and architecture schools concentrate on landing a position with a firm that has many commercial and industrial concerns. But we were lucky to find a young and creative architect interested in applying his talent to the area of home remodeling. At this point, we had truly made the transition to a design/build firm.

How It Works

Although every customer is different, and some are more interested in the design while others are more interested in cost, we generally proceed as follows:

Our first step is to meet with a customer. Here we function less as remodelers or designers and more as interpreters. During this "feeling-out" stage we listen very closely to the clients as they talk about the things most important to them— their home and family. In this consultation we

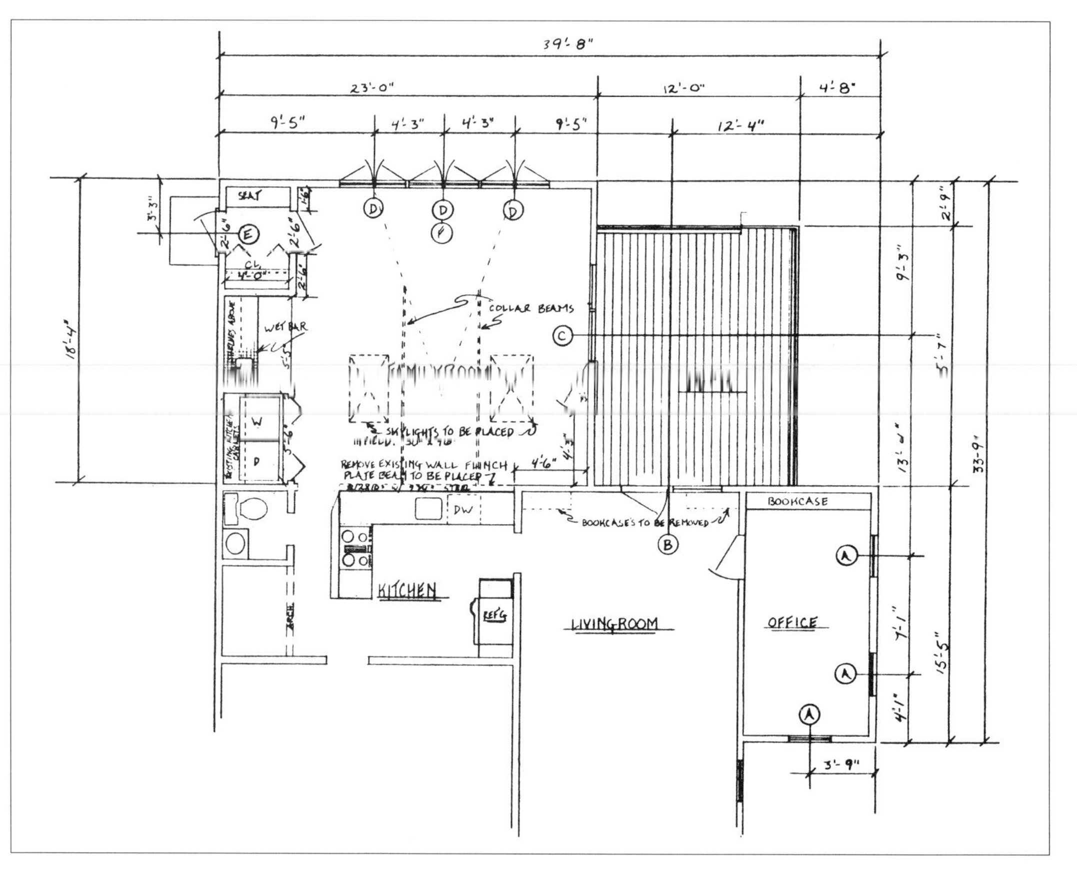

After the first meeting with the clients, Bolken Builders' designer-in-residence visits the site, and takes a Polaroid snapshot (above). Elevations and floorplans (at right) are drawn back at the office.

Bolken Builders Design and Planning Guide

What type of project(s) are you thinking about

- __ Kitchen
- __ Bathroom
- __ Addition
- __ Master bedroom suite
- __ Sunroom/Greenhouse
- __ Second-story addition
- __ Familyroom
- __ Interior renovation
- __ Replacement window(s)
- __ Repacement door(s)
- __ Skylight(s)
- __ Other__________

Does the owner have a plot plan__________

Septic__________ City sewer__________ Well__________ City water__________

How many bathrooms__________ How many bedrooms__________

Aproximate dimensions of addition __________

Can the existing heating system handle the increased demand__________

What type of heat: Existing__________ Desired__________

Do any services need relocating__________ Size of electric panel__________

Window manufacturer__________ Window area, large or small __________

Window material: Wood______ Alum______ Clad______ Vinyl Clad ______

Window type: Casement____ Double-hung____ Sliding____ Circle top____ Bay____

Flooring: Wood____ Tile____ No-Wax____ Ken-tile____ Carpeting____ Other____

If tile: Ceramic____ Marble____ Other____ Size(s)______

Area: Floor______ Wall______ Shower area______ Whirlpool surround______

Door(s): French ______ Slider______ Patio______ Single______ Other______

Access from what room(s) __________ Open to outside,deck, patio, balcony__________

Skylight(s): Fixed____ Venting____ Motorized____ In what room(s) __________

Ceiling: Cathedral____ Tray____ Flat____ Beamed____ High ceiling ______

Lighting: Recessed____ Wall-mounted____ Track____ Ceiling fan(s)____ Other____

Figure 1. *Early on, a remodeler should clarify the clients' objectives. Bolken Builders uses a form that covers items such as window treatments, flooring, and roofing, as well as lighting and decorative elements.*

DESIGN/BUILD PROPOSAL
Design Phase Only

Bolken Builders, Inc.

Page ________ of ________

Date ______________ 19 ______

We herewith submit Design/Build proposal at the sole request and order of:

NAME ADDRESS, CITY, STATE & ZIP CODE		Home Phone
		Office Phone
		Office Phone

hereinafter referred to as owner, for plans and specifications of projected work to be performed according to the following terms:

1. Contractor agrees to provide working plans and specifications for Company's own use in constructing proposed__________

   __________________ Type of Project

   at __________________ Job Address

2. Plans to consist of ☐ Foundation Plan ☐ Floor Plan(s) ☐ Elevations ☐ Section ☐ Perspective

   ☐ Other __________________

3. Proposed budget range for construction of the above under the presently understood scope is $ __________

   to $ __________, inclusive of __________ Major items included

   ____

   Major items not included: __________________

4. If the scope of either the proposed construction work or design phase work is changed after the signing of this agreement, the design fee will be changed and the above proposed construction budget range will be void.

   Any alteration or deviation from above terms involving extra costs will be executed only upon written order from owner or his authorized agent and will become an extra charge over the below agreed amount. No statement, arrangement or understanding, expressed or implied not contained herein will be recognized.

   The plans and specifications described in this proposal are intended for construction by this company and no liability or responsibility is assumed for construction by any other company using these plans and specifications. Said plans and specifications become the sole property of the owner upon full payment of the below agreed amount.

   We propose to furnish the above complete in accordance with the above terms for the sum of__________

   __________ dollars ($ __________). Payment to be made in full upon acceptance of this agreement.

You, the buyer, may cancel this transaction at any time prior to midnight of the third business day after the date of this transaction. See the attached notice of cancellation form for an explanation of this right.

Contractor's Acceptance

Design work to be started on or before__________19 ____

and be substantially completed on or before__________19 ____
assuming present scope of work.

Company Representative __________ Lic No.

Accepted By Contractor __________ Lic. No.

This proposal may be withdrawn if not accepted within ____ days.

Owner's Acceptance

The foregoing terms, specifications and conditions are satisfactory and are hereby agreed to. You are authorized to do the work as specified and payment will be made as outlined above. The owner upon signing this agreement represents and warrants that he is the owner of the aforesaid premises and that he has read this agreement.

Owner__________ Date

Owner__________ Date

Figure 2. *The design/build proposal makes very clear what kinds of plans the customers can expect: foundation, floor, elevation, perspective, and sections if necessary. It also provides a budget range for the proposed project.*

learn about the family's personality, habits, and lifestyle. We also find out if the potential clients are a potential heartache to us.

Often the clients will come equipped with magazine clippings to give us an idea of what they want. Clients' requests can be pretty zany. We had one client who requested that the skylight in a new addition be located so that the moonlight would reflect on a certain spot. The fact that the moon isn't always in the same place didn't seem to matter — until the job was done and the moon wasn't in the right place that week. We ended up moving the skylight.

But generally we try to steer the customers toward reasonable expectations. We supply them with our in-house design booklet, "How To Get The Most Out Of Remodeling Your House," which includes checklists and suggestions to help them formalize their ideas. We also have the customers complete a questionnaire (see Figure 1) to help them clarify their objectives. It covers everything from roofing, flooring, and window treatments to some personal questions regarding room traffic.

After the consultation we prepare a "ballpark" estimate of the project and a preliminary cost of the plans themselves. Once we get a verbal go-ahead to draw up plans, our architectural designer visits the site and takes a number of photos. He also records detailed measurements of the structure and the area to be remodeled.

We then work up preliminary plans — sometimes producing two to three different design alternatives, which we present to the clients. If they like what they see, we go over the specifics involved in our design/build proposal (see Figure 2). The proposal makes clear to the clients what they will be receiving: foundation plans, floorplans, elevations, structural sections, and detailed schematics if necessary.

Kitchens and bathrooms often require such detail. The proposal will also provide a budget range for the construction based on these plans.

It is at this point that we ask for full payment for the *design* aspect of the remodeling. We also aim for a complete understanding of the clients' goals. This is a time for keen listening, for the clients' feedback at this point will be instrumental in generating a set of plans that should satisfy their objectives. Since no actual construction has taken place yet, careful attention to the clients' message can eliminate costly delays and mistakes. Listen hard and ask the right questions. If we plan to use any nonstandard materials that might affect the total cost (but increase the quality), now is the time to explain why. Hopefully the clients will understand and approve.

After a couple of weeks, usually after some additional phone consultation with the clients, we produce final plans and drawings that incorporate their suggestions. We then present these to the clients. If they are happy with the presentation, this is the final rendering. There are times, however, when additional fine-tuning may be needed. Again, time and patience are the operative watchwords here. We try to avoid thinking of customer satisfaction in terms of just one job or this one client we are working with. We're working on the best form of marketing there is — the satisfied customers' testimonial.

Once the clients sign off on the plans we get into the production phase of the project.

Billing For Design

Unlike architects, we don't charge a fee based on a percentage of the estimated final project. Our fee structure is based on the project's complexity. Some remodeling projects are very simple to design, and we feel these should be priced accordingly.

Although all clients have the option of taking the design out to bid (once they have paid for the design), about 85 percent of our design customers have us do the remodeling. We find that the willingness to pay a design fee is evidence of a serious commitment to the project. The fee also takes care of those who would have us design for free and walk away to another builder who might build more cheaply. However, we do refund the design fee upon the first scheduled payment for the actual construction of the project.

Benefits of Design/Build

A full-fledged design/build team can provide the clients with virtually all the services generally accorded to an architect during the design phase, such as developing the outlines, schematics, preliminary drawings, and blueprints.

There are several immediate benefits that a design/build team has over a separate architect and contractor:

- *Control*. A design/build team supervises the whole remodeling job. Hiring, scheduling, and interacting with a horde of subcontractors is almost always left to the builder or remodeler, rather than the designer. Over the years, we've learned which subcontractors we can trust to turn out various phases of the remodeling project on time, on budget, and on a consistent basis.
- *Quality*. As builders, we are on-site from design to finish — to ensure adherence to plans, codes, and quality workmanship in construction.
- *Flexibility*. Design/builders don't seem to get in the type of bureaucratic logjams that the "traditional" approach to high-end home remodeling often results in. The design/builder can alter plans during construction without having to consult another professional, thereby avoiding added downtime.
- *Savings*. Additional architectural consultations also cost money. But even if additional consultations are not required, architectural fees can eat up a remodeling budget. Fees for architects here in Fairfield County vary from 8% to 15% of building costs, but can be higher. Our design process is billed at a much lower rate — usually less than 5%. If the design fee is refunded, as it is for most of our clients, the cost of the design process is covered under overhead.

"Disadvantages" of Design/Build

Perhaps the hardest part of being a design/builder in the remodeling field is that the approach is unfamiliar to most consumers. Generally, consumer magazines play down the clear advantages of the design/build approach and treat the designs offered by a design/builder as pedestrian. If the design/build service is not understood by the customer, he will not pay for it. So the biggest challenge is educating the consumer about the design/build idea.

A secondary problem lies in the standards that we have set for ourselves. As design/builders, we have more than just our reputation for quality workmanship at stake, we now have to maintain a standard for design excellence as well.

Without a doubt, every remodeling project is an opportunity to improve the existing home. A good design will carefully harmonize an addition with the original house and create a well-proportioned and hopefully, better-looking house.

Therefore, we like to discourage clients from cutting corners for the sake of economics when the result detracts from the house architecturally. A poorly designed addition is never a wise investment and should always be avoided.

We believe that the design/build approach, an approach that harkens back to the long-gone days of the "Master Builder," is a viable solution for most remodeling projects. But the burden of proof lies with the design/builder whose challenge is to design and build in ways that enhance the home — most families' largest investment. ■

Matthew DeTroy is president and director of Bolken Builders, a design/build team in Fairfield County, Conn.

Design Fees for Remodelers

by John Cargill

To avoid long hours without pay, ask for a design retainer after presenting rough drawings and a ballpark estimate

There is nothing more frustrating to a design/build remodeling contractor than spending 15 or 20 hours with potential clients — coaxing budget information from them, suggesting approaches and products, developing drawings, and generally holding their hands as we're expected to do — only to be told "Sorry, but my husband and I have decided to build a new house instead," or "We've decided to have Contractor B do the work you drew up because his price is $1,100 less." All you can do is smile weakly and suppress the impulse to say, "Lady, I'd have given you $500 two weeks ago to know this."

New Approach

In the last few years, I have been using a system that eliminates some of this risk. It's helped me sign nearly 80% of the clients I work with on a design, and pays me for part or all of the time I've put in on the other 20%.

Here's how it works. Once I've spent just enough time with potential clients to get an idea of their budget and the changes they want made, I prepare rough drawings and a ballpark estimate (I guarantee that it will be within 10% of the eventual price). Based on this, I ask them for a retainer of three hundred to several thousand dollars so I can proceed with a detailed estimate and a full set of plans and specs. The retainer is applied to the final draw of the construction contract if I end up doing the work, but it stays in my pocket if they drop the project or go elsewhere.

The beauty of this approach is that in presenting myself as a professional, I get treated like one, right from the start. It also puts me in control of the process. I was a remodeler for quite a while before I realized how important it is to be in control. My business life was built around what others asked of me. If the clients wanted the estimate a certain way, I gave it to them. If they wanted to see hard-to-find finish materials early in the planning process, I delivered. I gave them anything I thought would inspire their confidence, including detailed, original plans that I spent hours on. Sound familiar? When all this time leads to a signed contract, it's worth it. But when you see the plans that you sweated over and never got a dime for being built by another contractor, it feels a lot different.

Remodeler John Cargill, at right, discusses design options with two potential clients. After he produces rough drawings and a ballpark estimate, he asks for a retainer before progressing with a full set of plans and detailed estimate.

Initial Contact

The first step in using this approach is to carefully qualify your leads on the phone. A lot of remodelers respond to almost any inquiry by making an appointment with the caller. But it's important to realize that you're committing a *minimum* of ten hours of your time in sketching, estimating, drive time, and subsequent appointments by doing that. Spending between 10 and 30 minutes on the phone asking the right questions can save you that time if the caller just isn't ready, is working with an unrealistic budget, or is just shopping. I also use this initial contact to explain how our company handles design/build and our policy on retainer fees.

Assuming that this "lead" is a good one, I set an appointment time that allows me at least one to two hours with the clients. If you have a showroom you might want to make the initial appointment there, but I prefer the job site. I think people feel more comfortable talking about their personal habits and budget expectations (information you have to get in order to produce a good design) on their own turf. Also, it's a lot easier to talk about changes when you're standing in the "before" picture and can measure for fixtures, appliances, etc.

At this first meeting, I walk through the house with the clients looking at the scope of work. Much of this visit is spent listening to what the clients need and want. I keep careful notes and update them if new decisions are made in the course of the appointment. I also keep a running total in my head of approximate costs. This helps me to ensure that the clients' budget matches what the work will really cost. If I see things getting out of hand, I'll let them know a ballpark figure right away so I don't have to hit them with what amounts to a huge letdown later on.

You really can't proceed effectively without a clear notion of what the clients can afford. With some clients, I find it most effective to ask outright "What is your budget?" But with others, you have to probe more cautiously. Some clients hear these questions as "How much do you think this project is going to cost?" Since my clients are typically 30% to 40% under when it comes to guessing what things will cost, it's important not to mistake their "estimate" for their budget. Try to walk away with as clear an understanding of their real budget as possible.

Once I've felt the situation out and explained design/build again, I tell the clients that I will need to take some measurements before I leave. I then set up a second appointment to approve the sketches and budget on which I will base the final estimate. I schedule this appointment on the spot, and write it down in my book with the clients watching. I try to keep the time fairly short between these two meetings, although I have to give myself enough time to squeeze in the rough estimate and drawings—about four to six hours of work—along with all the other demands of my schedule.

If the house is vacant, I arrange to come back when I can check things out without interruptions. Otherwise I do the measuring during my first appointment. I use a checklist to jog my memory. A combination of photos and graph paper help me record the details. Some of the key locations and measurements I take are:

- Height of floor from grade
- Slope of grade away from house
- Style, size, and location of existing windows, including height above floor
- Style, size, and location of existing doors
- Ceiling height
- Size and location of radiators/ducts
- Location of main electrical service/sub panels
- Existing roof pitch/covering
- Width and type of siding
- Location of existing water lines and sewer
- Location of heating/cooling systems and possible tie-in points
- Location of possible obstacles such as gas, electric meters, hose bibs, well, septic, etc.

Two other things I keep an eye out for that can trip you up later are code violations on the existing house, and how local zoning requirements affect both these and the proposed changes.

Ballpark Estimate and Drawings

The size of the job affects how long you need for a preliminary estimate, but most estimates take me three to four hours. This is only a "ballpark" figure in the sense that the scope of the work is not yet precisely defined. However, I use accurate labor and material rates on the preliminary plan and do an actual takeoff. Where quantities are not precisely known, I round off high.

For subs, I use figures based on going rates. That's usually close enough. For example, in my area, it costs between $100 and $125 to relocate an hvac opening; $350 to $450 to paint walls, ceiling, trim in an average room; an electrical service upgrade runs $900 to $1,200. I typically use the higher of the two prices to cover the "I forgots," and then add my markup.

In fact, I tell my clients, "With a ballpark estimate, I would rather give you a high number to work down from than give you a lower one and keep adding to it." They appreciate this. When you add on to the price or say that something will be extra, clients get the feeling that there's no cap, and they get anxious.

Drawings

Although I try not to spend more than two or three hours on sketches at this stage, I also don't want to leave too much to the imagination. First, I work up a 1/4-inch scaled drawing of the existing floor plan with stuff that I know will be torn out drawn a bit lighter. With this drawing taped to the drafting table, I roll out 12-inch tracing paper over the scaled floor plan. Using a soft-lead 7mm pencil and scale rule, I do quick, free-hand "sketches" of the new work. When I see a dead end, I just roll out more paper, and start again until I get it right.

Once you have the basics sketched on tracing paper, you have to make a judgment call about how much further you want to go at this stage. I've shown clients everything from the bare bones tracing paper sketch to neatly drawn, color renditions. Time, client temperament, and a sense of how much more this plan will have to evolve before the clients are satisfied will help you decide what to present.

Second Appointment

For this appointment you have to take off your design hat and put on your sales hat (hat changing comes naturally to small contractors). First, I present the sketches and ballpark estimate, explaining that this is a custom plan fitted to their budget, needs, and desires. I listen and watch carefully when I'm presenting the drawings. Clients don't necessarily speak up if the design isn't what they had imagined, or if they haven't been clear about their budget with me and the one I'm presenting is too high. If I sense they're uncomfortable, I remind them that there is flexibility at this point and get right into exploring alternatives with them. Often the clients do more designing than we give them credit for.

On most of my jobs, I can make the necessary adjustments to the rough sketches and the ballpark estimate with the clients at this second appointment. And in most cases, they are pleased with what I've presented and have developed some trust in me.

The next step is to ask for the retainer. I explain again that they are paying for my time as a professional to develop a complete set of plans and a price that is within their budget, and that the money will be applied to the final payment if they proceed with

Cargill Design/Remodel

John R. Cargill IV
Certified General Contractor

We hereby express our interest in pursuing a project at the above address. In consideration of that interest we engage Cargill Remodeling to define and clarify a project consistent with our design requests and budget expectations.

We understand that a designer in the office of Cargill Remodeling will produce a floor plan, preliminary specifications, and a budget range.

The development fee will be applied towards a construction contract, and should we elect not to proceed, the fee is nonrefundable.

Project: ______________________

Target budget: ______________________

John R. Cargill IV

Date: ______________________

Development fee: ______________________

Special note: Because the development fee does not reflect the full cost associated with the definition of the project, all sketches and specifications shall be retained by Cargill Remodeling.

The author uses this preliminary contract to collect a retainer fee that covers his costs in developing finished drawings, specs, and prices.

construction. I also give them an idea of how I arrived at the ballpark estimate. This reassures them that I have done my homework, and that the final price will be within 10% of the ballpark estimate (if it's not, I return the retainer to them).

I don't have a set retainer amount. It depends upon the clients and the work. The larger your company appears to the clients, and the more established you are, the easier it is to ask for a substantial retainer—$1,000 to $2,000. However, starting out with a small retainer—$300 or so—is less risky, and it's better than none at all.

The contract that spells out the retainer is just one page (see sample contract, previous page) and is quite simple. I remind the clients that this isn't the construction contract, but just a go-ahead for me to develop a full set of plans and specs, and an estimate. Then we read through it together.

Once they've signed, I go right ahead with developing the final estimate and plans. It's important to remember that you don't have the job yet. That's also the reason for not leaving the budget or sketches behind, which the retainer contract makes clear.

Sound simple? Applying new approaches to your business is never easy. But this is one that works for companies of almost any size. I started using it when my volume was just $131,000 annually (I should do $325,000 this year), but I picked it up from a firm that does over $3 million.

Like any new business approach, you should check with your attorney to be sure that state laws allow you to collect a retainer for design work. And you should remember that this method isn't foolproof. Asking for a retainer can run a client off to one of your competitors who doesn't require a retainer. But the few times this has happened, I've wondered whether I'm losing a qualified client or a price-shopper who just wanted to pick my brain.

You can also get a retainer, put in all the work, and still lose the client. Even if you receive $1,000 retainer, "your" client may go with a competitor who is willing to come in $1,500 below your price in order to keep his guys working. That's a $500 savings for the clients even after forfeiting the retainer. But I find it doesn't happen very often. And when it does, there's not much you can say but "that's life." ■

John R. Cargill IV has owned his own design/remodel firm in Columbus, Ga., since 1980.

Contracts for Design/Build

For best results, use a contract that makes builder and architect partners, not adversaries

by Calvin Goldsmith

Last autumn, our construction company was nearing completion of a substantial new addition and renovation when we ran into a problem. The architect had specified a light fixture we thought inappropriate, and it occupied a highly noticeable spot in the new addition. This posed a dilemma for us, since we work frequently and well with the architect, and we were on great terms with the homeowner. We did not want to insult either of them, nor did we wish to buy the fixture, put it up, and then take it down again, especially since it was an expensive, nonreturnable, special-order fixture.

Builders often find themselves in this position. The architect has made a questionable call, and the builder is left with the decision of how to handle it. Unfortunately, a builder in this position is often tempted to play the hero and tell the owners about the architect's mistake, and how he, the builder, will fix it. This approach usually advances the builder's standing at the expense of the architect. At best, you may sacrifice a good source of future job leads. At worst, you may get sucked into a raging whirlpool of litigation.

We wanted neither. So the light fixture went up, the owner blanched, the light fixture came down, and it was replaced. It now sits in storage at our expense. At the time this seemed to be a small price to pay to preserve our relationships with the architect and the owner. But in retrospect, what we needed was better communication with the architect. Then we might have felt freer to question the architect's decision without fear of reprisal, and the light fixture could have stayed in its dark, dusty corner at the electrical supply house.

In an effort to improve our service to customers in situations like this, I set out to explore ways builders and architects could better work together. I began by examining the basis for the traditional architect/ builder alliance, and I interviewed a number of architects to get their input on the best way to structure a relationship. The result — a team approach — is founded largely on trust and communication. But since trust is a nebulous entity in a court of law, the team approach also provides a clear-cut method for handling disputes and resolving differences.

Traditional Contract

The architect has traditionally been the owner's representative, often taking full control of the construction itself. The contract architects have used is AIA Document A201, "General Conditions of the Contract for Construction." This document is based on the principle of inclusion — that is, include everything you think could ever happen in any construction project. It is long, complex, and certain to scare the hell out of any owner wanting to build a 2,500-square-foot custom colonial. But more importantly, it is written by AIA lawyers to favor the architect, often at the expense of the builder.

A201 puts the architect in control of everything and makes the contractor an adversary to be controlled. For example, Clause 3.18.1 under "Indemnification" states that the contractor shall indemnify the owner, architect, and their agents from any damages caused by the construction to the extent that the damage was caused by negligence of the contractor or subcontractors, "regardless of whether or not such claim, damage, loss, or expense is caused in part by a party indemnified hereunder." In other words, the architect is indemnified — that is, protected from liability — even if he is partly responsible for the problem.

Another example of the document granting unfair protection to the architect at the builder's expense is under "Execution, Correlation, and Intent." Item 1.2.3 ends with the statement: "Performance by the Contractor shall be required only to the extent consistent with the Contract Documents and reasonably inferable from them as being necessary to produce the intended results." This last clause could be interpreted to mean "if the architect left anything out, it is the contractor's responsibility to include it." This not only makes it difficult to bid jobs tightly, but puts an undue burden on the contractor.

Architect Mollie Moran, of Bourgeois-Moran and Associates in Dedham, Mass., points out that the contracts the AIA uses, as well as the training architects go through and the AIA specs, all put the architect and builder in an adversarial relationship. As a matter of practice, owners are led to believe they *need* a representative to act as bulldog and make the builder tow the line. This connotes a lack of respect for the builder as a professional, and puts the architect in the police business rather than in the consulting and design business he or she should be in. Moran and other architects I interviewed feel this archaic attitude must be replaced with a spirit of cooperation.

Recipe for a Team Approach

The architects I spoke with emphasized several important ingredients in the recipe for a team approach.

Early builder involvement. Most architects agreed that the earlier the builder is involved, the better. The builder is the one member of the team who is best able to provide accurate cost estimates.

Peter Thomas of Lincoln Architects in Lincoln, Mass., for example, has made this policy part of his regular business procedure. Rather than opening a project up to bid, Thomas likes to bring a good builder on at the very beginning of a project to get an early estimate based on a preliminary design. This assures him that his design is within the realm of the budget, and keeps him from having to do extensive changes during the expensive working drawing stage. The more complex the design, the more valuable is the early involvement of the builder. Thomas strongly advises the owners that they'll get the best deal by negotiating a price with one builder selected at the beginning of the project.

Without early builder involvement, a project can easily go awry before it ever gets out the door. Last winter, for example, our company bid on a project alongside two other builders. The bids all came in within 5% of each other, and approximately 50% above the budget the owner gave the architect. One problem was that the architect specified numerous flitchplates. He had little understanding of the costs he was running up for the owner. We submitted the bid with LVL beams as an option, with a corresponding reduction in price of almost $4,000. This was a small part of the project, but it's characteristic of the problem. Had we been involved early on, both the owner and the architect would have benefited.

Realistic expectations. According to architect Dave Glassman of Glassman Associates in Needham, Mass., setting realistic expectations is the architect's number one job. The architect should hold the owners' hands and explain the likely track the new construction or renovation will take.

For example, to be responsible, an architect should say, "We will take every precaution to minimize dust and give you back your kitchen quickly. But to be truthful, there will be dust and your routine will be interfered with, so please expect it. This is, after all, a major renovation project." Contrast this statement with an architect telling his client that he will stay on top of the builder and ensure that there will be no dust and only minimal disruption of daily routines.

Glassman believes these overly high expectations set the builder up for failure. Getting the owner to face reality as early as possible makes the rest of the process much smoother.

Mutual respect. Any type of team approach will depend heavily on the support each party can offer the others. The architect must take the opportunity to praise the builder to the owners, and the builder must do the same for the architect.

Think about it from the owner's standpoint. When are the owners more likely to feel confident that they are getting the best job possible? When the architect and builder support each other and communicate problems in an unbiased and constructive manner? Or when the architect and builder are locked in the all-too-familiar pattern of mutual blame and finger-pointing?

But while most architects will agree in theory that trust is of paramount importance, different architects have different ideas about how this trust is achieved. Not all the architects I interviewed were ready to advocate an equal relationship with a builder, especially regarding the issue of money.

The Money Issue

In the traditional model, the architect is the owner's advocate when it comes to money. In a team approach, the architect functions more as an adviser to recommend initial disbursement schedules, percentage of completion, and retainage.

Glassman sees the architect's role as a facilitator in these matters. In this capacity, he thinks the architect must balance the builder's need for sufficient cash flow with the owner's need for a level of comfort about money paid out. He stressed, however, that he is an adviser; the money decision rests with the owner, and ultimately this is the builder's concern.

To handle this, our company always presents a fair disbursement schedule up front with our contract. This ensures that our need for a reliable cash flow is met. We are much more sensitive than any of the other parties involved about who has to get paid how much and when.

An architect has a much different perspective, and his "advice," however casual, might not square with a builder's. Glassman, for example, feels the last 5% to 10% of the project is the most difficult, and he always recommends a modest (5% to 10%) retainage to the owner on those items where it is appropriate. I feel strongly, however, that retainage is often perceived by the owners as money for the taking if they can come up with enough complaints about the construction. It creates a monster out of an otherwise perfectly reasonable client. People hate to think they are giving up something they don't have to. I would never agree to more than 5% retainage, and then only very reluctantly as a last resort and only if we are working with an architect we know and trust.

Team Structure

The desirability of an architect/builder/owner team came through loud and clear from all the architects I interviewed. They all recommended this as a necessary condition for a successful, enjoyable project. However, not everyone agreed on the actual structure of this team.

The team approach can take many forms. A team can be a loose association, or it can be a builder with an architect on staff, or vice versa. It can also be a partnership between builder and architect.

In its simplest form, a team can be defined by a simple understanding between a builder and an architect to work together when necessary. Whenever either secures a client who also needs the other's services, that person agrees to call the other in. In this case, contracts for the architect and the builder will be with the owner, and will be separate from each other.

Our company is very comfortable with the simplicity of this approach. It brings the team together when there is a need, without prolonging the relationship after the need has been satisfied.

A more formal partnership, on the other hand, is an easier sell to the owner. You can offer one-stop shopping every step of the way. However, as the survivor of a couple of partnerships, I urge caution. Don't become a partner without weighing the pros and cons. These relationships are extremely hard to sustain. If you do intend to join a partnership, make sure it is a relationship built on the true needs of all parties involved. This need should be spelled out in a formal agreement that clearly defines shared responsibilities and benefits.

Fair Documents

Above all, make sure you understand the project documents you sign. Ideally, you should make a point to write any contract you sign, but because we aren't all lawyers, this isn't always possible. You may have to find a friendly contract written elsewhere.

Another AIA document, A191, "Standard Form of Agreements Between Owner and Design/Builder," is a good place to begin. This document governs the relationship between the owner and a single party contracting to both design and build the house. The agreement is written in two parts: The first is for the preliminary design and budgeting phase, the second for the final design and construction phase. In many cases this contract is far more appropriate for a team approach to design and construction than any other document; A191 is easier to read and less detailed. And while every contract tends to favor whoever writes it, in this case the favoritism also works for the builder, since the builder and architect are virtually indistinguishable.

A191 is best suited to a team that exists as a single business entity — a builder and architect that have a formal partnership. If you are not part of such a partnership, you may still want to use this document as a basis for your contract with the owner. You contract with the owner for your services, and the architect contracts for his services. Hire a lawyer who is knowledgeable in construction law to walk through the document with you, omitting items relating to design, unless your firm actually does the design as well.

By using A191, you are not a party to a contract with an architect, and consequently, you are not submitting yourself to the architect's formal control of the project. You and the architect work together to satisfy the owner's need, but each is responsible for his own work, and is legally answerable to the owner, who controls the purse strings.

The owner may request that the architect help interpret the degree of completion you claim on your disbursement request. You probably cannot escape this. It is the owner's right to do so. However, you have not agreed that the architect is a party to your contract, so his control is only in influencing the owner, which, if you have a good team, should not cause problems.

At the very least, make sure your contract includes an arbitration clause, so you have a clear-cut method for resolving differences without resorting to full-blown litigation. Arbitration may be painful at times, but it will almost always prove cheaper than going to court. An arbitration clause generally says that each party agrees to be bound by the findings of an arbitration hearing, alleviating the need for a judge or jury. A more detailed version of this clause appears in both AIA documents mentioned above. ■

Calvin Goldsmith is a former contractor and is currently the business manager of Tornesello Construction Company in Bedford, Mass.

The Perils of Design/Build

Contractors who make the leap into design/build face an array of legal obstacles in many states

by Carl Hagstrom and Toby Anderson

The move from building or remodeling to design/build is a natural progression. Most contractors don't hesitate to add a finished basement or a deck to a customer's prepared plans. Similarly, few hesitate to design a remodeled bath or small kitchen. From there, the next step may be designing a major addition or even a whole house from scratch. Such growth may proceed smoothly and seem natural, but there are many pitfalls that a contractor would be wise to consider.

When does a builder or remodeler who does a little design work become a "designer"? Can a builder/designer legally advertise design services? Can a designer who is not an architect get insurance against "errors and omissions" in design? And can he charge separately for design services?

While these questions are central to the livelihood of the design/builder, few have clear answers. The laws governing design vary greatly from state to state. In some states, builders who advertise and perform design services are on solid legal ground; in others they are technically breaking the law and are subject to criminal penalties. But even in the stricter states, the laws are generally unenforced. The issue does get raised in civil cases over insurance payments or contract disputes, but even here the guidelines are ambiguous.

So those who venture into design/build may face considerable risks, and those who don't may be losing out competitively. What's a reasonable course of action for the small contractor to follow?

National Summary of State Architectural Licensing Laws

FOR BUILDERS, REMODELERS AND DESIGNERS

NAHB Remodelors Council

American Institute of Building Design

State Laws on the Practice of Architecture

Every state has its own laws defining the practice of architecture, which spell out who may do what. These laws are summarized in the booklet, *National Summary of State Architectural Licensing Laws*, available for $20 from the Remodelors Council of the National Association of Home Builders (800/368-5242), and the American Institute of Building Design (800/366-2423). Some of these laws are actively enforced and others are not, so it's important to determine if your area enforces them. Your state builders association can be a good source of information.

Most states define "architecture" quite broadly, including everything from structural design through supervision of construction contracts, as well as the design of additions and other remodeling. Some include "aesthetic" design, environmental analysis, or "utilization of space." Literally interpreted, many of these laws would apply to interior decorating, siding on a garage, or installing a closet rod.

Typically, the laws allow only licensed architects to provide these services. Builders who violate these laws are subject to criminal penalties. In Pennsylvania, for example, if you prepare preliminary sketches or floor plans, you're subject to a maximum sentence of up to 30 days in prison and a $500 fine.

Fortunately, many of the laws are limited in scope. For example, 32 states exempt design of single-family residences from the architectural licensing requirement. Some of the laws are limited to larger structures, or ones intended for public use.

But many design laws are not so limited and, if enforced, would severely restrict the rights of nonarchitects to design. To a large extent, these broad-based statutes are ignored by builders, architects, and law enforcement officials. This puts the burden on the builder to decide whether he wants to take a chance by assuming, like everyone else in the business, that he need not pay attention to the strict language of the law.

If you're thinking of getting a friendly architect to put his seal on your plans, the lawmakers have thought of that, too. It's specifically banned by many state laws, and in most states architects are subject to disciplinary action or penalties if they place their seal on any plan not prepared by them or under their direct supervision.

While legislative activity and legal challenges have heated up a bit in the current building slump, there is little interest in "policing" the architectural state laws. This is certainly true here in Pennsylvania, which has some of the nation's strictest architectural statutes. So if the laws are largely ignored by those who are charged with enforcing them, what difference do they make?

The main way these laws come into play is in civil suits, most commonly in contract disputes. For example, let's say you're having trouble getting the last check, and you sue your client for the contract price. If he hires an aggressive lawyer, he might claim that your contract is unenforceable because it is illegal. In fact, agreements to violate the law, such as practicing "architecture" without a license, are usually not considered valid in court.

You, in turn, might argue that everybody ignores the law regulating design services, so the contract wasn't really illegal. Such cases can go either way depending on the specific state laws, contract language, and judge's interpretation.

Design Insurance

Another way the law could come back to haunt you is if your building were to suffer a structural failure resulting in personal injury or property damage. If that happened, your first phone call would likely be to your insurance agent, not to your lawyer.

Does your insurance cover your design services, and does it make any difference whether the services were technically illegal (because you're not a licensed architect)? None of the insurance agents we contacted in researching this article were able to answer this one definitively.

In some cases, they said, the insurance company would simply pay; in others, the answer would come only from a court after a lawsuit was brought against the insurance company. At issue is whether the "general liability" coverage most builders carry will cover the "professional liability" incurred by a design error. Most builders' policies do not specifically *exclude* such professional liability. But they don't specifically include it either.

For example, let's suppose a structural failure was caused purely by the design, not by any error in construction — for instance, a main beam was undersized and caused damage. Your first resort is your general liability insurance, which protects you against ordinary "negligence." This covers any mistakes you make in your construction of a building (or any *claims* in a lawsuit that you made mistakes, even if you say you didn't).

But most insurance companies will tell you that your policy was not intended to cover "professional liability" or "malpractice" — such as design errors. They didn't add in that risk when they figured up your premium, since few insurance companies recognize that most builders and remodelers do design work.

When such cases have been litigated, courts have generally ruled against the insurance company, unless professional liability was specifically excluded from coverage. The courts have argued that an insurance company should know enough to clearly state what its policy does and doesn't cover.

A few policies do clearly exclude professional liability. But many policy forms don't mention it at all. If you write your agent a letter asking about this, his answer will probably be that you are not covered for design liability, since it is not the insurance company's intent to provide that. At best, the agent might admit that a court would probably rule in your favor since there is no exclusion in the policy for design services.

If you want to buy insurance that will definitely cover your design services, your insurer might sell you an "errors and omissions" policy similar to the policies that architects have. More likely, you'd need to shop around for a company willing to underwrite such a policy. Your premiums may be higher and your limits lower than for a licensed engineer or architect, who often pay well in excess of $10,000 per year. The best strategy may be to purchase insurance through a trade association such as the American Institute of Building Designers.

Minimize Your Risk

Depending on your state laws, the design/build picture may appear pretty grim. Despite the lack of clear answers, however, there are ways to minimize your risk if you live in a state with restrictive laws.

The first and easiest strategy is to avoid the appearance of practicing architecture. If you advertise design services, don't use the word "architectural" or anything like it. Many state laws specifically bar use of the terms "architect" or "architectural" by nonlicensed persons describing their services. In fact, in 1991 the state architectural board in New Jersey attempted to ban the use of the word "design" by a remodeling contractor.

Second, it is best to state clearly that your design services are offered only in conjunction with your building services — that is, you are not offering to design anything unless you'll also be the builder. Some courts have decided that the licensing regulations don't apply to a designer who also builds the projects. Also, your design activities are less likely to come under legal challenge if you're the builder on the job.

When you're pricing the project, don't charge separately for "design." Instead, you could charge an hourly rate for "preconstruction services," or find some other way to get paid for your time. If the client is willing to sign a contract for the job before the detailed plans are drawn up, you could schedule periodic payments based on percentages of completion — which would include the "preconstruction" phase.

Sometimes your plans may find their way into the hands of another builder, exposing you to potential liability. Let's say a couple, who signed a contract and proceeded through the "preconstruction" phase with your company, decided to spend their money on divorce lawyers instead of building a house. They later resolved their differences, and decided to build the house after all — in another state. You can't keep them from taking the plans; after all, they've paid you and the plans belong to them. To guard against such situations, you should put a rubber stamp on all your plans stating that they are intended only for your use as the builder. This won't necessarily protect you from all risk, but it will help and the cost is low.

If the project is large enough, consider bringing in a licensed engineer or architect to "supervise" your design work, and put his seal on the plans. This may comply with your state licensing law, and would also give you a stronger position if an unhappy client sues you for problems with the design. It's interesting to note that many state laws don't address the fact that many professional engineers, who are not licensed architects, are performing work that falls within the definition of "architecture."

Associations For Designers

The American Institute of Building Design (AIBD) is a nationally recognized association of building designers. It was formed in California in 1950, is active in 46 states, and has over 900 members. Most of the members specialize in custom residential and small commercial projects. According to its literature, "it is the responsibility of the association to encourage members to comply with state statutes and to design their projects within the guidelines as set forth by prevailing building codes and engineering practices."

AIBD accepts members based on field experience and the ability to produce professional plans. It offers its members many services, including access to professional liability insurance, and an exam that can qualify them as Certified Professional Building Designers. For more information, contact AIBD at 800/366-2423.

Other organizations that offer training and certification for designers include the American Society of Interior Designers (ASID), 608 Massachusetts Ave., NE, Washington, DC 20002-6006; 202/546-3480; and the National Kitchen & Bath Association (NKBA), 687 Willow Grove St., Hackettstown, NJ 07840; 908/852-0033. ASID's program focuses on building interiors and NKBA's program focuses on kitchen and bath design. — *C.H & T.A.*

Rating The States

In this chart, each state has been listed in one of three categories according to how restrictive the state's statutes are regarding the design of *single-family residences*. The chart is intended only as a general guideline for builders and designers. For more detailed information, contact the NAHB Remodelors Council or the American Institute of Building Design for a copy of the *National Summary of State Architectural Licensing Laws*.

Very Few Restrictions		Some Restrictions	Many Restrictions
Alabama	Nebraska	California	Hawaii
Alaska	New Hampshire	Connecticut	Maine
Arizona	North Carolina	Delware	Nevada
Arkansas	Oklahoma	Idaho	New Jersey
Colorado	Oregon	Indiana	New York
Florida	Rhode Island	Michigan	North Dakota
Georgia	South Carolina	New Mexico	Pennsylvania
Illinois	South Dakota	Ohio	Tennessee
Iowa	Pennsylvania	Utah	
Kansas	Tennessee		
Kentucky	Texas		
Louisiana	Vermont		
Maryland	Virginia		
Massachusetts	Washington		
Missouri	West Virginia		
Minnesota	Wisconsin		
Mississippi	Wyoming		
Montana			

Safety Not the Issue

Arguments against builders having the right to offer design/build services often contend that the state architectural laws are in place to ensure public safety.

In most locales, however, plans are reviewed for code compliance before the building department issues a building permit. Many municipalities retain engineers to perform this review, and it's the building inspector's job to see that new buildings comply with a code that was written with the public safety in mind. For most single-family residences, it would seem that this procedure is adequate with regard to safety. For a structurally complex custom home, review by an engineer for structural soundness would certainly do the job.

Furthermore, virtually every state, with the possible exceptions of Hawaii and Tennessee, lets an individual design his own home, which could later be offered for sale. If public safety were really at issue, would the law permit this?

Look Before You Leap

As your building or remodeling business grows and you take on more design work, keep in mind the problems you may face. If you perform only enough design work to avoid misunderstandings with your customer, you may have little to worry about. If, on the other hand, you're considering an active design/build operation, you may be headed for some surprises.

While the intent of this article is to help alert builders to the risks of offering design services, it's important to note that it is not meant to be a legal guide. Contact an attorney before you make any final decisions.

Do your homework and know where you stand before you decide on a new venture. At least you'll know where the pitfalls are, and that can be a big help in keeping you out of them. ■

Carl Hagstrom is a design/builder in Montrose, Pa. Toby Anderson is a retired lawyer, turned writer, also in Montrose.

Chapter **14 • Risk Management**

Softening the Insurance Crunch

by Richard E. Schmidt

Minimize your insurance costs by selecting the right coverages and deductibles, examining your premium calculations, and shopping around

Look, no one likes having to pay for insurance. But if you intend to lead a prudent financial life as a contractor in the field, you must come to grips with this beast. Otherwise, insurance costs can eat you alive.

For example, say you bid a job based on such-and-such an insurance rate. If your renewal policy then comes in at triple that cost, it can be very hazardous to your company's economic health. But good management of your insurance program can help.

There are several ways to reduce insurance costs. The easiest and quickest is to eliminate unnecessary coverages.

Unfortunately, determining what is unnecessary can be difficult — and you may not find a willing partner in your agent. As good a person as yours may be, agents are capitalists, too, and tend to have a genuine bias for insurance over self-insurance.

Assuming Risk

Of course,"self-insurance" really has nothing to do with insurance. It simply means that you have cancelled or deleted coverage, and are assuming the risk yourself. You should only assume risk in one or both of the following situations:

- The risk is very small or remote.
- The cost of potential claims can be readily absorbed as a business expense.

If neither of these situations applies, risk assumption (sometimes called "going bare") is not for you.

There are other ways to reduce costs, however. If you feel squeamish about going without coverage, how about increasing the size of your deductible (or, in the case of many liability policies, adding a deductible provision to your policy)?

Be guided by the following principles when selecting the size of the deductible:

- If you have a relatively high frequency of smaller claims, set the deductible high enough to absorb these costs yourself.
- Use high deductibles in those cases where your chance of a claim is low, but where the asset you are insuring must be protected.

In the first instance, let me say that small, frequent claims drive insurance companies crazy, particularly when the cost of claim adjustment exceeds the amount of the damage.

More important, any insurer worth its salt will jack up your premium at the next renewal in order to cover these claims. On top of claim costs, they'll add your agent's commission, their profit, as well as some dollars to cover taxes on the premium. Therefore, assuming some degree of risk at the low end of the scale almost always makes sense.

The second principle takes a little more thought, however. In fire insurance, for example, most companies will discount your premium at least 15% if you agree to absorb the first $1,000 of any claim. All that's left to do is make a comparison between the amount of premium saved and the number of claims you're going to have to pay yourself.

Say the fire insurance premium on your building is $10,000, and you're carrying the standard $100 deductible. Increase the deductible to $1,000, and your premium will drop by $1,500. Now you ask yourself, how many claims do I have each year, and what would be the "action" in the $900 area of risk that I have just assumed (the difference between $100 and $1,000)?

Look back at the last three to five years. If you feel past patterns will continue, compare your premium savings with your anticipated out-of-pocket expenses on any claims under the higher deductible. If you come out ahead,

increasing your deductible is prudent. Ask your agent for a similar analysis of every coverage in your insurance program.

I am constantly amazed at the number of insurance buyers who carry standard, low deductibles and yet have never had a claim.

Playing with deductibles is one of the quickest ways to adjust premiums downward — assuming that you go about it properly by acknowledging the increased risk.

The Ratings Game

What I am about to suggest is that you embark on a more long-term project and get to know something about how premiums are calculated. Think of the following exercises the way you would when breaking down bid numbers.

Most insurance premiums are based upon a rate: a unit cost multiplied by a volume of something (usually payroll, revenue, number of items insured, etc.). This means your premium varies with quantity, or how much you do of a given thing.

The more you do, the more you pay — fair enough.

Behind the facade of your insurance rate is an elaborate mathematical process, the complexity of which would do Rube Goldberg proud. Happily, there are only two major things about insurance rating (the way your premium is calculated) that you will need to know:

- First, if you qualify for a particular rate based upon size or hazard of your operations, your premiums may be influenced, in part, by your past claims. The calculation is called *experience rating* because your loss experience is a major determinant of your current rates.
- Second, the rate, which already takes into account your chances of producing a claim, can be affected by the judgement of the insurer's underwriter. That's the terrible person who passes over your application and decides how much insurance to give you, as well as how much you must pay for it. He can rate you up or down, subject to certain limits, almost as he pleases.

Experience rate. If you qualify to have your individual claim history influence your rating, ask to see the claims and amounts that went into the calculation.

Many times, the amounts used are overstated because of poor communication or clerical error, and you may be in a position to get an inaccurate claim eliminated or reduced (increasing your chance for a discount based on reduced claim amounts).

This is particularly true for workers compensation, where data-entry errors are so prevalent that almost all calculations of experience modifications have some flaws.

Schedule/judgement rates. The theory behind this concept is that there are characteristics about your operation which may increase or reduce claim frequency (number) or severity (size).

For example, if you have installed a safety program, you probably deserve some consideration for rate reduction. After all, you're a better "risk."

Once again, if you don't ask your agent about the availability of this judgement discount, you may never receive it. Conversely, if you are being surcharged for an adverse characteristic (housekeeping, for example), most plans provide reasons why surcharges are applied, and you should be informed.

Why? If the charge is based on alterable physical conditions, you can have the charge removed by correcting the situation (depending upon the cost/benefit ratio, which is premium saved vs. cost of the correction). Again, however, you have to ask.

Knowledge of the rating process is not something that is bound to excite you. After all, if you wanted to know this much about insurance, you would have gone into the business. But insurance is too big a cost these days for any business owner to remain ignorant about it. Still, I am struck by the number of my own clients who have no idea how premiums are arrived at; there's almost a feeling of resignation about it (in addition to anger and frustration, of course).

My point is that you should look at the rating sheet and try to understand the basic mathematics involved — the "hows" as opposed to the "whys."

If your agent can't or won't show you these figurings (known as "rating data" in the business), then you probably should be dealing with someone else. How can an agent sell you something that not even he or she can explain properly?

Getting Religion on Safety

Let's go back to experience rating for a moment (please don't tell me you've already forgotten!), the system that translates your claim history into insurance costs.

It's very simple. The fewer workers compensation, automobile, and general liability claims you have, the lower your premiums tend to be.

So how do we go about having fewer claims?

I'm not going to insult you by trying to sound like an inspector from OSHA or your state labor department, but it is gospel that the soundest long-term way to control insurance costs is to reduce the size and number of claim-producing accidents. It would probably surprise you that insurance companies are not nearly as dismayed by truly catastrophic, six-figure claims as they are with the nickel-and-dime variety. In fact, all experience-rating schemes limit the amount that you can be charged for a large claim.

Here are, however, some bottomline guidelines:

- If you're not consciously devoting some time solely to the safety of your operations, you are not doing enough.
- Sloppy premises or work sites produce claims — period. In addition, insurance company inspectors get turned off, and they invariably go back to the local office and tell the underwriter what they saw. If you get the coverage at all, you're probably going to pay more than you should.
- Anyone allowed to drive a company vehicle should have his or her motor vehicle record checked prior to getting behind the wheel. This seems basic and common-sensical, but I actually had a client who hired a driver who had lost his license due to traffic citations. The driver went out and was involved in a fairly serious accident. Try defending that one in court.
- The surest way to safe operations (and reduced premiums) is to hire capable people. Slacking off in the hiring process, particularly for workers compensation costs, can be deadly.

Playing The Market

One last thing you can do to control and lower insurance costs is to bring the competitive forces of the marketplace to bear — that is, you go to bid.

You can do this in one of two ways: Either have your current agent submit you to a number of his companies, or get bids from a number of local agents. I tend to favor the second method, because you get to see all of the quotations that way.

You can also (ahem!) hire an outside consultant, but in truth, you may not be paying enough insurance bucks to justify the cost.

While relations with your insurance agent are based on trust and good faith, qualities which should not be held lightly, the fact remains that you are in a competitive bid business; your insurance agent should be, too. To check this, test the market at least every five years.

And that's how to save money on insur-

Don't Be Penny-Wise And Insurance-Foolish

The most important, yet least understood, aspect of insurance is coverage. You may be paying too much for insurance, but improper coverage is really asking for it.

My experience has been that if you don't buy the broadest protection, you will meet trouble down the road. Your liability coverages should be written on a comprehensive form and any property insurance should be written for the perils of "all risks of physical loss."

Maybe you'll dodge the bullet and luck out, but failing to plug all the gaps is, to my way of thinking, playing Russian roulette with your business.

Specifically, your agent should be giving you the following coverages:

General liability. The comprehensive form including the Broad Form Endorsement is a must (some states use the Commercial General Liability policy, which packages everything you need).

Auto insurance. Use symbol "1" for the Comprehensive Liability coverage (covers all business automobiles basically) and symbol "2" for physical damage insurance (assures automatic coverage for new automobiles).

Workers compensation. Include Voluntary Compensation and make sure that coverage applies automatically in any state.

Property and equipment. Nothing short of "all risk" coverage will do.

Job or bid specs. Have your agent review the specs for each and every job in order to tell you if there are any special insurance purchases which have to be made. No exceptions.

I'm sure that none of you is delighted by the prospect of paying more for insurance, but consider this: You will be hurt more by the uncovered claim than you will by paying too much in premiums. Get the right coverage first; cut costs second. — *R.S.*

ance: cut coverages, raise deductibles, understand premium calculation, become a better risk, and use the forces of competition.

I don't claim to have exhausted the possibilities, but this at least gives you some things to work on.

So what are you waiting for? ■

Richard E. Schmidt is a risk management consultant in Binghamton, N.Y. This article first appeared in RSI Magazine *(Edgell Communications, Cleveland, Ohio).*

Fighting Workers Comp And Winning

To avoid overpayment, keep good records and learn how to accurately calculate workers comp premiums

by Richard Cooley

As a small contractor, it's important that I accurately predict my expenses. I can do that now. But I used to have one expense that was about as predictable as the weather: workers compensation.

Like many contractors, I get audited every year by my insurance carrier. The auditor used to go through my books, prepare the audit, have me sign it, and a few weeks later I would be surprised with a bill larger than anticipated. Often the unexpected charges were premiums charged against wages paid to subcontractors whom I couldn't show were insured, and who were therefore considered by my insurer to be risks under my policy. I tried to learn the rules so that I'd know how to avoid these charges (or so I'd at least know which subs I was going to end up covering); but no matter how much I tried to learn about the system, I couldn't predict my charges.

Both my insurance agent and my carrier told me it was simple. (I work in New York State, but most of the basics in this article apply in most states.) All I had to do, they said, was multiply my uninsured payroll (employees and uninsured subcontractors) by the rates set by the rating board for each trade, and I would have my liability. They also advised me to keep it safe and simple — use only subs that carry their own workers comp, and get documentation to show it. That way, they said, I would be liable only for my own employees.

I tried to do this, but still ended up with unexpectedly high bills. Usually they would be $500 or $600 higher than expected. Though I hated writing the check, I would figure it was an amount I could live with. I figured I couldn't afford the time to haggle over these charges with the insurance company, especially since it seemed I would lose anyway.

After my 1990 compensation audit, however, I was given a bill for around $3,700 in unpaid premiums. I was expecting a bill for $600. With $3,100 at stake, I decided to take the time to haggle. It ended up taking nine months, but I got all but $600 of the bill erased.

Running Down the Facts

I started with a number of phone calls to my agent, my carrier, the state compensation board, the state insurance rating board, and the state office of consumer affairs. I told them I suspected I was being charged for covering people I shouldn't really be charged for, and tried to get them to clarify the rules for me. They all proved better at referring me elsewhere than answering my questions.

Things didn't start moving until I hired a lawyer. Suddenly the various bureaucracies involved seemed to take me more seriously: Here was someone who wasn't going to go away. A lot of the workers comp regulations are gray, and I suspect what often decides an inquiry is how big a fuss you are willing to make.

Of course, the lawyer was more experienced in querying those involved in the dispute about the critical issues. He was able to put his finger on the regulations and rules that were being overlooked or misinterpreted by my carrier, and which were resulting in my overcharges. Specifically, he found that:

- I was incorrectly charged $900 for covering the employees of an uninsured, but incorporated, plumbing and hvac sub I often used. Under New York State law, incorporated companies are required to carry their own comp insurance, and other companies cannot be charged for it.

- I was incorrectly charged $1,600 for covering subs who had no employees — a mason, a drywall taper, a trim carpenter, and a painter, all of whom I use often, but who also work for others and who are independent contractors. New York law does not require that I provide comp insurance for such sole proprietors who have no employees.
- I was incorrectly charged another $400 for charges relating to work done on a shop I was building on my own property. Apparently, a complicated interplay of insurance law, at least in New York, has it that a contractor does not have to pay workers comp for work on a property he owns — presumably other insurance already covers it.
- Finally, when these overcharges were corrected, I was relieved of another $150 or so that was basically service charges connected with the other overcharges.

Getting over $3,000 in charges canceled improved my mood. What made me feel even better was having a firm grip on the rules involved and why I had been overcharged, so that I could prevent these overcharges from happening again.

CERTIFICATE OF INSURANCE

SET TAB STOPS AT ARROWS

ISSUE DATE (MM/DD/YY) 1/1/91

PRODUCER

Acme Insurance Agency
P.O. Box 1000
Burlington, VT 05402

THIS CERTIFICATE IS ISSUED AS A MATTER OF INFORMATION ONLY AND CONFERS NO RIGHTS UPON THE CERTIFICATE HOLDER. THIS CERTIFICATE DOES NOT AMEND, EXTEND OR ALTER THE COVERAGE AFFORDED BY THE POLICIES BELOW.

INSURED

Rocky's Drilling & Blasting
P.O. Box 1000
Burlington, VT 05402

COMPANIES AFFORDING COVERAGE	
COMPANY LETTER A	Ace Insurance Company
COMPANY LETTER B	
COMPANY LETTER C	
COMPANY LETTER D	Ace Insurance Company
COMPANY LETTER E	

COVERAGES

THIS IS TO CERTIFY THAT POLICIES OF INSURANCE LISTED BELOW HAVE BEEN ISSUED TO THE INSURED NAMED ABOVE FOR THE POLICY PERIOD INDICATED. NOTWITHSTANDING ANY REQUIREMENT, TERM OR CONDITION OF ANY CONTRACT OR OTHER DOCUMENT WITH RESPECT TO WHICH THIS CERTIFICATE MAY BE ISSUED OR MAY PERTAIN, THE INSURANCE AFFORDED BY THE POLICIES DESCRIBED HEREIN IS SUBJECT TO ALL THE TERMS, EXCLUSIONS, AND CONDITIONS OF SUCH POLICIES.

CO LTR	TYPE OF INSURANCE	POLICY NUMBER	POLICY EFFECTIVE DATE (MM/DD/YY)	POLICY EXPIRATION DATE (MM/DD/YY)	LIABILITY LIMITS IN THOUSANDS	EACH OCCURRENCE	AGGREGATE
A	GENERAL LIABILITY X COMPREHENSIVE FORM X PREMISES/OPERATIONS X UNDERGROUND EXPLOSION & COLLAPSE HAZARD X PRODUCTS/COMPLETED OPERATIONS X CONTRACTUAL X INDEPENDENT CONTRACTORS X BROAD FORM PROPERTY DAMAGE X PERSONAL INJURY	L1M0111011	1/1/91	1/1/92	BODILY INJURY	$	$
					PROPERTY DAMAGE	$	$
					BI & PD COMBINED	$ 100	$ 100
					PERSONAL INJURY		$
	AUTOMOBILE LIABILITY ANY AUTO ALL OWNED AUTOS (PRIV. PASS.) ALL OWNED AUTOS (OTHER THAN PRIV. PASS.) HIRED AUTOS NON-OWNED AUTOS GARAGE LIABILITY				BODILY INJURY (PER PERSON)	$	
					BODILY INJURY (PER ACCIDENT)	$	
					PROPERTY DAMAGE	$	
					BI & PD COMBINED	$	
	EXCESS LIABILITY UMBRELLA FORM OTHER THAN UMBRELLA FORM				BI & PD COMBINED	$	$
D	WORKERS' COMPENSATION AND EMPLOYERS' LIABILITY	M2N1222122	1/1/91	1/1/92	STATUTORY $ 100 (EACH ACCIDENT) $ 500 (DISEASE-POLICY LIMIT) $ 100 (DISEASE-EACH EMPLOYEE)		
	OTHER						

DESCRIPTION OF OPERATIONS/LOCATIONS/VEHICLES/SPECIAL ITEMS

CERTIFICATE HOLDER

Tri-State Construction
123 Meadow St.
Pleasantville, VT 22222

CANCELLATION

SHOULD ANY OF THE ABOVE DESCRIBED POLICIES BE CANCELLED BEFORE THE EXPIRATION DATE THEREOF, THE ISSUING COMPANY WILL ENDEAVOR TO MAIL 30 DAYS WRITTEN NOTICE TO THE CERTIFICATE HOLDER NAMED TO THE LEFT, BUT FAILURE TO MAIL SUCH NOTICE SHALL IMPOSE NO OBLIGATION OR LIABILITY OF ANY KIND UPON THE COMPANY, ITS AGENTS OR REPRESENTATIVES.

AUTHORIZED REPRESENTATIVE Patricia Gilroy

Subs with employees should provide you with certificates of workers comp insurance. Otherwise, you risk paying higher rates to cover the sub.

It Pays to Know Your Stuff

I now keep my books on labor and sub charges very carefully, tracking the charges for which I am truly liable. This, and knowing the areas where I have been overcharged before, helps tremendously when the auditor comes around for his yearly visit. Dealing with this audit visit properly, I believe, is a key to preventing future problems.

When my auditor comes around, I show him both my own payroll records and the ledger on which all my chargeable subcontractor payments are listed. This ledger does not include sole proprietors working alone, and it does not include subs for whom I have proof of insurance. It does show what part of each payment is for labor, so that material costs don't get included in the basis for my charges.

The auditor then compiles the chargeable wages straight from this ledger and my payroll. He might also flip through my checkbook to check the payments there against the ledger. Anything that doesn't match, we hash over. So far, with my better recordkeeping and knowledge of the rules, I've won all my small disputes. I've had two audits since the year I was overcharged, and haven't had any further problems.

What if the auditor does find a problem? If he disagrees with my listed chargeable wages, I'll tell him, "Fine" — he can turn in his audit to the carrier, but I won't sign it. Here I'm gambling a bit on how the audit game works: My auditor works piecework, and, I'm guessing, doesn't get paid for a completed audit unless I sign it. My persistence and confidence in my numbers, I like to think, has made it clear to the auditor he won't leave with my signature unless I totally agree with the audit.

I think this may be one reason he's been willing to take my objections into account.

Doing Your Homework

To deal with auditors with this sort of confidence, of course, you have to be on top of your numbers and have everything well-documented, so that if the auditor questions something, you can show him why you're right. This means having the documentation and financial records to back yourself up.

Substantiate those subs. If you don't want to have a sub's wages included in the chargeable wages, you need to be able to show either that the sub carries workers comp insurance, or that he or she is an independent contractor who works alone.

For subs with employees, this means getting copies of their certificates of workers comp insurance. I won't contract with or pay a sub unless I have a current copy of his certificate, or unless his price and/or work is so good I'm willing to carry him for the length of the job in question.

For subs working alone, your problem will be proving that a) they are working alone, and b) they are truly independent contractors and not employees. Establishing these conditions with absolute certainty is often difficult, but there are a few crucial pieces of documentation that will usually convince your auditor. One is a certificate of general liability insurance showing the sub either has a "dba" (doing business as) or is incorporated. If he's incorporated, then he is an employee of his own company and must, by law, carry his own compensation policy.

Another way to show that a sub is independent is by doing business only under individual contracts for each job, which describe the work to be done and a payment schedule. It helps if subs use their own tools, set their own hours, and are self-directed. According to the New York compensation board, it also helps your case if your subs supply their own materials.

I also make sure to get my bills from subs broken down into labor and materials; I won't pay a sub's bill unless it's itemized like this.

All this has paid off. I've saved thousands of dollars. And I now know how to predict my charges and avoid being overcharged, so I can set aside the right amount of money before the bill comes at the end of the year.

Many contractors are overpaying. Most auditors know little about what to look for. Your best protection is to understand the rules, and thus your own liability, and impress upon the auditor and your carrier that you won't agree to anything more than the proper charges.

If that doesn't work, call your lawyer. For about $400, mine saved me thousands. ■

Richard Cooley is a general contractor in Schenectady, N.Y.

Am I Covered?

The wording and coverages of liability policies vary. Make sure you understand yours before you need to make a claim.

by Patricia A. Ayars

Unfortunately, many contractors don't check their liability insurance coverage until there is an injury or property damage on the job. When the contractor does contact his insurer, he is often surprised to find he's not covered. In this era of high jury awards for personal injuries, lack of coverage could mean the end of a good contracting business. So it's important to understand the basics of your liability coverage. Let's start by looking at the standard coverage available, using three basic questions as a framework:

- Who caused the injury or damage?
- How did the injury or damage happen?
- Who or what was injured or damaged?

Standard Coverage

The standard liability policy states that it will: "pay on behalf of the insured all sums which the insured shall become legally obligated to pay because of bodily injury or physical damage to which this insurance applies, caused by an occurrence." What does this actually mean? By using our three questions, we can make sense of this insurance language.

Who Caused the Injury?

The policy states that it will pay those sums that "the insured shall become legally obligated to pay." The key terms here are "the insured" and "legally obligated to pay." The term "insured" is defined in the policy.

The "insured" includes you, the contractor buying the policy, who is known as the "named insured." If the contractor is an individual, the term "named insured" covers the contractor and perhaps a spouse. If the policy is bought by a contracting partnership, the policy will cover the partnership itself and perhaps the directors or officers of the corporation. The term "insured" also includes employees who are performing certain acts for you, such as operating your equipment.

The policy states it will pay for those sums that the insured is "legally obligated to pay." Under agency law, an employer is generally liable for the acts of an employee who is acting within the scope of his employment. Consequently, you may be legally obligated to pay for those injuries and damages caused by your employees while the employees are working on the job. The policy would cover these injuries or damages. However, an employer is usually not obligated to pay for the acts of an independent contractor. Consequently, the liability policy would not generally cover injury or damage caused by an independent contractor working on your job.

If you want to have the actions of additional parties covered under your policy, you need to have these parties named as insured on the policy. This type of coverage is available and is called "extended coverage."

How Did the Injury Or Damage Happen?

The policy states that it will cover "an occurrence." Simply put, an "occurrence" is an accident. An accident is anything that happens or is the result of that which is unanticipated. In other words, an accident is an unintended, sudden and unexpected event.

All liability policies exclude coverage of "intentional damages or injury," a term carefully defined by law. The definition of intentional damage or injury goes beyond the intention to perform an action which subsequently happened to cause an injury or damage. It means, rather, that one not only intended to perform an act but that the action was also intended to cause injury or damage, or was substantially certain to cause injury or damage. Thus, if your employee intentionally drops a board on someone's head or drops a board from a second story window knowing that it is pretty certain that the board will strike someone below, the action is "intentional" and the results will not be covered by the policy. In contrast, if the employee stacks boards on the second story, anticipating that the stack is stable, and a board slips from the pile and strikes a passerby, the injury would probably be covered.

The policy will also state that it excludes certain types of events. These items are found in the "exclusions" section of the policy. There are four common exclusions that are of particular interest to the contractor:

Collapse hazard. The policy generally excludes the "collapse hazard." This exclusion means that there is no coverage for structural property damage or the collapse of or structural injury to a building or structure if (1) the damage is caused by you or your employees and (2) the damage is caused by the grading of land, excavating, burrowing, filling, backfilling, tunneling, pile driving, coffer dam work, or caisson work, or the moving, shoring, underpinning, raising or demolition of any building or structure, or the removal or rebuilding of a structural support. Incidental coverage is available for the "collapse hazard."

Underground property damage hazard. Policies also generally exclude the "underground property damage hazard." The policy thus does not cover property damage to wires, conduits, pipes, mains, sewers, tanks, tunnels, or any similar structure that is located beneath the surface of the ground or water if (1) it is caused by you or your employees and (2) is caused by and occurs during the use of mechanical equipment for the purpose of grading land, paving, excavating, drilling, burrowing, filling, backfilling, or pile driving. Coverage for this type of damage is available through incidental coverage.

Completed operations hazard. Policies also generally exclude the "completed operations hazard." Under this exclusion, there is no coverage for bodily injury or property damage that occurs on the construction site after construction has been completed or abandoned. Construction is deemed completed when one of the following events occur:

- When all operations to be performed under your contract have been completed;
- When all operations that you are to perform on the site have been completed;
- When the portion of the site on which the injury or damage occurs has been turned over to the owner or user and that

portion of the site is being used for its intended purpose.

Even if further service, maintenance or warranty work is required on the site, the project will still be deemed to be completed for coverage purposes.

Violation of law. Policies also usually do not cover those injuries or damages arising from violations of statutes or ordinances, such as OSHA or state safety codes.

Who or What Was Injured or Damaged?

Most liability policies do not cover injuries to your employees if the injuries are covered by workers compensation. Thus, if your employee falls or is injured by another employee, the liability coverage usually does not extend to those injuries.

In addition, your own property is usually not covered by the liability policy. Property or equipment that you are renting, using, borrowing, or is in your care, custody, or control is also excluded. Many things can be under your care, custody, or control. For example, if you have exclusive access to a building and damage occurs to that building, you may not be covered by liability insurance. Items that are not in your control, like tools of employees that you cannot borrow except with express permission, will be covered. It is important that you have adequate property or equipment insurance to cover the excluded items.

The liability policy usually does not cover damage or injury occurring during the use, loading or unloading, operation, ownership, and maintenance of your on-road vehicles, like automobiles or trucks. Your vehicles include those that you own, operate, or rent. The exclusion is so broad that injuries to an independent contractor occurring during the loading or unloading of your vehicles may not be covered. The on-road vehicle exclusion does not apply to vehicles parked on the site if the vehicles are owned by your employees or third parties, and the vehicles are not being used in the course of construction.

Conclusion

The wording and coverages of liability policies vary. Using this guide can help you understand your coverage, but you should read your policy carefully. And if you are uncertain as to what is covered under your liability policy, discuss the policy with your insurance agent. ■

Patricia A. Ayars, Esq., is an attorney with Perlstein & Ayars of Glastonbury, Conn.

Chapter 15 • Computers

Making Computers Earn Their Keep: Four Case Studies

Mention computers to a contractor and the response may range from "Don't waste my time with that hype!" to "I can't imagine doing business without them."

In covering computers for the last several years, we've discovered that both responses have merit. Yes, the computer field is rife with unrealistic promises about how the new electronic brains will save you loads of money, spit out flawless estimates instantly, and whip your business into organizational perfection — all while you're out fishing. And a number of buyers who bought those claims wholesale have been disappointed. But a much larger number of contractors who had more realistic expectations, and who have invested the time and money needed for training, now find their electronic tools indispensable.

For many who have worked at it, computers have enabled them to reach levels of financial control, office organization, and business growth they could not envision achieving otherwise. And almost every user, even the guy who uses his computer only as a glorified typewriter for form letters and contracts, attests to the professional image that computers foster. For that function alone, a modestly priced system can pay for itself.

To help give you a realistic idea of what you can expect from computers, we've sought out four small to mid-sized contractors who have brought computers — both IBM and Macintosh — into their businesses for varying uses and with varying levels of success. Here's what they've learned from taking the leap.

Seth Melchert: Running the One-Man Office

Company: Master Builders; Oakland, Calif.
Market: Residential remodeling.
Number of Employees: At least four in the field; no office help.
Volume: $300,000 - $400,000.
Computer Hardware: IBM-compatible 286 with 40MB hard drive, modem, mouse, and color monitor. Dot-matrix printer.
Computer Software: Lotus 1-2-3, Wordperfect 5.0, Quicken 4.0.

Like many small residential remodelers, I'm not only the owner of my business, I'm also the accountant, estimator, secretary, and even foreman at times. I don't want to get much larger, because I can operate my business thoughtfully at this size and produce work I'm proud of. But it requires me to be very organized and to have systems in place that I'm completely comfortable with. My computer is at the center of this organization.

I shopped around for about six months before buying my computer—I knew they weren't going to go away. I didn't really know what I'd be using it for, but I had a sense it was the way of the future. I also knew I needed to step into the world of "real" accounting very soon, and I figured I might as well do it with a computer.

Hardware

I went to a number of seminars, spoke with friends, and looked around the stores before I bought a basic, IBM-286-compatible on sale. I did splurge on a top-of-the-line Sony color monitor with a graphics card, figuring if I was going to spend a lot of hours in front of this thing, I wanted it to be as pleasant as possible.

I never agonized much about the Mac versus PC question; I figured I couldn't go wrong either way. The cost of Macs was one major factor against them. Now that PCs have Windows software and Macs have better prices and more business software, I wonder if the debate won't just fade way.

I also bought an IBM Proprinter II, a 9-pin dot matrix printer that's inexpensive and does my work just fine. In all I spent about $3,000—in 1987. The only hardware change I've made is to add a modem and a mouse.

Software

It's true that your greatest investment in a computer system is not the hardware or the software, but rather the time you take to learn how to use it. I'm glad that I bought my computer when my business was smaller and I had the time to learn and practice. I was also fortunate to have a neighbor who makes his living setting up computer systems and tutoring their users. Without the help of someone experienced in computers it would have been much more difficult.

All of my software is standard, off-the-shelf stuff: Lotus 1-2-3 for a spreadsheet, Wordperfect 5.0 for word processing, Quicken 4.0 for my general ledger, Procomm for modem use, and some utilities. I've looked at a number of large, expensive integrated programs for contractors, but as a remodeler, I really need something that I control and change to suit different situations.

Word processing. I use Wordperfect more than anything else. I write bids, invoices, memos, contracts, correspondence, and business forms on it. I see three great advantages of using a computer here:

- First is presentation quality. My bids are detailed, extensive, and professional, with a nice cover page and sections like Scope of Work, Work Not Included, Logistics, Subs, Schedule, and References. Even when my bids are turned down I receive very positive comments and offers to pass my name on.
- Second is the filing capability. I can refer back and forth between documents —

Lumber On-Line

An important part of my computer use is a unique service provided by my local lumber company, Truitt & White, in Berkeley, California. They have created a special program called "Modem to Lodem," or M2L, that provides builders direct access to the store computer via modem. Through it, I can look at a list of every item they carry in stock, what quantity is on hand, the list price, and my discounted price. I can also look at all the purchases made on my account in recent months.

At the heart of the program is the quote entry. This allows me to write up a quote, just as one of their clerks would, which I can then use in a bid, knowing the prices will be honored. Later I can release it to the store to be filled and delivered. While I could always go through the salesmen to get quotes for jobs, M2L allows me to do my bidding at night and on weekends when the store is closed but I'm not busy putting out fires in the field.

Using it requires a modem, but it's relatively easy. As I plow through an estimate on Lotus and come upon the framing line, I simply transfer to M2L and do the take-off. As soon as I complete the list, I print up the quote and return to the spreadsheet with the total.

Another feature I find handy is the Keyword Scan. When I enter a word or description, it calls up every product that is related to it or contains that word. For example, if I enter "hinge," I will see everything from a 35mm Forstner bit to the hundred or so varieties of hinges in stock.

With M2L, I can also play "what if" with different materials. It's easy to find out which is cheaper for closet shelving, No. 2 Pine, or ABX plywood. I can look around for materials I hadn't even thought of; it seems as if every month there is a new line of primed casing available that I didn't know about before.

But best of all, there is no shuffling through old invoices to find out the (sort-of) recent price of an item. For accuracy in bidding, M2L is unbeatable. — S.M.

notes to myself, bids, contracts, schedules, correspondence, employee lists, phone numbers—all within seconds, and create still another document from them if I choose.

- Third is being able to create and access standard forms and documents instantly. I have standard invoice forms, job cost summaries, letterheads, fax cover sheets, time sheets, punch lists, and so forth.

Number crunching. I use two programs for keeping track of the business: Quicken and Lotus 1-2-3. Quicken is a simple and inexpensive bookkeeping package that works great for me. I enter every check I write into the general ledger, and assign it to its account and subaccount. It will then give me a balance sheet, profit and loss statement, and cross-referenced summaries of how much I've paid in a given account, to whom, and when. I can even set budget totals and compare these to what I actually spent. I do not use Quicken for writing checks, but it has that capability too.

I use Lotus 1-2-3 for estimating and job costing. Lotus is a complex and powerful program. I use only a fraction of its capability, but it's easy to learn enough to start using it.

I've built a Lotus spreadsheet that is suited to my needs for estimating remodeling. It includes a dozen sections, from General Requirements to Finishes. Within each section I build the particulars that a specific job will need. While a retaining wall job will have little carpentry and no millwork, an attic addition will have a large carpentry section with stairs, closets, floor beams, siding, etc.

Each section is split in two halves: one for labor and one for lump-sum costs of materials and subs. I select the crew members who will do the job and the hourly costs of their wages and burden; then I estimate the hours to perform the particular work. If I choose to estimate a task by the lineal foot or square foot, I can do this too. Each line item and section gets subtotaled, and then the whole sheet gets totaled. At the bottom I add a percentage margin for profit and overhead.

For job costing I use an identical spreadsheet next to the one I used for bidding. At the end of every week, I punch in all purchases and all worker hours spent on each task. Having the two spreadsheets side by side lets me instantly compare my costs to my bid. I can keep track of the total job cost to date, and see how far ahead or behind the payment schedule is running.

This spreadsheet system suits me precisely because it's so flexible. Rather than having to enter a fixed unit cost, I can see how a 50¢ per hour raise for someone affects the overall job, or what might happen if I reassign framing responsibilities to an apprentice instead of a journeyman. However, because I have customized it to the way I operate, it probably wouldn't work well in a larger setting where it would have to be taught to an office staff.

Originally I used Lotus for payroll also. Now I use a payroll service — it's inexpensive, fast, and most important, breaks out workers comp categories and automatically figures my tax payments and reports. These features should save me several thousand dollars a year; contractors with a lot of variety in their work should look into a service.

Other software. I have a calculator program, Pop Calc, resident in memory that I can pull up while in any program. I use it frequently and I highly recommend it. I have never tried Critical Path scheduling programs, but I have seen them demonstrated, and this will be my next purchase.

I originally had the idea of using Autocad to develop a design/build capability in my office. I found that it is extremely complicated, takes a lot of time to learn, and is simply not for me. I also tried a simpler CAD program, the IMSI Designer. It is easy to use and fast, but it lacks accuracy and I haven't used it much.

When I'm asked by other contractors about the benefits of "computerizing" I admit that I don't know anymore what it's like to be without one — I've developed my whole business around it. As a result, I think of it in the same way I do the telephone and my pickup truck: They've gone beyond tools to being simply "how I do things." □

Paul Eldrenkamp: Rx for Growth Pains

Company: Byggmeister Inc.; Brighton, Mass.
Market: 85% residential remodeling, 15% commercial remodeling.
Number of Employees: Three (down from six a year ago).
Volume: $550,000, 20 jobs.
Computer Hardware: Macintosh SE, 20 MB hard drive,4 MB RAM.
Computer Software: Microsoft Excel, Microsoft Word, Quicken.

I bought my computer so I could have more up-to-date financial information about my business. In addition, I felt it would help me present a more professional image to prospective clients. It has given me both those things, and their value has been even greater than I expected.

Finding the Right One

I knew I needed a computer for a while before I bought one. Finally I assigned an employee to do research and make recommendations. On his advice, in March 1989 I bought an Apple Macintosh SE with a 20 megabyte internal hard drive and one megabyte of memory. Within ten months I added another three megabytes of memory so I could run more than one program at once. I also bought an Apple Imagewriter II (dot matrix) printer.

Though I paid lip service to the standard advice of choosing software first, then a computer to run it, I confess I never looked closely at IBM-compatible software, I was so wowed by the friendly interface of the Macintosh. Although the IBM world offers a wider variety of small construction software, I find that to be of limited importance, for two reasons: First, the software gap is narrowing, and there is now a variety of good packages for the Macintosh. Second, it is much easier to learn a wide variety of tasks on the Mac.

Software

My first major software purchase was Microsoft Office, a package that included Microsoft Word (a word processing program), Microsoft Excel (a spreadsheet program), Powerpoint (for presentations), and Microsoft Mail (for intercomputer communications). I don't use Mail or Powerpoint, but the package cost less than buying Word and Excel separately.

Word and Excel have become vital parts of my business operations. I use Word for all my correspondence and business documents. With a little study I have learned to use Excel — a powerful but initially difficult program — for effective, accurate estimating, job-costing, and lead-tracking.

Other software I use frequently includes Acta, an outliner, and InTouch. InTouch is a desk accessory for tracking names, addresses, phone numbers, and contacts — a computerized rotary file. It's easy to use and effective, but not ideal; I'd like, but haven't yet found, a powerful program to track the status of leads and proposals.

False Starts

My biggest problem has been finding good estimating and accounting software.

Rough estimating. One of my first software purchases was the Turtle Creek collection of estimating and accounting software: MacNail (a set of "templates" that runs on Excel) and HyperEstimator, which runs on Hypercard, a general-use program that comes free with every Macintosh. Although these were easy to learn and use, I was disappointed with the way they handled overhead and profit, resulting in unrealistically low estimates. (In fairness I should add that since then, I've heard and read that MacNail's updates have gotten better and better.)

To get estimates more in line with the way we do business, I decided to write my own templates for Excel. This involved learning the niceties of Excel, which took some time.

But I now have an estimating program tailored to my own business and way of estimating.

Balancing the books. Finding a suitable accounting package also proved difficult. But it taught me some important lessons.

I do my own books, with quarterly assistance from a tax accountant, so I wanted something that would let me do them on computer. After dropping MacNail, I bought an expensive and powerful accounting program called Computerized Classic Accounting. It turned out to have too many capabilities for me to learn and sort through effectively, given my limited knowledge of accounting.

I finally decided to start at the bottom. I had been doing my personal checkbook functions with Intuit's Quicken, an inexpensive checkbook program. It lacks true accounting features, but is easy to use. So one morning I transferred my current business ledger to Quicken, and I've been using it ever since.

This will amuse accounting sophisticates. But using Quicken as my first computer "accounting" package has provided real advantages. For one, it's easy: It takes me only five minutes a day to keep my records current. And it's flexible enough to get useful information from. With a little trial and error, I set up my chart of accounts in a way that enables me to get vital information about where my money is going quickly and regularly. For more sophisticated calculations of profit/loss ratios and other important percentages, the program allows me to export data to the more powerful Excel.

I've since concluded that a good strategy for choosing software is to start off cheap and easy, then develop increasingly advanced — and expensive — capabilities as your expertise grows. With Quicken, for instance, I could do adequate record-keeping immediately and easily, and in the process get a sense of what capabilities I still needed. I was able to do useful work and learn at the same time, rather than having to learn a complete system first — a luxury most small business people can't afford.

But now I have advanced from Quicken to MYOB (Mind Your Own Business), another inexpensive financial software package, but one that comes closer to a true accounting package. Eventually I plan to move up to the more expensive accounting program I bought initially.

How It's Changed the Way I Work

A computer initiates a sort of vicious cycle of increasing business sophistication, but higher administrative costs. Budgeting, financial planning, and job-cost accounting are all things that can be, but rarely are, done manually. With a computer you can do these tasks quickly and accurately. When you do, you get a better handle on your true cost of doing business—which turns out to be much higher than you suspected.

So you raise prices. Which means you get fewer jobs. So you start using the computer to enhance your image to attract enough work to cover what you now realize is your real overhead. Which keeps rising, because while you were once in the field driving nails for minimum wage because of the lousy bids you submitted, you're now doing sales and administration for minimum wage (which is more than you were paying yourself to do them before), and you've replaced yourself in the field with a $16-an-hour lead carpenter who wants vacation and holiday pay.

Eventually, of course, you have to lay your carpenters off, because you can't compete against the bids from builders who don't have computers and don't know how much money they're losing. That's how buying a computer changes the way you work.

This exaggeration makes a point. People think computers save time by taking care of business matters faster. But the reverse is true. You actually spend more time on business, because as getting organized becomes more possible, you suddenly see its importance. I never did job-costing before, for instance, but I do now because it's easy. Maybe my costs went up, but so did my awareness, and I can now charge accordingly. Fortunately, your value to clients goes up along with your prices, and if your marketing is effective—which can be helped greatly by computers—you'll get the business anyway.

My $3,600 computer has helped me grow from a small company doing a few jobs a year to a small company doing 26 jobs a year and $750,000 worth of business. That's called "no-man's land," that area of business volume between $500,000 and $1,500,000 where it's supposedly impossible to prosper. But I am. And there's no way I could do it without a computer and have any other life at all. □

Stephen King: Mastering the Numbers

Company: Common Vision; Hamden, Conn.
Market: Design/Build residential remodeling.
Number of Employees: Six in the field; part-time bookkeeper.
Volume: $800,000 - $900,000 each of the last five years.
Computer Hardware: IBM-compatible 286 with 40MB hard drive, math co-processor, and color monitor. Wide-carriage dot-matrix printer.
Computer Software: Master Builder 5.0, Quattro Pro, Word Perfect 5.1, Act! 2.0, Generic Cadd, and a custom marketing database.

In order to run my remodeling company effectively it's critical that I have up-to-date, accurate information—bank balances, job costs, accounts payable and receivable, etc. —at my fingertips. In the early years of my business I'd spend weeks at the end of a year trying to sort out what had happened. Did we really make a profit, and why? I hated bookkeeping and office work.

When I heard there was a computer software program that could give me systems to stay organized daily, I decided to take the plunge. I knew nothing about computers, but I imagined them to be the miracle I needed to survive and grow. Perhaps because I knew so little, I didn't bother to study the situation; instead I bought a computer on impulse.

Buying A System

I turned to another small business for help, run by a woman who provides office support of all kinds and is a rep for Master Builder software. She worked with a local computer dealer on the hardware side and came up with a complete package — an IBM-compatible, Epson 286 with a 20MB hard drive, a monochrome monitor, a basic Epson printer, and the Master Builder integrated software. The best news was that we could do all of this on a lease. This eased cash flow and removed the stumbling block of having to put out $7,000 or $8,000 for a product I knew very little about.

We were up and running within a few weeks with the help of training visits from our rep, but it was still tough. Since we were starting midway into our fiscal year, we had to enter the past six months' data. We learned the programs by following the screen prompts, and using the help windows and program manual. Even with superb telephone support from Master Builder, it took much of the year—and a lot of voided entries and journal transactions—to get comfortable.

But the frustrations of learning were worth it. At the end of the year I had easy access to our accounting numbers, could rapidly turn out accurate estimates, and was tracking our job costs daily. During the first year of owning a computer, I doubled the company's volume without hiring extra office help. I did have to steal a lot of time from my family during that first year of learning, but for the last couple of years, the computer has more than made up for the lost time.

Upgrading

I view upgrading as a necessity, since computer technology and one's business needs are always changing. Within the first year, we upgraded to a NEC Multisync color monitor (easier on the eyes), an Epson LQ2550 wide-carriage printer (to print out spreadsheets), a math co-processor (for a CAD program), and a 40MB hard drive (you can never have enough storage). I figure we'll be using a scanner (which can transfer drawings from paper to electronic computer files) in another year, as well as taking power and speed up a notch by networking with a 386 system.

As I grew more comfortable with our computer I also bought more software packages. I purchased Word Perfect 5.1 to do our invoicing, contract specs, and letters; even so, I barely utilize its capabilities.

I also bought Quattro Pro, a spreadsheet program that will do more sophisticated financial reporting; Norton's Commander and Utilities for DOS management; a custom designed database for direct mail; and Generic Cadd for computer drawings. All together, between hardware, software, training, and consultants, we've spent about $18,000 in the last three years. This may sound like a lot until you realize that the computer has allowed me to avoid hiring another office person, and has eliminated the manual errors in bidding that used to frequently cost me my profit. All this for $6,000 a year is not bad.

How We Put It to Use

Master Builder is the core of our system. It's a business management system specifically designed for remodelers by two former remodelers in California. It includes accounting, job costing, estimating, scheduling, and other utilities that interface with each other — an integrated program. The current version is an improvement on an already good system; it prevents you from making common entry errors and has an outstanding manual.

Integrated system. On Monday, we do payroll, Wednesday is general ledger and accounts payable and receivable, and Thursday we enter invoices for job costing. Russ, my bookkeeper, spends less than 15 hours a week in the office running a company that has eight employees and does an average of $830,000 in business a year.

In the first year, I used the "Budgeting & Bids" capability of Master Builder for estimates. This is basically an electronic version of stick bidding. But as we got all of our data from past jobs entered, we switched over to "Power Take Off," which uses these numbers for unit calculations. Either way, the program eliminates all the careless errors that compound in a bid when you are presenting four or five refinements of a design/build estimate to a client.

Spreadsheet. I use Quattro Pro for custom financial reports or to enlarge Master Builder reports (they are limited to an $8^1/_2$x11 format). These aren't a necessity, but I feel a lot more comfortable if I'm able to track my company from a lot of different angles. This spreadsheet program allows me to find in minutes answers that would take dozens of hours of calculating. It will also produce this data in all kinds of graphs.

For instance, to use Quattro Pro for budgeting, I take my profit/loss statement categories and compare projected numbers to actual figures each month. But unlike some builders, I don't like the spreadsheet for standard estimating, job costing, etc. That's what Master Builder does quickly and without complication.

Custom database. In our slow Northeast economy, we've turned to direct mail marketing to help stimulate job leads. We had a consultant design a database to handle our list of 9,000 potential clients. The program has a number of features, such as an automatic search for duplicate entries, that save us a surprising amount of money.

Job leads software. I also use Act! 2.0, a personal contacts and job leads database. I enter in job leads with all pertinent information, and can track communications with our clients from week to week. It prints reports that show which marketing and advertising methods worked best, based on numbers of leads generated and dollars spent. Act! also categorizes and tracks the types of jobs we've done — bath, kitchen, additions — throughout the year.

CAD program. This is an area where I'm still a little disappointed. As a design/build firm, I do a lot of drawing, and most of it is still by hand. There has been some improvement in lower cost CAD programs in the past few years, but they're still not user-friendly, nor do they contain drawing templates for common items like Pella and Marvin windows. I'm hoping those will be available in another year.

I still feel somewhat computer illiterate, but I'm very comfortable with the systems we have in place. They offer greater accuracy than I had before, access to information that allows me to actively steer my business, and an efficiency that allows me to keep office overhead way down. □

John Call: Mac Marketing Drums up Business

Company: Home Energy Consultants; Haverhill, Mass.
Market: Energy-efficient design/build.
Number of Employees: Three.
Volume: $133,000 last year, down from $327,000 in 1988.
Computer Hardware: Mac IICX with 40MB hard drive, 5 megabytes RAM, color monitor, letter-quality printer.
Computer Software: Microsoft Works (integrated package), Claris CAD, Pagemaker.

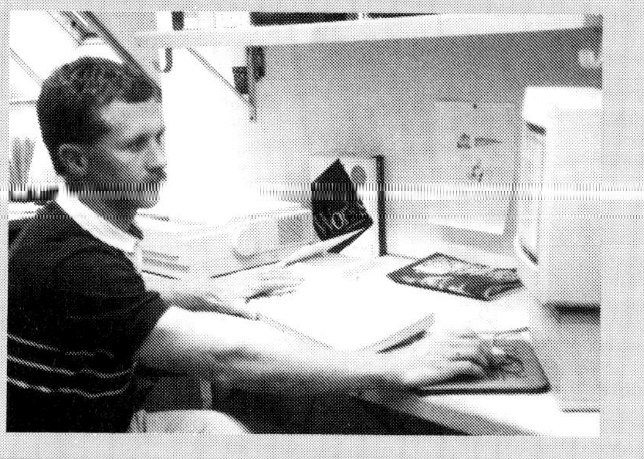

When our sales fell almost in half in early 1989, I knew we needed to start doing something different. During the 1980s, we had grown from an insulation and replacement window company to a remodeling company centered on energy-efficient remodeling, with a few spec houses mixed in. We were running three crews, one each for weatherization, insulation, and carpentry.

But in early 1989, our business fell dramatically. The recession brought a general slowdown, and Massachusetts gutted the low-income weatherization program that provided much of our work. High land prices drove us out of the spec-build market. And we had two homes we couldn't sell. The markets we had been succeeding in were largely drying up.

A Marketing Strategy

We realized we needed a marketing plan and marketing materials — something beyond the yellow page advertising and occasional mailing that had taken us through the boom. We decided to concentrate on the remodeling market, specifically on energy-efficient additions and solar spaces, and to develop appropriate marketing literature to emphasize our experience in those areas. We wanted to be able to produce these materials ourselves, so that we could revise them often, and so that we could tailor presentation materials to particular clients.

These needs, and a conversation with Bill Smith of Coastal Computer of Portland, Maine, convinced us to buy a Macintosh. We figured the Mac's desktop publishing capabilities would give us the flexibility we needed in developing our marketing materials.

So in August 1989 we bought a Mac and a letter-quality printer. For software, we bought Microsoft Works, which combines a word processor with a database, a spreadsheet, and a telecommunications module. We also bought Claris CAD, a sophisticated drawing program; an estimating program; and Aldus Pagemaker, a page layout program.

Early — and Lasting — Success

We got to work right away. We used the word processor to create the first copy of our newsletter. With the database we compiled a client list. We used the spreadsheet to develop an inventory of equipment and stock. We then put this information together into a loan proposal to establish a credit line, which we got.

Excited over these quick results, we put more energy into our presentation and marketing materials, hoping to woo customers as successfully as we had the bank. We used the Mac not only for the newsletter, but to develop more professional-looking letters, memos, and proposals.

We soon began incorporating drawings from our CAD program into these materials. These professional-looking drawings have become an important part of client presentations, and also help greatly with our own planning, and with change orders once the job is under way. If you have to make a change and get approval, you just go back to the office, make the change, and print it. These drafting capabilities alone make the computer worthwhile.

But it is the newsletter, generated on Pagemaker, that has brought the most return in terms of sales. It goes out to the past, present, and potential clients we keep on our mailing lists, which are easy to print out. It includes information about our company and stories about the energy situation and home energy efficiency. In addition to informing and maintaining contact with clients, the newsletter forces us to constantly reevaluate our advertising and marketing programs, and to define both ourselves and the market we are trying to sell to and build for.

It has paid off. We just completed a $47,000 addition that is directly attributable to the newsletter. We're now putting together a presentation book, complete with photos and drawings of our previous jobs.

We see our Mac as a valuable tool box. The marketing tools it gives us are valuable during the most crucial time, when business is slow. We can use it not only to plan and take care of correspondence for existing business, but to generate new business. And although we haven't gotten far in these areas yet, we plan to make use of the computer for estimating and accounting in the near future.

Why sit around and wait for the phone to ring? I think our Mac is the best piece of equipment to have in a sluggish economy. ■

How Do I Use My Computer?

by Guy Hermann

To get the most from your computer, you'll need training. Here's where to look for help.

Computers, despite outward appearances, are not like TVs—you can't just buy one on sale, plug it in, and expect to use it. You will need to learn a whole new set of skills and lingo, just as if you were starting out as an apprentice carpenter. You may often be tempted to toss the whole rig in the dumpster.

The most common mistake someone with a new computer makes is to not ask for help early on. Help is available—and you will need it. Below are a few places where you can get it.

The Manual and Tutorial

Learning the lingo is often the hardest part of getting started. This is why you should read your manuals, with special attention to the glossaries. If your computer or software comes with an on-disk "tutorial," it too can be helpful. Run it several times.

The People You Paid

The stores, mail-order companies, and software companies from which you bought your computer and software owe you support along with their products.

The store —either the salesperson or someone on the technical staff— is the place to turn to with the first week's questions: "Why doesn't my printer print?" "What does 'subdirectory' mean?"

If you purchased via mail-order, you should be able to get the same support. The best computer mail-order companies provide outstanding technical advice by telephone. Call them often. You paid for the help when you bought the computer or software. Some software publishers also have good technical phone support. I won't buy from companies if they don't offer support.

Classes and Users Groups

In most areas, local high schools and community colleges offer courses on IBM, Macintosh, and Apple II computers and dozens of software packages. These classes are cheap, readily available, and often as good as or better than more expensive "computer schools."

There are users groups for every imaginable computer and software package. These groups typically meet monthly to swap experiences and help each other solve problems. They're also good places to pick up free and inexpensive software. Also, big groups have subgroups for specific interest areas. The Boston Computer Society (617/964-2547), for instance, has a construction subgroup that meets monthly and publishes its own (primarily IBM-oriented) newsletter.

Finding a users group can be tricky. Your dealer can sometimes help, as can electronic bulletin board services like Compuserve (more on those later). Some computer and software companies maintain lists of users groups across the country. Apple (800/538-9696), for instance, will tell you the name of the Apple users group nearest you. In addition, *Computer Shopper* (800/274-6384), a

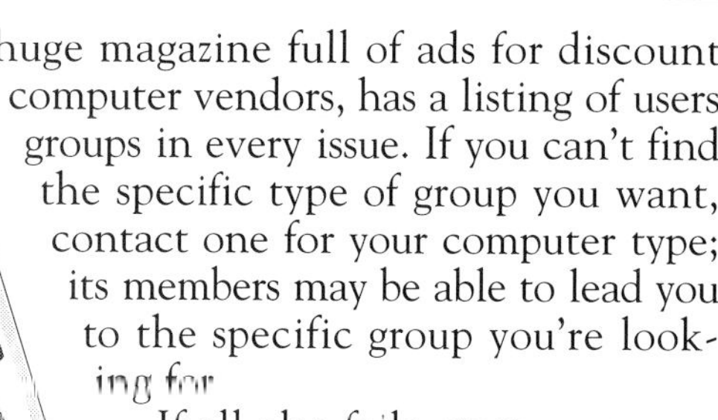

huge magazine full of ads for discount computer vendors, has a listing of users groups in every issue. If you can't find the specific type of group you want, contact one for your computer type; its members may be able to lead you to the specific group you're looking for.

If all else fails, start your own users group. Invite several other computerized builders (or CAD users, or whatever your special interest is) to your office to see and discuss how you do things. Next month, go to another office. You may expose your inexperience, but you'll learn a great deal. Apple has a free information packet on starting a users group that is also helpful.

Electronic Bulletin Boards

One of the best ways to get together with and learn from other computer users is through an electronic bulletin board service, also known as an on-line computer information service. The biggest and best is Compuserve (800/848-8990), though there are several other major services with their own faithful followers. You hook up to these with your computer and a modem, usually for around 20¢ a minute.

"Going on-line" gives you access to an incredible range of services: encyclopedias, news wire reports, financial and economic information, and huge informational databases. You can also converse, either through "live" forums or by leaving and picking up messages, with other users in any area of interest, ranging from aviation to war games, and for every computer type and all major software. These forums can be great places to get help with your hardware and software problems.

The National Association of Home Builders has a new service, available through Compuserve, called NAHB Net, which has a database of NAHB updates and information on construction-related governmental and regulatory subjects such as wetlands. (For more information, call NAHB at 800/368-5242.) Unfortunately, other than this, I couldn't find a builder's forum in any of the major services. Perhaps someone out there would like to start one?

Publications

Good books can help immensely. Two books for beginners are *The Little Mac Book*, by Robin Williams, and John Dvorak's *PC Crash Course and Survival Guide*. Both are short but explain the essential terminology and commands you need to use your computer.

To keep up with new products and techniques, you should subscribe to at least one of the major computer magazines. The best are *PC Magazine*, *PC World*, and *PC Computing* for PCs, and *MacUser* and *MacWorld* for the Mac. My favorites are *PC World* and *MacUser*. I find them more specific about features and more critical of shortcomings than their competitors. They are also less technical and easier to read. *PC Computing* is likewise good for beginning PC users.

Newsletters can be much more specific than general purpose magazines, but they cost more. One of the best is *Macintosh Construction Forum* (P.O. Box 1272, Sandpoint, ID 83864), published by Craig Savage. Check with your software companies for other construction-specific newsletters.

Consultants

A good consultant can save you loads of time with choosing hardware and software, and can do everything from designing and setting up your system to creating an estimating system tailored to your needs. A consultant may seem expensive at first, but if you are busy, the time and headaches saved will make up for the money spent. The best way to find a good one is to ask around — ask your dealer, other computer users, and any of the other sources mentioned in this article. Then ask the consultant a lot of questions. If the answers make your eyes glaze over, call someone else. Apple has an active consulting program called *Apple Consultant Relations* (800/538-9696). PC consultants can be found in the yellow pages. Be sure to check references before making any commitments. ■

Guy Hermann runs a computer consulting company called Clear Thinking, in Mystic, Conn.

Computers—What Good Are They Anyway?

by Martin King

Resist the urge to buy until you weigh the benefits and the cost

The computer is not a labor-saving electronic device — it is an irresistible social force. If you don't recognize that fact you are a contractor at risk.

Contractors' problems take a variety of shapes, but they eventually boil down to one thing: time. Whether you survive in this profession you love and pursue from dawn to dark is largely decided by how you spend your time. No matter how large or small your company is, each day gives you precisely the same amount of time it gives to every other contractor in the world. And it's never enough.

That's where computers come in. In the seconds it takes you to read this sentence, a computer performs thousands of operations. Surely anything that fast will save a lot of time! How can you survive without one? That's the message blowing from every corner. Owning a computer is a credential that certifies you as a modern, technically proficient business person. But think about this: General Motors, Eastman-Kodak, Burlington Industries and a thousand other firms grew into giants without the benefit of a single computer.

Okay, let's be realistic. There is probably no way to resist the pressure. Who wants to be out of step with the world? You are going to get a computer. But if you're a full-time contractor trying to squeeze fifteen productive hours out of a twelve-hour day, the worst thing you can do is spend several thousand dollars and a hundred hours of precious time learning to perform automatically something you should not be doing in the first place. From a business standpoint the net effect of a computer can be equivalent to breaking your leg.

The best way to handle the decision may be to buy the cheapest computer you can find—and store it in the closet. Permanently.

This may seem a strange prologue for a discussion of computer applications for contractors, but you have to be hard-nosed about decisions as important as this. Contractors usually know enough about construction; they stumble for lack of business savvy. As a business person, your computer decision hinges on this: "Are the dollar benefits of this investment greater than any other expenditure I could make?" Until you answer that question with a reasonable degree of certainty you will not be making a business decision. The answer requires that you have a good idea of the dollar costs and benefits a computer is likely to generate.

The purpose of this discussion is to help you evaluate those costs and benefits. We're turning on the cold water first, because it will be safer in the event you should stop reading early.

The Real Costs

When you need construction equipment you balance its price against its expected performance, durability, and frequency of use. Then you either buy it, rent it, or secure it through the services of a subcontractor. The costs are clearly known.

With computers you will be dealing with less tangible costs. The hardware and software costs of a system are easily determined. The difficult things to quantify are the time it will take to learn how to use the system and the administrative time required to mesh it with your existing operation. These costs are usually underestimated. By a factor of ten.

We know a masonry contractor who does quite well. He paid $7,800 to a well-known manufacturer of computers and construction systems for a fully-integrated package: accounting, job cost, estimating, project management — the works. His wife had used a word processor and she was going to operate the system. After three years no part of it has come on line, and there is no sign it ever will. The $7,800 is not what hurts most. It is the 300-plus hours of his wasted time and the lost activities those hours represent. Also important: he probably will never try another one!

Do not underestimate the difficulty of comprehending the strange medium of electronic data processing. Just because computers are shown in all the magazines and every business seems to have one doesn't mean they are easy to use. Computer usage is like a foreign language, except that it is sternly unforgiving. Data entry is either precisely correct or entirely wrong. Many who rush into computer applications find it tough sledding.

It takes time to become proficient with any software, but not sporadic time. Like the field trades, computer skill requires continual use to develop and keep sharp. This means some person other than the boss will have to be dedicated to computer operation. It doesn't mean they can't do other things as well, but their duties should include regular keyboard time.

Also start out with a healthy respect for the time and effort it takes to change the way you do business. Your business may not have a procedures manual, but you do have procedures, even with an office staff of one. These are a matter of habit—you expect actions to follow a certain sequence, you expect to find documents in certain places. A strange computer program may require a lot of procedural changes all at once. That means administrative time will be spent in implementing, monitoring, and fine-tuning these changes.

Balancing the costs is the fact that a well-utilized computer system will be amortized over many years, probably ten or more. It can be a major business asset. So within reason, actual hardware and software costs don't matter that much. The major cost and most of the risk lie in the time spent training (computer jargon: "the learning curve") and integrating electronic data processing into your existing operations.

Along with the costs, we must examine how software will benefit your specific applications. Understand that computers don't solve problems — they only help with things you are already doing well. The following are some of the benefits a contractor can find useful.

In The Office

Correspondence. You can produce more accurate and more attractive letters. There is no time savings over an electric typewriter in the hands of a very good typist, but there is considerable time savings over a mediocre typist, since corrections are quickly made on the computer.

Form letters. You can store standard letters or paragraphs and combine them easily. You can individually address form letters quickly. You can set up a file to print a complete set of standard letters and forms for a new project.

Contracts and clauses. You can store a variety of contracts and clauses and combine them easily to fit specific conditions. You can also revise contracts quickly to reflect changes after negotiation.

Mailing lists. If you mail notices to suppliers, advertising letters to prospective clients, or reminders to old customers, the computer makes it easy to update your mailing lists and print mailing labels.

Office forms. You can compose forms for internal office or field use, and easily revise them to fit procedural changes and new conditions.

Production Applications

A variety of programs can be used for project management. They are based on data collected at the site. But if you are not already collecting and recording daily production data, you will find it difficult to establish procedures from scratch.

Software vendors may promise a marvelous array of reports and tables that let you peer into the soul of your business. What they seldom emphasize is the cost of collecting, entering, and updating data for those reports. How current do you want the information to be? Daily accuracy requires daily entry, weekly accuracy, weekly entry, and so on.

And how useful will the information be compared to the cost of capturing it? How much time do you have available to digest reports and act on them? Only information that leads to action is needed. The rest is a waste of time.

Also, using computer-generated reports introduces the question of accuracy. Since the records on which your report is based no longer have to be assembled the old-fashioned way (in a file folder), it may be difficult to check the back-up documents. Whenever any report shows a significant change, a question invariably comes to mind: "Is this thing accurate?" Data may not be easy to confirm on the spot.

On the brighter side, once you establish

procedures you will be committed to capturing and entering that data, and checking to be sure it is done consistently.

Project status. You must be able to track the status of work in progress. Every contractor has to produce a list of this type to plan the coming week's activities and control the cash. With a few ongoing projects it is easy to produce the project list manually. When the projects multiply beyond a certain number, a computer can handle them more efficiently. Progress payments and subcontractor draws must be in sync with the work performed, completed operations must be verified, various inspections scheduled. The accuracy of the tracking depends on timely data collection and processing.

Scheduling. Most contractors use some type of visual scheduling system, such as a large scheduling board with magnetic markers. You can't beat the clarity and efficiency of this tried-and-true method. Complex projects such as major condominium rehabs, however, require detailed charts which may be handled more efficiently by computer.

Job cost. A periodic analysis of individual job costs is essential, if only to correct the bidding of future jobs. If your jobs are not too large, too numerous, or too complex, this can be performed manually as jobs come up for billing. Invoices and payroll records are close at hand and their accuracy is easy to verify.

However, if you want to track job cost in order to control the project while it's running, the requirements change drastically. Using invoices and payroll records incurs a lag in reporting time. Late invoices don't appear, and change orders may not be billed by subs until after the job is over. That's why large projects have a way of looking good until the very end, when suddenly they turn into losers.

Job-cost programs that run from the general ledger tend to inform you when it's too late to bird-dog the trouble spots or bill for the extras. Instead of running job cost from the general ledger, you can use a computer program that handles it separately. This can give you accurate, up-to-date information. Again, the frequency with which the data are collected and entered determines the accuracy of the report. But it also represents a cost.

Estimating Applications

Anyone who has labored into the night calculating wall areas and cubic yards of concrete has surely asked himself "Why can't a computer do this?" It is a logical application. Computers can perform area and quantity calculations instantly. They are also beautifully equipped to store price data and flash it out at your command. They can apply predetermined ratios to estimate material quantities based on the overall dimensions of a structure.

Unfortunately, computers can't estimate. The multitudes of variables that bear on cost, such as site conditions, structural irregularities, quantity, travel distance, new products, and weather, etc. remain the human estimator's business. Sharp estimators may increase their efficiency with a computer. Estimators who don't know their business won't learn it from the computer.

Every computerized estimating system reduces flexibility to some degree. Technological speed requires a standard format; the automobile that can stop at any street corner is both more flexible but less swift than the airliner, which stops only at airports. The same principle applies to computers. The faster you wish to travel, the less precision you can apply to the finished product.

There are several approaches to computerized construction estimating. One approach lets you enter the overall dimensions of a structure, the types of materials, and other cost factors. The program then applies built-in ratios to calculate the material quantity, costs, and labor requirements. This might be called a *whole structure approach.*

The *systems approach* lets you select various preassembled system prices for different types of walls, floors, foundations, roofs, and so on. A floor system might include the joists, bridging, headers, subfloor, finished floor, nails, and adhesive in a single unit price. You enter the dimensions for each major system and the built-in ratios provide a total price.

A third method lets you select the items of work to be performed on each part of the structure in a *room-by-room approach.* Items are called up from a built-in inventory of material and labor costs. This method can work with either stick-

or unit-prices, permitting flexibility in listing different items of work for different parts of this structure.

There are variations and combinations of these systems. There may also be useful by-products from the computerized estimate, such as trade summaries, material lists, work specifications, invoices, and other reports derived from the estimating data. Some programs have special material take-off forms that simplify the process.

In general, you are likely to enjoy more success using a system that matches your present estimating method. If you get a good fit, computerized estimating can cut estimating time substantially and provide a superior product.

Work Specifications

After you are awarded the contract you probably put together some sort of work specification to tell everyone what to do. However, the work specification can also be a powerful sales tool. Detailed specs submitted along with your contract can make the package more attractive—people find it easier to buy when they know what they're getting.

Accounting

There are a plethora of computer programs for tracking payroll, accounts receivable and payable, and general ledger. Accounting is a discipline that requires precise habits and the ability to sit still for long stretches of time. Contractors usually detest it, so it remains a mystery to me why one would want to perform his own bookkeeping on a computer. The cost in lost productive hours is prohibitive to anyone but an accountant or comptroller. Leave this work to an accounting professional and let him evaluate accounting software since he will be the one who lives with the choice. ■

Martin King publishes Clearscope, *an estimating program for remodeling and rehab contractors. He is also president of Martin Churchill Assoc., Inc., in Arlington, Va., which specializes in investigating and consulting on structural damage, and appraising reconstruction costs.*

Chapter 16 • Conflict Resolution

Mediation: Peaceful Solutions to Client Disputes

A third-party mediator can help you keep small conflicts from growing into major lawsuits

by Larry Hayden and Herb Schwartz

Building and remodeling breed strong emotions, and strong emotions breed disputes. As consultant Linda Case says in her remodeling seminars, "There will come a time in every remodeling project when the client will be very, very angry with you."

Construction disputes may be inevitable. But if you plan for conflict, rather than hope to avoid it, it can strengthen rather than threaten the owner-contractor relationship. Mediation is perhaps the best way to plan for conflict, by providing an amicable way to resolve it. Unfortunately, few contractors know how mediation works and what it offers.

Maintaining Focus

Mediation means asking a neutral third party to intervene between two conflicting parties to help them settle a dispute. The mediator does not dictate the solution; he or she simply helps the two parties find a mutually agreeable settlement.

As a nonconfrontational, voluntary means to resolve a dispute, mediation offers great advantages over arbitration or court. The mediator improves communication, keeps things focused, and helps the owner and contractor address the root questions of why there is a dispute and how it can be settled. This helps them avoid getting sidetracked into other disputes and constantly raising the stakes, as so often happens when lawyers run the show in the arenas of arbitration and court. In mediation, it is the two parties, not a third party, who control things.

Perhaps most importantly, mediation is the only dispute-resolution method that requires a mutually satisfactory result. And it works: According to the American Arbitration Association, 90% of all cases submitted to mediation are successfully resolved.

Mediator Larry Hayden, at center, helps client and contractor hash things out with minimal cost and distress. The mediator can only suggest solutions, not dictate them. Nevertheless, 90% of mediated cases end with both parties agreeing on a settlement.

The beauty of the mediation process is that both sides receive positive affirmation of the results. Without the neutral mediator, almost any resolution will leave one party feeling victorious and the other

Getting To Mediation: Some Essential Tools

Mediation is a tool, much as a hammer or a change order is, but you need to know how to use it—how to actually get a client into mediation. To do so, we recommend that you incorporate two elements in the way you do business: a Mediation Clause in your contract, and a Mediation Agreement that allows you to set aside disputes during the course of construction (if their size doesn't warrant mediation at the time they occur, but you want to keep the job moving) so that they can be mediated later.

Mediation Clause

The parties to this construction agreement establish this dispute resolution section to resolve any misunderstanding, concern, dispute, or question about this agreement and/or the related construction work, in a prompt, efficient, and cost-effective framework.

Mediation: *In the event either party desires an independent, neutral third party to act as a mediator, they may call any independent mediator or mediation organization mutually acceptable to both parties to request a nonbinding, confidential mediation.*

Put It in Writing: Mediation Contract Language

Although the Mediation Clause shown here draws on 57 combined years of legal, business, and construction experience, it may not be appropriate for your construction contract. For example, we have intentionally omitted an "attorney's fee" provision—a common provision stating that the loser has to pay attorney's fees resulting from any disputes. We feel that requiring each side to pay its own legal and arbitration costs encourages mediation of disputes. Others might rather have the "protection." You should check with your attorney before including these clauses in your own contract.

MEDIATION AGREEMENT

This Agreement Is Between:

Contractor: ______________________ Owner ______________________

______________________ Owner ______________________

Contractor and Owner hereby appoint Damage Control Mediation Service as mediator of their dispute. Both parties have read, understood, and agree to be bound by the Mediation Rules appearing on page 2 of this document.

The Issues To Be Resolved Are As Follows:

____

____

____

____

____

____

Contractor:

____

By ______________________

____
Owner

____
Owner

Date: ______________________

Mediation Fee Deposited: $ ______________________

1

itching to get even. This residual negative feeling causes subsequent disputes to get off track and moves the parties in the direction of lawyers.

Excising The Cancer

If you intervene early with something like skin cancer, it's an in-and-out office procedure. But if you wait, it may require major surgery or even kill you. Disputes are the same: Intervene early and you can be in and out quickly. But if you wait, you may end up in arbitration or court.

We feel that arbitration and litigation are surgical solutions for disputes that have been too long ignored. To operate requires two or more lawyers, their paralegals, investigators and experts, three arbitrators or a judge, etc. It also means long delays, hours of painful hearings, and incredible expense. The process leaves personal scars and can bankrupt a business.

Of course, some situations do require the use of lawyers and a judge or arbitrators. *But*

MEDIATION AGREEMENT

NONBINDING MEDIATION RULES

1. **Nature of Mediation**

Damage Control Mediation Service (DCMS) is the mediator appointed by the parties to facilitate negotiations between them. Mediation is an agreement-reaching process in which DCMS assists the parties in reaching an agreement in an informed manner. DCMS has no power to decide disputed issues. (This is not binding arbitration.) DCMS acts solely as a facilitator between parties who choose to preserve their relationship. DCMS may communicate separately with an individual party, but DCMS shall remain impartial and shall not champion the interest of either party.

The objective of this mediation is to assist each party in quickly and inexpensively reaching their own best agreement. DCMS has an equal obligation to work on behalf of each party. DCMS will not give legal advice.

2. **Absolute Confidentiality**

DCMS and the parties shall conduct the mediation as a strictly confidential exchange of information including records, reports, statements, or other documents. None of the statements made, opinions expressed, admissions of responsibility, agreements to pay, and/or unsigned mediated agreements are admissible in any subsequent dispute-resolution process. The parties shall sign a state law-required confidentiality agreement which will establish confidentiality of this document and all other communications related to this mediation.

DCMS shall not be called to testify concerning this mediation or to provide any materials in any other dispute resolution process. The mediation process is a settlement negotiation that is intended to result in a signed agreement.

3. **Fee Schedule**

a. Filing/administration (nonrefundable)................$100.00 ($50 per party)
b. Mediation sessions (per hour/per party)......$100.00 (per party per hour)

When the parties execute their Agreement to Nonbinding Mediation, on the reverse hereof, they shall **each** send a deposit of $250.00 to cover filing/administration, and the first two hours of mediation. If the mediation is completed in less than two hours, a pro rata refund of the initial deposit shall be returned to the parties. After completion of the first two hours of mediation, DCMS and the parties will estimate the remaining time to complete the process. The parties shall deposit the estimated fee with DCMS at that time subject to a pro rata adjustment at the conclusion of the mediation.

—2—

You may also want to include a standard arbitration clause, such as that supplied by the American Arbitration Association, specifying that any disputes not settled by mediation will be submitted to binding arbitration.

Agreeing to Disagree: The Mediation Agreement

The mediation agreement is a form that can be used to begin the mediation process. It also lets you "sign up" for mediation but set aside until later any small disputes that are not worth the cost to mediate right away. While this may seem like procrastination, it is actually much better than "settling" a disagreement unsatisfactorily and causing resentment that can sabotage the relationship. This is usually the result when even a minor dispute is settled unsatisfactorily; at least one party will feel like a loser and want to "get even" later on, so that even a minor problem can later assume epic proportions. When each party gets validation from a neutral third party that they have reached a fair settlement, this resentment doesn't crop up.

The mediation agreement does not in itself settle disputes; it merely lists the things on which you agree to disagree. This keeps the project moving forward while setting up an amicable way to settle things at the end. It also gets everyone thinking mediation from the beginning.

The mediation agreement at left is one offered by our mediation company, Damage Control Mediation Service. Other agreements could appoint predetermined mediators (as does this one) or specify that the parties could later choose a mediator together.

—*L. H. & H. S.*

these are exceptions. In our opinion, you make a big mistake when you take any construction dispute to arbitration or court without first going to mediation. Construction disputes are too complex to settle efficiently in court. Going through the rules of evidence in court or the battle of expert witnesses in arbitration is at best a zero-sum game. But if you intervene early with mediation, you can address the gut issues and resolve the dispute quickly and to everyone's satisfaction.

Our personal experience suggests that 95% of disputes involve honest people who believe their opponents are crazy, malicious, stupid, or all of the above. It is these unreasonable assumptions that some lawyers prey on, and on which they base their approach.

In reality, all contractors want a profit and a referral out of every project. Every homeowner wants the job done well with minimum hassle and no legal entanglements. Contractors and owners frustrated with their inability to communicate often go to arbitration or court hoping to make things right again, only to be disappointed. In doing so, they lose any chance of getting what they originally wanted, which is a fair settlement that leaves everyone satisfied, if not exactly happy. Mediation, on the other hand, because it requires a mutually satisfactory result, offers a way to resolve a dispute and probably even maintain a positive relationship afterward.

Two Approaches

To get an idea of the difference between a mediated dispute and one that goes to arbitration, consider the following two cases that we handled. In one case, Larry Hayden served as one of three arbitrators, in the other he acted as sole mediator. In both cases, the amount in controversy at the outset was $7,000. The names and some of the facts have been changed for the usual reasons.

The arbitration approach. In one case, a contractor named Bill asked Larry to arbitrate a case that had been pending 18 months. The client was upset because of some problems with the stucco finish. Bill had proposed ways to fix the problem, but was frustrated because the client would not give him a final punch list so he could finish the job and deliver the final bill, which was for $7,000. Bill wanted a written punch list, because he feared he would have to go back repeatedly to fix an ever-increasing list of problems he felt he wasn't responsible for, or that he would be asked to meet unrealistic expectations.

For months, Bill and his attorney had been unsuccessfully trying to get the homeowner into arbitration. So Bill finally decided to exercise his arbitration clause. He knew he had made mistakes on the job, but he thought arbitration would settle things quickly and fairly and get the irrational homeowner off his back.

Unfortunately, once arbitration papers were served, the homeowner counterclaimed for $158,000. (The original contract was for $135,000.) Once the client received his summons to arbitration, he contacted a lawyer, who recommended that he have an engineer "expert witness" look the place over. The engineer drafted a long list of code violations and work that didn't precisely meet the specs. Suddenly Bill, instead of seeking a final $7,000 payment, was fighting for the life of his company.

So what happened? The arbitrators had a prehearing to gather basic information and then made a site visit. At the visit, the homeowner's engineer expert pointed out construction defects and code violations *ad nauseum*.

At the arbitration hearing itself, the contractor, being the plaintiff, went first. He presented his case in about half an hour. The homeowner's rebuttal took eight hours, during which he cited documents filling four 3-inch binders.

After many hours of debate, the arbitrators awarded $40,000 to the homeowner, with each side to bear its own attorney fees and half of the arbitration costs. For the homeowner, these fees added up to $34,000, leaving a net award of $6,000. The contractor, meanwhile, is today pursuing his subcontractors and his own insurance company for approximately one third to one half of the $40,000 award against him.

Who won here? Clearly, the lawyers, the expert witnesses, and the arbitrators, who were all paid fees and costs.

The mediated approach. Once again, the amount in dispute in our second case was $7,000. At issue was the cost of shoring up foundation diggings when unseasonably heavy rains threatened to collapse them. The contractor felt justified in asking for the money because in California, where this took place, it wasn't reasonable to expect rainfall at that time of year. However, the owner didn't feel he should have to pay for it.

The parties met with Larry for a total of three and a half hours. It was understood that nothing said would be used anywhere else, and that the mediator could not be called as a witness by either party in any subsequent litigation if the parties couldn't reach an agreement.

The parties met together in joint sessions, and then each met with Larry in private. Discussions focused on the diggings and on the key point that emerged during the mediation — the time factor involved. The contract had called for the job to be completed by a certain date so that a special event could be held at the home. The contractor felt "under the gun" to get the job done on time, and so spent the $7,000 to keep things moving. The contractor never felt he had the option to stop the project to save the money. As it turned out, the owner felt the contractor should have given him the option of slowing the job to save the money; but he never told the contractor that, so the contractor did what he had to do to stick to the schedule.

When these misunderstandings came out, the two sides began to see what had happened and moved toward a mutual understanding. Subsequent joint sessions showed that a resolution was possible. They finally settled, agreeing that the client would pay the contractor $5,000. The mediation cost each side $350.

Conclusion

The homeowner hires the contractor to make an intangible idea—a new house or remodel—a three-dimensional reality. Often, the reality fails to meet expectations. Disputes arise. The smart contractor anticipates these and uses them as constructively as possible. Mediation, used as a tool in the construction project, is the state of the art when it comes to dealing with disputes. If successful, it can actually turn a dispute into a referral. ■

Contractor Larry Hayden, president of Federal Building Company, and attorney Herb Schwartz jointly operate Damage Control Mediation Service in Oakland, Calif., which resolves construction disputes.

Settling Disputes out of Court

by Bob Merz

Do everything you can to avoid conflicts in the first place or to resolve them in a friendly fashion

Sooner or later you will likely be involved in a customer-contractor dispute that is tough to resolve. As a mediator and arbitrator for the last 15 years, I've tried to help settle over 1,900 such construction disputes. Most of these don't stem from bad intentions on anyone's part, but instead from a mismatch of expectations between the client and contractor.

When this happens, the first thing to do is to try to work something out with the client. This starts with the acknowledgement that, even though your work might be of good quality, a problem still exists because a certain part of it isn't to your customer's satisfaction.

Let the clients know what you can't or won't do, as well as which ideas simply won't work. It might be the case, for example, that a proposed repair will make something even less aesthetically pleasing — the cure they want might be worse than the disease. Point this out in as helpful a way as possible, and try to find some other solution.

Don't be afraid to ask your clients for suggestions. They may have an idea saved as a fallback position that could help resolve the situation.

When Things Get Serious

If after a sincere effort, however, you aren't able to find a compromise, the stage is set for serious conflict and a possible lawsuit. Once you get to this point, you should keep two things firmly in mind:

First, prepare your position (or defense, if you will) on all issues as if you are ultimately going to court. Ask for all complaints and repair or warranty requests in writing. Acknowledge their receipt, as well as any offers or corrective action you make, in writing. Contact subcontractors and suppliers to verify run number, dye lots, and other important product information. Inform manufacturer's representatives of any possible product flaws involved. Check with distributors to see if other contractors have had similar problems with a material. Document your findings as you go.

Second, at all costs, avoid going to court. You will rarely gain by doing so. Due to its complexity, construction is an industry in which disputes are quite expensive and laborious to settle in court. And win or lose, it doesn't help to have it said around town that you're being sued.

Find An Alternative

If you can't work it out cordially and you don't want to go to court, what's left? Fortunately, the abundance of lawsuits over the past ten years or so has spawned a whole generation of "alternative dispute resolution" methods. These methods generally rely on assistance from a neutral third party, in the form of a mediator or arbitrator. Being private and confidential, these methods are less likely to generate adverse publicity.

In mediation, the third party tries to facilitate communication, untangle misunderstandings, and suggest compromise solutions; the mediator does not "rule" on the issue, but simply acts as a facilitator for negotiation.

In arbitration, both parties agree beforehand that they will present their arguments to the arbitrator and then accept whatever solution the arbitrator specifies. Generally, an arbitrator's ruling is as final and binding as a court ruling.

Mediation

On the surface, mediation may seem a waste of time — if you couldn't work it out before, how is another person going to make a difference? But a mediator can help you focus on the real issues and steer you away from nonproductive areas of discussion. The mediator should, of course, be well versed in the means and methods of your industry.

Since one of the rules of a mediation is that nothing said there can be used against you later, both parties will feel free to tell all, admit mistakes, or consider solutions that might be used against them if said outside of mediation. Finally, many people tend to be more reasonable when presenting their arguments and requests to a third party.

Sometimes, of course, mediators are unable to find a middle ground, and the dispute remains unresolved. At that point, the only forum short of court is arbitration.

Making It Final

Arbitration is a "quasi-judicial" process that generally has the force of law. Courts will reconsider a previously arbitrated case only in the most unusual circumstances, such as when it is found that the arbitrator had a conflict of interest. Arbitrators, like mediators, should be chosen for their experience in the industry and their experience as arbitrators.

As in court, you'll do best by supplying documentation that the work done and materials used are up to industry standards, and that you have made good-faith efforts to satisfy the customer within reason. The arbitrator will be less interested in questions of character than in facts.

Avoiding Disputes to Start With

There are several things you can do to help reduce the chance of unresolvable disputes and court battles.

Include an arbitration clause in your contract. A typical wording for this clause might be:

> "All claims or disputes arising out of or relating to this agreement, or breach thereof, shall be decided by binding arbitration in accordance with applicable state statute."

Check with a local attorney for the proper language in your state. Such a clause gives you a known end to any disputes that arise, and virtually eliminates the possibility of court battles. Many home warranty programs such as Home Owners Warranty provide arbitration as part of the warranty. You can also find either mediators or arbitrators through the Better Business Bureau in most areas, or by looking in the yellow pages.

Outline a procedure for callbacks. Laying out the specifics of where and how complaints are handled will increase your efficiency in handling them, and will give you more credibility in the eyes of both the client and any third party called upon to settle the dispute.

Specify the standards of performance on which you base your work. Phrases such as "to the standards of the industry" or "workmanlike manner" are as likely to breed arguments as to resolve them. You might consider using and specifying in your contract a more exact standard, such as the NAHB Remodelors Council's recently updated Quality Standards for the Professional Remodeler.

Provide a written warranty. This will assure customers that you intend to do quality work and increase your credibility in case of a dispute. Be careful not to make yourself responsible for things outside of your control or to promise to correct "all defects," lest you end up bearing responsibility for material defects beyond your control. (The manufacturer should be responsible for those, providing you took reasonable care in specifying and selecting them.) ■

Bob Merz is president of Construction Arbitration Associates Ltd., a private service in Atlanta, Ga.

How to Prepare for Small Claims

You don't need a lawyer, just a well-documented presentation, to win your case in small claims court

by Lorie M. McCollum-Lombard

Consider this scenario: you're a successful floor finisher and have just completed a project that consisted of sanding and resurfacing a client's dining room floor. You've made the floor gleam, packed up your equipment, and even finished the job on schedule. But when you request payment the client expresses dissatisfaction. Portions of the floor were gouged, he says, and the surface appears to have bubble spots. The client refuses to pay the bill, which — labor and materials included — you've set at $575. How can you collect payment without hiring a lawyer? By taking the matter to small claims court.

Quite often, a small contractor is in the position of seeking relatively small monetary damages from a client, supplier, or co-worker. Rather than requiring such cases to go through the civil docket court procedures, which can be a long and expensive undertaking, most states have an informal process that allows the average lay person (often without aid of an attorney) to pursue legal action to resolve such claims. This avenue of relief is small claims court.

Keeping Things Simple

While states vary in their precise filing requirements and court procedures for small claims court, these guidelines and rules are generally designed to keep things simple. All states, for instance, have monetary limits beyond which cases can't be heard in small claims court; in Connecticut, for example, the maximum is $2,500, while in Virginia it's $1,000. And most small claims courts encourage the parties to represent themselves. Thus, both plaintiff and defendant must be strong advocates for their own interests.

For the sake of brevity, this article will follow the small claims procedure for the state of Connecticut. To familiarize yourself with your state's procedures, call the small claims court there; court officials are usually quite helpful and willing to provide any information you need.

How to Succeed in Small Claims

The most basic directive for success in small claims court is to be prepared at all stages of the process. This means collecting and organizing all the relevant evidence and testimony you can. It also means knowing what's expected of you in court. If you adhere to the following basic steps and prepare yourself well, you stand a good chance of winning.

The first step is to determine which court has jurisdiction over the location in which the incident occurred. Contact the small claims court there and discuss your claim with the clerk to make sure, first of all, that the dispute can be handled through small claims. Find out the monetary limits for damages sought, and have the clerk send you any information you'll need to file a claim. Filing a claim will probably involve submitting a "letter of intent" and a claim form with the court, which will then deliver those documents to the defendant.

Once the letter of intent and claim form have been sent to the defendant and a court date has been set, organize your court presentation, beginning with any evidence you've accumulated. Evidence might include copies of work orders, correspondence, requests for payment, canceled checks, or photographs. Photographs work particularly well to graphically support a claim.

You should also try to find witnesses who can support your claim. If a witness can't attend the court proceeding, ask the witness to write a letter stating his or her testimony so that you can read it in court. (Check the court rules on this; some courts may not allow testimony from persons not present.) Be sure you know what the witness's testimony will be before entering the courtroom.

At the actual presentation, be brief and to the point. A judge will hear your case. The plaintiff — the person seeking damages — will be heard first, followed by the defendant. If you're the plaintiff, usually the judge will simply want you to describe (and document with your supporting evidence) what happened. Referring to an outline will help things go smoothly. Also, have your documents neatly organized in chronological order to facilitate review by you, the defendant, and the judge. Neat organization will get your story across more clearly and effectively. Always give the opposing party the opportunity to ask questions, and respond in a professional manner. It is to both parties' benefit to follow court procedures and to not speak out of turn.

Finally, make sure your documentation includes something — such as estimates, supply bills, or work orders — to corroborate the amount of damages you are claiming, and be prepared to explain why you are requesting this amount.

If You Are a Defendant

If you're a defendant, the guidelines above still apply. You respond to notices from the court as requested. If you fail to contest a claim within the specified time, for example, a judgment may be rendered in favor of the plaintiff. Judgment may also be made against you if you respond that you are contesting the claim but do not attend the hearing.

Prepare your version of events so you can explain it at the hearing. Be organized and bring whatever documentation you can. During the plaintiff's presentation, write down questions and make note of where the testimony may be contradicted. Following the plaintiff's presentation, you will have the opportunity to present and document your side of the story and to question any witnesses.

Judgment

Once the judge has reviewed all the submitted evidence and testimony, he or she will render a judgment. In certain cases, the judgment may be delivered at the court proceedings; in others, the parties will be notified through the mail a short time afterward.

Small claims court isn't perfect. But it does provide a relatively simple and effective way to settle the sort of small legal disputes that occasionally pester contractors. Use it judiciously and you'll have a fair chance of getting justice. ■

Lorie McCollum-Lombard works in the real-estate research department at Fleet Bank, N.A., in Hartford, Conn.

Don't Sue, Arbitrate!

Arbitration is somewhat faster and less expensive than going to court. But you must understand the rules if you hope to get satisfaction.

by Perry Safran and Carolyn Annis

Construction projects begin with contractual agreements and too often, these days, end in courtroom disputes. In the past ten to fifteen years the number of such disputes has multiplied many times over.

The arbitration system was designed to handle this increased load, and to more quickly settle construction disagreements. This method of third-party intervention offers several advantages. It can be less formal, less expensive, and less time-consuming than the court system. For example, in some large cities it can take seven to ten years just to get into court. This extended wait can cause you to lose at the bank even if you win when you finally get to the courtroom. Arbitration, on the other hand, can take as little as a few months. But before you can make it work for you, you must know the rules.

The rules, in this case, are the Construction Industry Arbitration Rules, which are available in pamphlet form from the American Arbitration Association (140 West 51st St., New York, NY 10020-1203; 212/484-4000). They have been simplified so the reading is not tedious and it is well worth your time to read them.

It's always better to settle your dispute amicably if possible; but if you must use arbitration, the following tips should help you.

The Arbitration Clause

First, check your contract document for an arbitration clause. AIA documents and other industry-standard contracts contain these, but don't assume the provision has been included. Many owners, or their agents, will use these standard contracts, but stack the odds in their favor by adding or subtracting pertinent clauses. (This advice applies not only to arbitration clauses, but also to other important contract clauses such as liquidated damages and interest-bearing clauses.)

Make certain you understand the entire arbitration clause, including *all* of its terms and their implications. Consult your attorney about unfamiliar language. You may be in a hurry to sign the contract and begin working, but take the time *before* signing to fully understand how your disputes will be settled if they do occur. No one likes to begin a relationship with a discussion about future arguments. But, it's much easier to lay the ground rules in the beginning when the association is still rosy.

Filing The Claim

When filing your claim, answer, or counter claim for arbitration, provide as many details as possible and identify any unusual, industry-specific details that could influence your case. For example, an owner may be unhappy with material you used on the job. Provide any product literature, photographs, or other evidence indicating that the item you used is standard and recommended by the industry for the job.

A detailed claim serves a dual purpose: first, it satisfies the technical requirement of sufficiently notifying the other party of your position (Rule 7) and second, it provides the AAA Regional office with an adequate background of the case so that an appropriate list of arbitrators can be selected.

While contractors can competently speak the language of their trades, the climate of the industry today forces them to learn the legal language and communicate in these terms as well.

How well you document the day-to-day events of the job will often be the pivotal point in your case. Develop standard procedures to document change orders, payments, deliveries, etc. If events occur that spell future conflicts, document them well: for example, with photographs, measurements, detailed notes. This paperwork is just as important to the job as are your tools and equipment.

Contractors often feel that they can best represent themselves because of their expertise in the field. But you should consider consulting a lawyer to at least draw up your claim. This will help you to understand the type of information the arbitrators will be looking for. The claim should not be verbose, just clear and concise. Remember, always stick to the facts.

Discuss Your List

After filing your claim, the Regional Arbitration Director will provide you with a list of potential arbitrators. When you receive it, call the Director and discuss the list. You need to find out the rationale for each candidate (usually 10 to 15 possible arbitrators are chosen from a list of from 500 to 2,500 names). The answer you get can give you valuable insight into how a third party viewed your case. It will also demonstrate whether you provided enough information for a third party to understand the principle issues of your claim.

The Arbitrator

From the list you receive, you will choose three to four names. Consider your choices carefully. Although your first reaction might be to choose a panel composed entirely of industry-related individuals, consider having a lawyer on the panel. All arbitrators, whether lawyers or laypeople, are trying to reach an equitable agreement. If your case will be heard by more than one arbitrator, an attorney/panelist can provide guidance and direction for the lay panelist. He can also be helpful in conducting the hearings. You may want to avoid choosing panelists within your trade. They tend to be less sympathetic to your predicament than a business owner in another trade who has perhaps been in your shoes in a similar suit. In general, your competitors will tend to judge your work more harshly.

The list of three to four names you choose will be compared to the opposing party's list. If you've both chosen a particular person, that person will arbitrate your case. If no names coincide, the Regional Director will then choose an arbitrator you'll both have to live with. Usually, only one arbitrator is chosen, but if the case appears complex, the Director may choose more than one.

Cost

The arbitration system is not free. When deciding the issue of payments, you may want to waive the "one-free-day" rule and pay a little extra. A few extra hundred dollars can be money well spent to get the most knowledgeable and intelligent arbitrator(s).

Generally, the person filing the claim is required to make a deposit if the case looks like it will get complicated and run on. In any case, when judgement is finalized, fees are usually split 50/50. But don't assume, be certain to check out the arbitration clause in your contract. It may specify a different method, and you don't want to get caught holding the bag because you failed to read the fine print in the arbitration clause.

One way to keep costs down is to keep the process simple. Although some people go to great expense by renting fancy boardrooms, and catering fancy lunches, you needn't do that. The rationale for arbitration is to avoid courtroom expense. It doesn't make sense to substitute other types of expenditures. Simple cases can even be held at the job site.

Prearbitration

The prearbitration conference is the first opportunity you will have to present your case to your arbitrator(s). Yet, if your arbitrator has not scheduled this conference, request that he does so. You will need to prepare an outline of your case to present at this meeting. Construct it as carefully as you did your initial claim. A written copy is not required, but it makes an impressive added touch. It will also relieve the arbitrator(s) of burdensome note-taking.

During this session you will discuss and agree on issues such as the guidelines for "discovery," which is how each party may get information from the other. Remember, volunteering information is a far more effective tactic than appearing to try to hide it. You will also set the times, dates, and locations of the hearings. The best plan is to use a low-cost central location. Simple cases can be decided in a day, but it is wise to schedule more days than you will need; it is always easier to cancel a hearing than to add extra dates. Although you should avoid several back-to-back sessions that will tire out your arbitrator, the Arbitration Association prefers that you schedule all hearings within a five-day period.

Once your dispute has reached the point of arbitration, you will have become acutely aware of the importance of documentation. All of the facts and information concerning the case will, at this point, be included in a claims/defense book. Your success or failure can ultimately hinge on whether this book is complete and contains the necessary paperwork to support your claims. Keep the need for this information in mind when you begin working on a contracted job. It will take time and effort to maintain this degree of paperwork, but think of it as an investment in legal insurance. Contracting parties never begin their relationship with the expectation of future conflicts. But later, when the honeymoon is over, they often wish they had started out with a clearer perspective. ■

Perry Safran is a North Carolina construction attorney. Both he and Carolyn Annis are associated with Proof Management consultants, Richmond, Va.

Chapter **17** • **Diversification**

Diversification and the Small Builder

by Chuck Moriarty

You can put your building talents to use in many ways—from home inspection to real estate management. But you need to find the right mix for profits and peace of mind.

I'm a principal in a small, well-diversified building company in Seattle, Wash. When my partner and I established our business in 1976, our sole purpose was to purchase, improve, and resell or rent single family residences within the city limits of Seattle. Our goal was simply to acquire enough rental property to provide ourselves with sufficient income to enjoy our "golden years" (you know—when we turned 40). We have since diversified into several related fields including property management, construction management, custom home building, design/ build, marketing real estate partnerships, home inspections, bookkeeping for small companies in related fields, and consulting for other contractors and subcontractors.

Obviously, some of these endeavors have fared better than others and what follows is my experience in the various fields. We have kept the ones that worked for us, and thrown out the ones that didn't. One must understand that our successes and failures are strictly a function of our personalities and strengths and I would encourage you to try any and all diversification opportunities that present themselves, even though they may have failed for our company.

Some of the areas we have explored have been natural extensions of our original business, whereas some have been things that we just wanted to try. But we have done them all with an eye towards simplicity of structure and operation. What we have never wanted was a top-heavy organization fueled by its subsidiary parts in a never-ending quest to pay for all the additional overhead.

My partner and I excel at sales and at developing long-term relationships. As a result, we have been particularly interested in businesses that depend heavily on these. We leave the mundane (but all-important) daily business operations to our general manager and staff.

Delving Into Real Estate

When we started our business, we were buying, rehabbing, and reselling or renting single-family homes. As the market tightened, we sold our inventory of rental homes and bought several larger buildings for rehab. Some of these we kept for rental income and long-term gain, some we sold on contract to produce interest income, and some we sold for cash to produce working capital. This end of our business is shaped by income tax issues and our long-term retirement goals.

A natural extension of this business was offering partnership opportunities to other investors. As the General Partner, we would acquire an interest in the building without using any of our own capital. We would also charge a management fee to the Partnership for managing the building.

It is surprisingly easy to come by investors for real-estate opportunities. We have never advertised for investors, but we have often been approached by people who

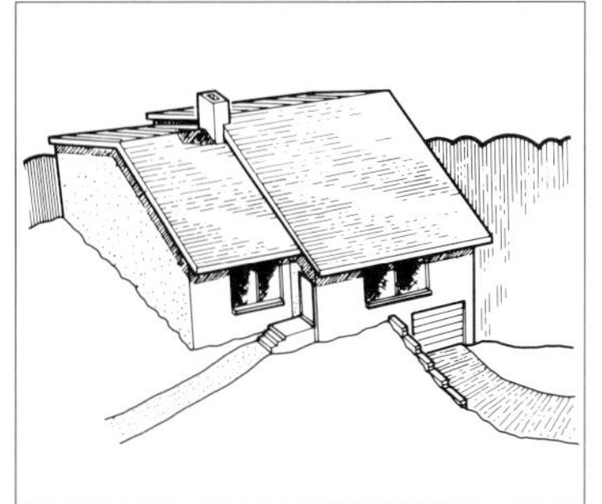

Before

After

Spec remodeling has paid off for both the contractor and the investors who put up all the cash (in return for 50% of the profits upon resale). This Seattle bungalow was transformed into a two-story contemporary and yielded these figures:

Purchase	$58,000
Rehab	55,000
Sales price	157,000
Sales cost	11,920
Net profit	32,080
50% to investor	16,040

= 29% return on investor's money

know what type of work we do. These people include past clients, friends, employees and many professional people we have met through our years of business.

Most people recognize the value of investing in real estate but don't want the hassle of maintaining and managing buildings. I'm one of those people myself, but I also love looking for property, working deals, and selling partnerships. Fortunately, my partner doesn't mind showing rental units, and our secretary does quite well in collecting the rents.

Our rental buildings are small, mixed-use or apartment buildings. Although smaller buildings probably take more time to manage (percentage-wise), they are easier to keep full and are much more liquid than larger complexes. We have found our best opportunities in the more marginal (affordable) type of apartments than in the higher- and lower-end units. Middle-income tenants just seem better at paying their rent on time.

Most of our mixed-use buildings are in neighborhood business districts. They typically have one to four street-level commercial spaces with apartments above. We spend a lot of time looking for buildings that are underutilized and undermanaged. The best ones are buildings where we can lease the commercial space to tenants who have money to spend on improving the space.

As we gained experience managing our own buildings, it was natural to begin managing rental properties for other owners. For 10% of the gross, we show apartments and collect the rent. If the tenants have problems, they call our office (they generally don't even know who the owner is). All advertising and maintenance costs are additional. Basically, we relieve the owner of all the headaches of owning rental property and provide a bookkeeping service. Depending on the owner's needs, we can pay underlying mortgages, taxes and insurance, or simply subtract our fee and send the proceeds to the owner. Additionally, we give the owner a monthly financial statement and a year-end profit and loss statement on their building.

Reluctant Remodelers

Since the housing market was exploding during the early years of our business, we were able to successfully market our rehabbed single-family units ourselves. With the aid of our site signs and advertising, our name became synonymous with quality remodeling work. Our first diversification opportunity presented itself when we somewhat reluctantly began accepting remodeling contract work. I say reluctantly because I still don't quite know how I ended up doing so much contract work and making so little money at it. But since our speculative business was extremely profitable, we were able to make many families happy by unintentionally subsidizing their remodeling projects.

It wasn't until we got out of single-family remodeling that I realized how poor this end of our business really was. Not having been active in my local homebuilders association, I was busy reinventing the wheel on pricing and markups. My accounting system would not give me the information I needed to accurately assign overhead costs related to the contracting side of my business, as opposed to the speculative side. So it was quite a shock when I finally realized how much this business was *costing* me. With that knowledge in hand, I quickly raised our prices — and our construction gross dropped from $450,000 to $80,000 in one year's time.

But at least I made money on that $80,000.

Now I knew how to make money in the contracting business, but unfortunately the contractors I was bidding against still didn't. So I was consistently 20%, 30%, even 50% over the competition. As a result, we lost most jobs that we bid on. But by this time, we were a lot smarter than when we had started out, and we had joined the local chapter of the National Association of Home Builders. Our membership gave us access to educational opportunities that we hadn't known even existed. Through the local chapter's Remodelors Council I had the pleasure of meeting many remodeling contractors in my area. It amazed me how open and friendly these contractors were and how willing most were to answer our questions.

My best educational opportunity came when I started attending the annual NAHB show. Participating in seminars and roundtable sessions brought me the friendship of remodeling contractors from across the country. These people have been my best resources for making my remodeling division a profitable venture.

Design/Build

Since I couldn't get a job at my price through the competitive bid process, we diversified into design/build remodeling. Every time I met someone who was into design/build I would grill him on how he did it and what his experiences were. It took several years to get good at it but now it comprises 95% of our residential work. We still do some bidding for a few select architectural firms, but we always find out who our competition is so we don't waste our time when there is no chance we will get the project.

There are a multitude of advantages to design/build, not the least of which is control. When a customer hires you to both design and build a project, you are the undisputed expert and the project usually just rolls along smoothly.

Another advantage is that you are getting paid to do the budget. When you bid on someone else's design, you have less than a 50% chance of getting the job, depending on how many bidders there are and whether the project even gets built. With design/build, you have a 100% chance of getting the job.

But the best part of design/build is that you can quit competing on price and begin selling value. I personally find it extremely rewarding to take a family's dreams and work through the design process to give them the project they want at a price they can afford — and at a price that I can make a fair profit on.

Design/build also seems less prone to lawsuits. This is probably because both you and the client have a chance to get to know each other before your crew moves in with them. And it's far easier to walk away from a project with a difficult client during the design stage than when it's half completed.

Tenant Improvements

We have also enjoyed tremendous success in the field of tenant improvements in retail stores and offices. Back in the days when I couldn't get any bid residential work, I started calling on building managers and architects looking for office and retail improvements. Here was a field in which I could be extremely competitive since I was at the smaller end of the scale in company size. The repeat business is also fantastic. Since most retailers upgrade their stores about every five years, you can enjoy a relatively stable work load by developing a good list of commercial clients.

Custom Home Nightmares

And then there's custom home building. The opportunity to build someone a custom home is something most of us don't want to pass up, probably because it satisfies our egos. I have built half a dozen homes, generally for people who didn't have the money to pay for them.

Actually, the biggest problem in custom homes is that the owner generally does not pay the architect enough to provide a complete set of drawings and specifications to accurately price the job from the site work to the towel bars. Since I don't want to hire a house designer, how can I build custom homes without the benefit of complete plans and specifications, yet still make money on them? Enter the concept of construction management. We will build your home at cost plus a predetermined fee. My one experience in this had a court date of July 1988.

This particular owner dragged me out to a subdivision and took me through a model home he liked. The house was priced at $150,000. Could I build him one like it with a "little better quality" for the same price if he furnished the lot? No problem, I stupidly replied. He started in on his drawings, we both looked at several lots, and finally he purchased one in a relatively complete development of $200,000 homes. We received several generations of drawings and by the time the actual building had started, the construction budget had grown to $190,000. The lot cost $50,000 so we were now in the $240,000 range, definitely at the top end of this particular subdivision.

While the foundation was being poured, I was trying to get the finish package together. We were having a difficult time specifying what was going into this house, so I went on a Sunday drive with the owners to see several examples of what they would like. By the end of the day I was very quietly sitting in the backseat thinking "How am I ever going to get out of this one?" All the homes they had shown me were in the $400,000-plus range and I knew we were in big trouble. While they were busy chattering about swim spas in the basement, I was trying to figure out how we were going to build this home without any lumber so we could use the framing budget for the interior finishes.

These people were anything but novices. The husband was a mortgage broker whom I had known for years and had done some business with. His wife was a real estate appraiser.

The following day I met with the husband and explained my fears of the cost overruns involved. He immediately recognized the concerns, but the house was his dream home — his response was: "Let's do our best on the budget and just build it." In order to save money, we would just set the fiberglass swim spa in the basement and he would hook it up in a couple of years when he could afford the dehumidification system required.

Of course when the hvac man arrived, it was only an additional $10,000 to install the system — so the owner decided to go ahead and do it. I reminded him of our tight budget but since the owner was paying all bills directly, I gave in and figured he knew what he was getting into.

To make a long story short, I think the house finally cost the owner a total of somewhere around $350,000. Who ultimately absorbed which costs was in the hands of the lawyers.

Profitable Partnerships

Over the past couple of years, we have begun to do some speculative projects in partnership with private parties. Because of the low interest rate banks are currently paying on CDs, we have found many people interested in financing the acquisition and rehabilitation of existing homes for resale.

The concept is simple. The investor puts up the cash, and we put up the expertise. To keep the acquisition costs low, we buy the homes on real estate contracts with short term cashouts. We retain 10% of the gross construction cost as an offset to our overhead for doing the project, and split the net proceeds 50/50. The investor is protected because he holds title to the property; we are protected through the lien provisions of our state law. The hard part is finding a property that has the potential for the return we expect. There has to be substantial profit in it or it simply isn't worth doing. We have found them, however. Over a typical six-month period, our investors have enjoyed returns of 14% to 60% on their money.

This system works well for both parties, but gets bogged down if you are borrowing money from a bank. The additional fees and interest charged by the bank make many of these projects strictly marginal.

Home Inspection

Over the years, some of our past clients have asked us to look at homes they were thinking of buying. For a nominal fee, we would go out and look over their prospective purchases. As these calls increased, we decided to formalize the business and actually started marketing home inspections. Because of the liability risk, we decided early on to keep this business small enough so that only the partners handled the actual inspections. We already earn our attorney a pretty good living so we didn't need the risk exposure of a hired inspector signing off on a crawlspace without actually inspecting it.

This said, I think home inspections are an excellent opportunity for the owner of a small company to generate some additional income without having to add to his overhead. If you run a company that does not require your constant presence on job sites, you could begin marketing for home inspections and create a nice little cash cow with just a little better organization of your day. The fees for home inspections seem to be very market sensitive so you would need to do a survey to find what the going rate is. In our area it averages about $200. Our housing market is relatively stable right now, so it is fairly easy to schedule the inspections to meet *our* needs. Our early marketing efforts were directed towards real-estate agents and we spent a lot of Sundays at open houses meeting them.

Direct your efforts at the top agents in your community and you can keep yourself busy during a recession. The best agents continue to sell year in and year out, just like the best remodelers continue to work right through the downturns in the economy.

There's Always Consulting

Over the years many subcontractors have asked for my advice and expertise on business management. I offered to help some of the people that I knew and particularly liked, but quickly found I had to start charging for my time. So another natural diversification was to do business consulting on a fee basis. Surprisingly enough, not only were people willing to pay me, they were much more likely to follow my advice than when I was advising them for free.

I have consistently found that the building industry is made up of expert craftspeople who are lousy businesspeople. To be successful does not mean running a million dollar business. It means making a *profit*. Far too many of us have absolutely no handle on the costs of doing business and can only afford our Porsches and Mercedes because we have such dynamic cash flow. When that flow is turned off, we suddenly find we can't make last month's payments with this month's retainers. The biggest hurdle I have is convincing contractors that it is *not* the gross dollar amount done yearly that makes their business successful.

As our business became fully computerized, we decided to turn our office into a profit center and started marketing bookkeeping services to other small companies. Although we do some retail shops, we concentrate on small subcontractor companies where the owners want to spend their time on activities other than bookkeeping. Of course this is also a natural extension of my consulting practice. Our contractor clients easily pay our fees through their increased profitability.

Over the years our business goals and strategies have changed and some years we emphasize some areas of our business more than others. The general economy, our hired expertise, and the partners' personal goals all play a part as we plan our future business strategies. We retain our ultimate goal (enjoying our golden years) but the road there is ever changing. Ours is a well-diversified company—which is ideal for us. Others, however, find it better to do just one thing and do it extremely well. We find it best to do many things, throwing out the unprofitable ones along the way, and retaining the profitable ones that interest us. Because of our diversification, our work load is extremely stable and we find that we are well-equipped to ride out the economy's ups and downs.

As I said in the beginning, our successes and failures are strictly a function of our personalities and strengths, and each person has to determine what works best for him. ■

Chuck Moriarty, a former chairman of NAHB's Remodelors Council, is president of Moriarty & Matzen, a Seattle-based general contracting firm specializing in residential and commercial remodeling.

The Bright Idea (A Cautionary Tale)

by Edward Behr

In August 1982, I was looking for a new building project and I called up a builder friend, Ralph. I wanted him to join forces with me on my bright idea—we should borrow a whole lot of money, find an abandoned 19th-century house, move it 100 miles to a fancy suburb, restore it perfectly with all the modern conveniences, and sell it for a fat profit. On spec.

Good idea...huh? We certainly thought so.

Our assumptions about the market were based on our dealings with restoration customers and on our observations of the local real-estate market. Practically the entire upper end of the real-estate market in the Northeast Kingdom region of Vermont, where we lived, was made up of affluent outsiders. They loved the New England countryside and were buying up the quaint old farmhouses, say 1850 and earlier, despite their decrepit conditions. We figured that if we could move a fully-restored example of this same product to a much higher-priced area and keep our costs near the comparatively low Northeast Kingdom levels, we should be left with a gorgeous profit. Images of Saab turbos danced in my head.

We had solid experience in single-family construction. Each of us had built several top-quality houses. We knew one another mostly by this reputation. And in rural New England we had plenty of experience with old buildings. Beyond that we had at least a minimum of actual restoration experience coupled with a basic knowledge of architectural history.

We were sure we knew our market, our method, and our profit. Financing too, seemed under control. I planned to mortgage my house and then leverage the rest using the partially-completed project as collateral. I was fearless.

A day or so after our first meeting we

"Just think, if we moved this antique house to an upscale town, and tastefully restored it, boy could we make a bundle..."

Before: *Hidden behind the weeds in the Northeast Kingdom region of Vermont, the 1850 house (above) featured a steep roof, pilasters, pediments, and moldings in the Greek Revival style.* **After:** *One hundred miles away and two years later, the house stands fully restored (below) near the shores of Lake Champlain. Note how the arched barn door was converted to an entryway.*

were on the road scouting out all the restoration candidates that might be for sale — cheap. We quickly focused on an abandoned farmhouse, suffering heavy moisture damage and in an impossibly loud and dusty location — opposite an active gravel pit and asphalt plant.

The Perfect House

The house was, in fact, for sale and we were the only interested would-be buyer. The floor system was so thoroughly rotten from the wet cellar, that there was nothing to jack. And we were the only ones prepared to completely disassemble the thing and reconstruct it.

Ralph is a very hard-working guy and he came up with cost estimates for the entire project overnight. I didn't even have time to procrastinate. The seller hadn't given the final word yet, but enthusiasm was running high. We expected to get double satisfaction from doing top-quality work, and excellent pay for our trouble.

We agreed on a price of $1,500, a questionable bargain (considering that moving and restoring it would cost more than the price of a new house) and about $1,000 more than we had hoped to pay for it.

The building itself was a large, clapboarded Connecticut Valley house, with a steep-pitched roof and strong decoration. It was built in 1851 by a blacksmith named Asa Hovey during the last gasp of the Greek Revival style, and it had all the requisite charm that people expect to find in a country house in Vermont.

Despite belonging to the Greek Revival period, the house was not a columned temple with its gable-end facing the road. Still, the details of pilasters, pediments, and moldings were all in the Greek style. The building had enormous appeal: It seemed that everyone we knew had idle dreams of rescuing this house. Instead of the traditional New England austerity, it combined a favorite New England form with rich decoration.

The long side of the rectangular building contained a nice front entry with sidelights, and addressed the road with a 48x20-foot wing extended to the right of the main block. Both were a story-and-three-quarters. There were about 2,200 square feet of finished space including a kitchen downstairs in the wing. Half the wing, though, was taken up by a carriage barn, which was entered through a fine, elliptical arch. This naturally became the garage.

The Not-So-Perfect Site

Once again Ralph and I hit the road, this time making a series of day trips to Chittenden County in northwestern Vermont, where we checked with real-estate agents and followed up leads from ads. Chittenden County, which is home to the city of Burlington, is a booming and especially attractive area, with impressive views of the Green Mountains to the east, and of Lake Champlain and the Adirondacks to the west.

We gradually focused on Charlotte, which is the most affluent town in the county and borders on Lake Champlain.

After weeks of scouring the countryside for a site, we settled on a five-acre piece. We found a good deal, the right price for that size and area. But we had made a major blunder. Our distinctly Connecticut Valley house with its steep-pitched roof had an entirely different feeling from the architecture of the warmer and wealthier Champlain Valley. And we foolishly wanted a piece of land that felt like the rolling, wooded, upper Connecticut Valley. So we chose a small, pleasant field, almost entirely enclosed by trees, and at the base of a hill, where you could see the Green Mountains only if you knew where to look for them — a location one real-estate agent described later as a "hole."

More than that, as outsiders, we later found that we had picked the wrong side of town. We were three miles inland from where we should have been. In a lakeside town the phrase "lake view" carries more weight in the homebuyers' market than we realized. We could have had a flat, barren five acres (ugly to us) with the right views, near the lake shore, at a similar cost. It would have sold our house a whole lot quicker. But we didn't know that then.

In October, with our eyes on a bright future, Ralph and I incorporated, which was not a bad idea. And when all the real-estate papers were signed, we set to work. With the labor of three men the old house in Waterford came down in three weeks. We carefully took it apart, labelled each piece, and neatly bundled related pieces together. Then the work of scraping, minor repair, and priming of most of the exterior was subbed out to a conscientious worker. We ordered sash and some reproduction doors from two nearby millworkers. Interior woodwork, clapboard installation, and miscellaneous exterior repair and replacement would be done at the new site. So far the restoration was going beautifully.

We made a decision, a good one, to put the old house on a new frame. The 1851 frame was seriously deteriorated, if not from the wet cellar, from two serious roof leaks. It was an up-and-down sawn frame, interesting to a builder, but not to a customer. We saved time by sticking to familiar, modern stick-built construction. Our crew knew it best and so did we. No one could tell the difference until they went into the cellar or the attic.

A Nice Winter Project

Over in Charlotte the initial site work was done and the foundation was poured. Our crew started work on November 7. At the end of two weeks the building was framed and most of the cedar shingles were on the roof. Then, when construction was going perfectly, we did an incredibly dumb thing: we dismissed most of the crew and decided we could do the rest of the carpentry by ourselves. A nice winter project! Five weeks later, Ralph pointed out to me what a mistake we had made. He also questioned our estimated costs. I felt like a man waking up from a dream.

Instead of completing the project for the original sum and selling for another 25% less any sales commission, we were looking at figures that showed costs 50% over budget. Eventually they ran even higher. We needed money right away and fortunately we found a bank that didn't blink at the necessary mortgage. Probably the worst part of our financial planning was that we were completely unaware of the frightening costs of loans, interest, and property taxes on a project gone out of control. By January we knew that we were building the most expensive house in the neighborhood by about $20,000. But we plugged ahead, still with too few workers. I still didn't understand how time-consuming it is to restore all the woodwork in a house, clapboards and all.

When construction was complete, the house was a meticulously finished piece of work — *and* four months late.

The Market, the Market

The house sat on the market for over a year. We had made another mistake. Naturally, we had respected the original floorplan as much as possible. We didn't think to find out what buyers might be looking for. We assumed they respected history as much as we did. Wrong. The Chittenden County high-end market was very suburban, and wanted conventional, affluent, suburban floorplans. We needed a very large living room instead of our small parlor, and a bigger dining room than we had. A master bedroom suite wouldn't have hurt, either. Our two upstairs bathrooms had unconventionally low ceilings. We could have compromised fairly easily with the nineteenth century to solve these problems, but we hadn't recognized them as problems.

Plowing Ahead

Then there was trouble with the adjoining landowner, who had sold us our property. He was also a builder and he was developing the acreage of the old farm that our piece had been part of. Like us, he was a first-time developer, and he needed the success of our project to help give his project a boost. But he dug out a big gravel pit right next to us in order to build his roads and driveways. His intentions were good; it was his deed that required five-acre lots and design-controlled houses. He even had to grade and seed his gravel pit — someday. But we had no way to get him to do it anytime soon and it kept getting bigger and deeper and kept eroding, month after month. This was really unforeseen. When we bought our land the site of the gravel pit was invisible behind trees and underbrush. Now it gaped at our historical house — and visitors were asking questions.

And yet another problem: Our brand-new well produced water tainted with sulfur and iron bacteria. Back in our part of Vermont no one had ever heard of such a thing. In Charlotte we had to install treatment equipment. Even at wholesale cost this was another burden, one we could ill-afford. And it was something else to explain to prospective buyers.

And unbelievably there was another problem that never occurred to us country boys — the town raised the grade of our road, effectively lowering the site of our house by 8 to 10 feet. This wasn't a case of there being an established grade. The grade was up to the judgement of the old-time road commissioner. And he made those decisions when he got around to doing the job, without benefit of a transit. He just gradually raised the road while we built the house.

The Quick Sale

During the bleak period of waiting for a buyer to appear, we heard a lot of advice. "It only takes one buyer" was the most popular phrase. Eventually I began to answer, "Yes, but it does take *one*." Potential buyers looked, often they came back twice or more and asked to meet with us to discuss the house. But no one made an offer. One or more of the home's shortcomings always held them back.

In October, five months after our May completion, we went back in and redid the floor plan, opening the wall between living room and dining room, and compensating the low ceiling in the upstairs bathrooms by moving walls and changing plumbing — a major expense.

Fortunately I sold my other house, the one mortgaged to finance the project at the start, which reduced my interest payments. Then, in November, I moved into the Charlotte house to further reduce my costs and put some appropriate furniture in it, since a furnished house is easier to sell. Gradually over the coming months the gravel pit was graded and landscaped. And we prepared a brochure of the house with before and after photos.

We had two real-estate agents. We let the first one go in frustration after four months. But both of them worked hard for us. The second agent kept his belief in the project, advertised the house, held repeated open houses, advised us not to make a big sacrifice on the price, and ultimately came up with two competing offers on the same day. Hooray! In September of 1984 the house was sold.

Was It Worth It?

In hard dollars-and-cents figures, we more or less broke even. But if you place a high value on time and lost opportunity, we lost around $30,000 — a sobering sum.

In retrospect, we were amazingly naive. We didn't know how to be hard-headed in the business sense. We didn't know that partnerships have a greater rate of failure than sole proprietorships, for example. We assumed that two heads were better than one, a combination of strengths rather than a combination of weaknesses. A couple of nice guys don't ask a lot of unpleasant questions about optimistic figures. And there are simple maxims that bear saying out loud now and then. One of them is "time is money." We came late to that one.

The problem was that we didn't really know what business we were in. We thought we were in the business of producing craftsmanship at a high aesthetic level. But we were actually in the business of real-estate development. If we had had our priorities straight, the real-estate development business might have offered us some opportunity for personal enjoyment of the craftsmanship and aesthetics that we sought—plus a fair profit.

The Moral of This Story?

When you get a bright idea and decide

to build on spec, keep in mind the lessons we learned:

- Know what business you are in.
- Remember that time is money.
- Know your market. Don't assume anything. Do research. (At least find several experienced people to ask.)
- Design your product to meet your market. Make it commercial, don't make it personal.
- Don't depend on other people's good will. Know how motivated they are to follow through. Be in control of the situation.
- If you choose a partnership, know how it may help or hurt your business.
- Buy land that will support your development. Especially in high-end construction, the only land that's a bargain is the land that will sell your development.

Hindsight is wonderful stuff, often expensive, and not very useful unless you can turn it into foresight. If you choose to build on spec, I wish you better luck than we had. ■

Edward Behr now lives in Peacham, Vt. and has retired from the construction business.

No Job Too Small

Rather than turning down small jobs, learn to make them profitable

For most remodelers, small jobs are a nuisance. While everyone defines a small job differently, something simple like replacing drywall after a roof leak, patching around a toilet, or replacing rotten deck joists still means two or three trips to the supplier, half a day wasted while the mud dries or the tile sets, and rounding up a sub or two who are willing to do a few hours worth of work.

Then there's all that overhead for a job that may yield only a few dollars in profit. For most companies, scheduling, billing, and contractual procedures are the same whether the job is worth $100 or $100,000. Yet on a small job, passing along that overhead to the customer may put you out of the bidding competition.

Here are two contractors who have found a way to make small jobs profitable. Their approaches are different in many ways. For instance, one emphasizes the quality of the people he hires to perform small jobs, while the other uses these jobs as a training ground for new employees.

But despite the differences, they have at least three things in common:

- *Each has created a separate division with streamlined scheduling and billing practices designed to cut the overhead on small jobs.*
- *Both add in liberal markups to their small jobs. This means they can't compete with the guy who operates from the back of his pickup. But the companies are providing professional service with guaranteed workmanship and fast delivery. They figure that's worth the extra cost.*
- *And finally, they have both changed their attitudes toward small jobs. No longer are these jobs a bother. Instead they are a productive, profitable part of the business.*

Call The Handyman

by Tom Swartz

About five years ago, we set a new company policy: no more small jobs. A look at our books for 1986 showed that of the 1,500 jobs completed that year, fewer than 300 earned 80% of our profits. The remaining 1,200-plus jobs were small ones, netting less than $2,000 each. But each one passed through the sales, production, and accounting departments, requiring the same kind of paperwork as the $100,000 additions. They were eating us up in overhead.

Often these jobs were even more troublesome than the big jobs. Scheduling was a hassle. It meant pulling our crews from the big projects to attend to the little ones. If we were busy, the small jobs had to wait, often for weeks, resulting in customer complaints. To top it all off, we were spending a fortune to bill all these people for jobs costing as little as $50.

Everyone agreed that the company's policy to turn down small jobs was a good one. Yet despite our intentions, month after month the small jobs kept appearing on our records. Our company has been around for more than 70 years and has built up a strong and loyal client base. We found it impossible to say no when these people asked us to come and hang a storm door, repair a roof leak, or even get a squirrel out of an attic. It was time to change our objective from eliminating the small jobs to finding a way to make them profitable.

Condo associations provide a steady source of small jobs, such as this storm door replacement.

Introducing the Handyman Service

With great fanfare, we introduced our Handyman Service in early 1988. We rented four booths at the local home show and parked one of our newly painted handyman vans right in the middle of the space. No one could miss the big, shiny, orange truck. We started getting calls the next day.

Our system for handling the calls is simple. Our receptionist first asks the caller exactly what type of service is needed. In many cases, people respond to our Handyman Service advertising by asking us to do larger jobs like additions or decks. These calls are routed to the sales department.

We define small jobs as those that can be completed by one person in a day or two. We don't attempt to classify them by dollar volume. (The customer may be asking us to hang a door, but the price of the door could range from $100 to more than $1,000).

If the job is meant for the Handyman Service, the receptionist gets the ball rolling by taking all the information required for scheduling and billing. We've designed a simple, four part form that we call the handyman ticket. This acts as invoice, work order, receipt of payment, and control copy. If you have ever had a service call on an appliance, you'll recognize the system.

Rough estimates are usually easy to provide since all service jobs are based on time and materials. The receptionist will tell the customer our basic charges, roughly how long the job will take, and what materials might be involved. The handyman will provide a final estimate to the customer before starting the work.

We charge $50, plus materials, for a service call which lasts less than an hour. Calls lasting more than an hour are billed at $32.50 per hour, plus materials. These prices may be considered high by some remodelers in our area. But we've found that most of our customers will pay them because they know we'll get the job done right and on schedule.

We've also developed a price list that includes about 200 standard items, such as cutting off a door, fixing roof leaks, and patching plaster. If the customer is looking for one of these services, the receptionist simply quotes the price from the most recent list.

The customer is then informed that upon completion of the job, the handyman will accept Master Card, Visa, a check, or cash. Thus the administrative expenses of billing and collecting are eliminated. We've found that 90% of our handyman jobs are collected before leaving the job site. The additional 10% are usually commercial jobs where payment policies vary.

Scheduling

After the receptionist takes the information, the form is given to our production supervisor, who passes it on to one of our four handymen. The handyman is then responsible for calling the customer and setting up a time to take a look at the job and provide a final estimate. Our goal is to get back to the customer and have the job completed within a week of the initial call. In some cases, a roof leak for instance, we try to get there within a day or two.

It is up to our handymen to give the final estimate, do the work, turn over the invoice, and collect the money. The handymen are educated about our pricing policies, including markups and overhead. They recognize that a minimum gross margin of 50% (achieved by marking up materials and labor by 100%) is necessary to cover the costs of the Handyman Service. They keep all they need to provide an estimate in the truck, including a copy of our price list, a schedule of basic material costs, and job contracts.

Once the handyman has taken a look at the job, he goes back to the truck, figures the price, and writes up a contract that's presented immediately. The customer must sign it and put down 50% on the spot. The handyman then schedules the job. We've found that we close on about 80% of the jobs we estimate this way.

The handyman gives the office the control copy of the handyman ticket so that we can track the jobs. Within 30 days of completion, the handyman receives a commission on each job that achieves the targeted gross margin. This has proven to be a good incentive to schedule more effectively and charge customers correctly.

Profile of a Handyman

Our handymen are different from the typical carpenter. For one thing, all of them have a lot of experience. Two of the four have had their own companies so they know how the business works. They also know how to sell our company and our services. Most importantly, they know how to deal with people. Not everyone can cope with customers peering over their shoulders while they work.

The handymen are probably more independent than other members of our field crew. They have to be good at setting their own schedules and working alone. Now and then, if the handyman end of the business is slow, they'll work with the other field staff on bigger jobs. But for the most part, the handymen are in and out of the office only long enough to pick up their work orders.

Making the Division Work

The Handyman Service has been part of our company for three years now. We still do as many small jobs as we used to, but every year we've gotten a little smarter, and these jobs are now earning higher profits. In 1990 we did about 2,000 jobs, generating $3 million in sales companywide. The Handyman Service earned about $600,000 of this.

There are other benefits from providing this kind of service. Our customers are happy. They know they can turn to us for any type of job, from building a crate to shipping a giant cactus across the country (yes, it's true), to redesigning the entire downstairs of their home. At the same time, our Handyman Service has generated lots of referrals for bigger jobs. □

Tom Swartz is president of J.J. Swartz Company, a 70-year-old remodeling firm with offices in Decatur and Bloomington, Ill.

J.J. SWARTZ CO.
GENERAL CONTRACTORS
2120 N. OAKLAND AVE
DECATUR, ILL.
62526
HOME BUILDING & REMODELING

INVOICE

Name: Joseph Green Date: 3/30/92
Address: 11 Walnut St.
Anytown, UT 11111
Phone #: 222-3333 Job #: H/123
Work #: 222-4444

Work to be completed:
Replace existing rear storm door with wood combo (style #1). Salvage and reinstall latch and closer. Add new weatherstrip. (Owner will paint.)

Amount due for above work $285.00

Additional Work Completed:
DOOR CLOSER MISSING PARTS — INSTALLED NEW ONE.

Special Instructions:

Labor: Hrs. Rate

Material: 1 EA. DOOR CLOSER $12.75

TOTAL DUE $297.75

METHOD OF PAYMENT:
X Check ____ Master Card ____ Visa
Acct. # ____ Expiration Date ____
Authorized Signature ____

To hold down small-job overhead, this simple, four-part form serves as an invoice, work order, receipt, and control form.

A Niche for Small Jobs

by Jim Walter

Like many remodeling firms, we got our start doing small jobs. And we quickly learned that you don't prosper on small jobs alone. Gradually, as our business grew, we turned our attention to larger, more profitable jobs like room additions and whole house remodeling. These now make up about 80% of our annual sales. But after 12 years in business, the small jobs still play an important role in our company.

In 1990 we did about 500 jobs, generating about $750,000 in sales. Of these, 480 were small jobs, grossing about $150,000. We have a full-time staff of 14 with two part-time employees. All of us, from the most experienced to the new hires, do small jobs now and then.

The author completed 480 small jobs last year and 20 large ones. All are tracked in the bins and on the erasable schedule board shown above. The goal is to complete small jobs within a week of the initial call.

Small Jobs, Big Perks

While many remodeling firms consider small jobs more work than they're worth, we've found that they offer several advantages. First, none of our crew needs to worry about getting laid off: There's always plenty of small jobs to provide us with a steady source of work. These jobs provide constant income that gets us through the slow times.

Also, we like the scheduling flexibility that small jobs give us. A two-day job fits neatly into the schedule when, for example, you're waiting for the electrician to finish the rough-in on an addition. At the same time, that flexibility lets us use the whole crew if we need to get a roof torn off and a second-story addition roughed in. Nothing pleases a customer more than the sight of a big crew working together to get a job done quickly.

Small jobs are also a good source of referrals. We usually pick up at least two referrals on every job we do (we ask for them). Many of these may not pan out, but those that do help us to expand our customer base.

Finally, small jobs are a service to our customers. They appreciate the fact that we show up on schedule and do a good job. And we give them a price up front and stick to it. That creates goodwill, which means more jobs down the line.

Managing the Small Jobs

Most of our small jobs come from two sources: condominium associations, and a local home center for which we do product installations. In the case of the associations, we're familiar with the projects, usually small repairs and maintenance, and we know right away what the job will entail. And since the home center reimburses us a fixed amount for each installation, all we have to do is schedule the work. This enables our production manager, Tom Walter, to estimate and schedule most of our small jobs right over the phone, thus cutting out overhead costs and increasing profit.

About 15% of our small jobs come from past customers or other sources. In these cases, the work is more of a service. We typically spend more time on these jobs since we take a look at the work before providing an estimate.

We've created a separate division that we call the Service Department to handle small jobs. Our goal in that department is to strip away as much overhead as possible. We define small jobs as anything under $10,000, but the majority fall in the under-$1,000 range. These jobs, as a rule, don't require subs and rarely need more than one or two workers to complete them. Except for the jobs assigned by the home center, all our estimating is based on time and materials.

We charge a higher markup on small jobs

than on bigger jobs since costs on small jobs can be disproportionate to profits, especially if things don't go smoothly. You can easily eat up the profits on something like hanging a storm door if you accidentally get the wrong size door.

Our fee is $30 per hour with a one-hour minimum per job. Labor is marked up 100% and materials by 20%, yielding an average profit margin of about 40% per job. On almost all small jobs, labor is the larger share of the cost.

We have a unit-pricing system which we update constantly. This gives us standard prices on things like reroofing (usually $75 per square, depending on pitch), siding replacement, ceramic tile installation, and drywall. We track our job costing and do all our billing on a computer.

We use a simple, standard work order that Tom fills out while talking on the phone to the customer. If Tom needs to see the job to give an estimate (if it's not on our unit price list, for instance), he drives to the site, meets with the customer, and mails out or delivers an estimate the next day. He typically collects a deposit of 20%.

Scheduling

We try to get to all the small jobs as quickly as possible, usually within a week of the initial call. In fact, it's one of our biggest selling points and something we're known for in this community. But scheduling can be tricky, especially when the weather isn't cooperating or one of the jobs takes longer than expected.

Once again, Tom is in the hot seat. He keeps an erasable board in his office with the names of all our employees and a roughed-in monthly schedule. When one of our crew members is between jobs or waiting for subs to get his work done, Tom greets him in the morning with a pile of small-job work orders. Tom has already made the appointments, so our person just has to show up and do the work.

It's a system that works well most of the time. But when there are problems, Tom will call the clients and reschedule. Then he'll contact our crew members and make some quick reassignments, often giving out work order information over the phone.

Who Does the Work

The best way we've found to train new employees is in the Service Department. They gain all kinds of technical experience in a wide variety of trades, and they get used to dealing with customers. It's also a quick and easy way to see where their abilities and weaknesses lie. Because the jobs are small, feedback from the customer is immediate.

Small jobs give our crew members a chance to do something different, to take a break from their everyday work. We don't provide bonuses on these jobs, since we've never come up with a fair method for quantifying them. But for the most part, our employees enjoy the change of pace and the responsibility of seeing a job through from start to finish. ■

Jim Walter and his partner, Bob Mager, are the founders of Lakeview Lumber & Construction, Inc., in Lake Barrington, Ill., a full-service remodeling company that specializes in second-story additions and other small-repair and light-commercial jobs.

Opportunities in Insurance Repair

by Walter Lumpp

Special skills and a commitment to service are needed to make good in this recession-proof market

Catastrophes show no favoritism. Fires, floods, storms, and other disasters create havoc with equal disregard for all people and places. Last year, such devastation cost the insurance industry an estimated $30 billion in payments to policyholders — who then sought builders to restore their damaged property.

A Market Largely Untapped

Insurance restoration forms a significant segment of the construction industry. Yet it remains a largely untapped market for most builders, who can't see beyond the charred remains or muddy mess of a disaster scene to the remodeling job ahead.

Dealing with the aftermath of a disaster requires special skills and flexibility. But it also provides a recession-proof source of

Insurance restoration jobs range from a porch that collapsed under a snow load (above), to a house damaged by a fallen tree (facing page, below).

work. Disasters don't care about the season of the year or the mood on Wall Street. For those who develop the necessary skills, insurance restoration offers a solid buffer against the uncertainties that are a constant part of the general construction industry.

My firm began specializing in insurance restoration 21 years ago, after frustrations with the high risks and slim profit margins in general construction almost convinced me to leave the industry. I investigated insurance restoration as an alternative and enjoyed the challenge of reaching out to a specialized market of insurance agents and adjusters. The profit margin in insurance restoration turned out to be more comfortable, and the work load did not depend on economic trends.

Two decades and 33 branch offices later, I am convinced that builders who enjoy challenges and being of service to people should consider insurance restoration. After all, once you've done the cleanup (which can be subbed out), insurance restoration is mostly remodeling. The other major differences relate to the delivery of service following a disaster and to marketing.

Disaster's Aftermath: Everything Is an Emergency

If you have any misgivings about being on call 24 hours a day, 365 days a year, forget insurance restoration. In this business, emergency calls are the rule, not the exception. Getting to the scene as soon as possible after a disaster is vital, because you often must start work immediately to minimize further damage.

Also vital is the ability to deal well with distraught people. Under the best circumstances, property owners will be distressed. When property loss is coupled with an injury or death, the trauma experienced by those facing the restoration decisions can be overwhelming.

Unfortunately, while dealing with the emotions of the property owners takes time, the job of salvaging the building often can't wait. Broken windows and doors must be boarded up. Standing water must be removed. Smoke odors should be treated as quickly as possible. In time, the owner will be able to appreciate your concern for protecting what is left of the property. (And insurance adjusters certainly will remember your quick response to their calls.) Initially, though, you may have to balance the urgency of salvaging the property with a sensitivity to the emotions of its owners. Make what progress you can without seeming pushy or aggressive.

Attention to owner emotions reaps rewards, however. In one of our jobs, an elderly widow whose house burned was distraught because the engagement ring on her dresser was buried beneath the rubble of her former bedroom. Despite her substantial loss, she could talk only about the missing ring.

One of our workers asked the woman where the dresser had stood. He then took a pan and whisk broom and spent two hours painstakingly sifting through the ashes until he retrieved the ring. Both he and the widow were overjoyed. Recovering that ring brought the trauma of the fire to an end. The woman could then focus on the job of rebuilding her house, and work was easier for everyone.

Marketing Your Services

Once disaster strikes, it's too late to start promoting your services. In insurance restoration work, you must market your business in advance so insurance adjusters know who to call when property damage occurs. Quality of service is as important as price in getting such referrals. Unless your prices are completely out of line, you will get calls once adjusters and agents know you give immediate and reliable attention to their clients. The calls will come from insurance agents, telephone adjusters, and field adjusters, but the majority of insurance restoration work comes from the field adjusters.

After you've taken on this type of job, you may find that the working relationship between the property owner, the adjuster, and the contractor can become complicated.

The adjuster is the mediator who must see to it that the property owner receives reasonable compensation while the insurance company pays no more than it is liable for under the policy. Although adjusters are paid by the insurance companies, it has been my experience that most of them honestly try to reach an agreement they believe is fair to both parties.

Between the adjuster and the property owner stands the contractor, whose loyalty is split. The contractor serves both the adjuster who asked for the estimate — though the two have no formal relationship — and the property owner, who will sign a formal agreement with the contractor once details of the restoration are settled. The property owner controls the current job; the adjuster controls future referrals.

Remaining neutral is the only way a contractor can deal fairly with both parties. You must give your estimate of the scope of repairs and then stay out of negotiations until the adjuster and property owner come to an agreement.

Estimating Costs and Time

The first step in estimating any insurance restoration job is to take pictures. This gives you a record of structural features and damage. It's also a good idea to have the adjuster on the scene while you estimate the scope of repairs so you can both agree on the extent of the restoration to be done.

If more than one contractor is preparing an estimate, both parties should walk the job together, agreeing on the scope of work with the adjuster. This will ensure that a fair comparison of the total price will be made.

Costs. The biggest difference in estimating for restoration work (as opposed to ordinary remodeling) is figuring the cost of the cleanup. Cleanup after fire or water damage can be difficult to assess, not to mention to perform; until you gain experience in this area, it might be wise to subcontract such work, including the estimate. You can find companies specializing in such work in the Yellow Pages under "Fire and Flood Damage" or "Water Damage." To avoid hiring your competitors, make sure you use companies specializing in cleanup only. As with any other subcontractor, it is possible to develop long and productive relationships with such companies.

When using subs for cleanup, however, make sure you clarify at the beginning just how far their cleanup will take you toward the point of being able to begin reconstruction. There is often a gray area between cleaning and rebuilding. For instance, some surfaces or structural components may need to be removed before rebuilding, and the cleanup sub may not include such work in the cleanup estimate. You need to identify these gray areas of work at the beginning.

The other estimating peculiarity in restoration work is hidden damage. What happens when damage becomes apparent after you've started work? In my experience, insurance companies almost always are willing to approve a changer order for hidden damage as long as you can explain why something originally thought to be salvageable must be replaced. You'll be most successful with change orders if you submit a very clean set of specifications for repairs and replacements with the bid.

Except for cleanup costs and hidden damage, the cost estimating process for insurance restoration work is identical to that for remodeling: You take the rooms apart, operation by operation, and calculate the cost to rebuild them. You can consult standard

estimating books or use computer software.

Obviously, the bigger the job, the more time a detailed estimate will take. However, if you are familiar with construction operations and costs, you should be able to deliver a reliable estimate within 24 to 48 hours. Always give the adjuster an adequate amount of time to review the estimate.

Time. Estimating how much time a restoration job will take can be tricky. Because insurance restoration is emergency work, you can't order supplies or schedule subcontractors in advance. Disaster restoration is often a game of "hurry up and wait" — you start a job immediately, only to encounter delays when an unusual piece of material is out of stock or the carpenters you normally use are already committed to another job. Be conservative in giving your completion date. That way your client will be less frustrated when the inevitable delay occurs.

Understanding Adjusters

There are several types of adjusters you may have to deal with when doing insurance repair work:

Agency Adjusters

The local agent's role in handling small insurance claims has changed rapidly in recent years. More and more agents are being given larger check or draft authority by various companies. The agency's authority will depend on its staff, experience, and longevity.

The agent may have a small draft authority for some insurance companies, while for others he may have no authority at all. This applies to two kinds of agents: The independent agent who writes on more than one company; and the company agent who writes for the company with which he is licensed to do business.

Many agencies write enough business within a local community to justify their own adjuster who settles many claims without relying on the insurance company for help. Again, the adjuster's authority will vary from one insurance company to another.

Occasionally an agency writes enough policies within a given community to require an adjuster from a particular company to be in-house. however. This is rare, however.

Insurance Company Adjusters

Here again, there are several types of adjusters. Many insurance companies have watts lines on which their claims are called in to one location. They are then screened according to size and location and assigned in one of the following ways:

Telephone adjusters never see the job, but simply request one or more estimates from the property owner. The adjuster then agrees to the scope of the repairs. When the check is issued, a proof of loss will be required from the property owner. A proof of loss is a signed and notarized document that verifies that the loss occurred and that the owner abided by the rules of the policy.

Field adjusters will visit the site and review the damages when the size of the loss makes this necessary. In these cases, the field adjuster will actually write an estimate of repairs when he inspects the property. Using this, he will compare it against a contractor's estimate or even close the claim without ever discussing the repairs with a contractor. It is then up to the property owner to hire a contractor to have the repairs done.

General adjusters will only handle the very large or complicated losses, frequently referred to as jumbo losses, and they specialize in specific areas, such as property damage. The general adjuster understands policy coverage and is skilled at dealing with clients and other members of the business community.

Independent Adjusters

Generally speaking, the independent adjuster contracts his work from the insurance companies. The independent adjuster must have experience in several areas (auto, property, workers comp, etc.). However, independent adjusters also have a general adjuster on staff who is extremely knowledgeable in specific areas of business.

Public Adjusters

The public adjuster has become increasingly popular in recent years. He represents the property owner in adjusting the claim and in many cases is paid through the proceeds of the claim. Public adjusters usually serve large metropolitan areas.

What To Expect

No matter what type of adjusters or agents you deal with, get to know them personally. Usually, they are local business people and have local business interests at heart. The key is to market the adjusters and agents and convince them that you can be counted on to produce an estimate and make repairs quickly and without problems. — *W.L.*

Writing The Estimate

Write the estimate so that each room requiring repairs is listed individually with accurate measurements of every dimension of the room. Make a line-by-line listing of replacements with materials and labor in separate columns on the right. This cost breakdown will be very helpful in establishing a budget and tracking job costs during the reconstruction of the job.

After the proposed repairs for one room have been completely listed, repeat the process in the next room. When the interior of the house has been completed, move to the exterior and do the same thing.

A properly itemized estimate will make it easier for the adjuster to compare one estimate against another, and against the coverage of the policy. Remember, the adjuster may not be a construction expert, so keep the estimate in lay terms. The adjuster must understand the work being done well enough to explain the estimate to his supervisors. Keep in mind that an adjuster's needs, in order of importance, are:

- competitive unit costs;
- detailed itemization of covered repairs;
- a quick response;
- timely completion;
- completion without complaint.

The property owner should also understand the estimate and know exactly which repairs are and are not going to be made. His needs are:

- a quick response;
- competence;
- empathetic treatment;
- quality workmanship;
- timely completion.

Champaign-Urbana News-Gazette

Heavy fire damage requires a quick response from the restoration contractor. Minimizing water and smoke damage will be a top priority.

As a contractor, never allow yourself to get into the crossfire of trying to settle a claim for either party. When you review a job, do so in a professional manner, listing the items in need of repair or replacement based on your expertise. Then allow the adjuster to settle the claim with the claimant.

Safety First

Use common sense when entering structures damaged by fire or other disasters. You may need to wear a respirator, since many household items leave toxic residues after a fire. Be sure floors are sound enough to carry your weight. Wear protective gloves and clothing during cleanup. Heavy shoes with steel shanks are a must to protect against exposed nails.

During cold weather, restoring heat immediately is a priority. You also will need fans and dehumidifiers if the structure has been exposed to water. And though you'll need electric power quickly, you should never use any damaged existing wiring until a licensed electrician inspects the system. A portable generator or current from a neighbor's house can serve as a temporary power source.

Getting Paid

Because insurance companies vary in their payment policies, restoration contractors must be in strong financial positions and have good relationships with their banks.

Many insurance companies pay policyholders 80% of an estimate as soon as an agreement is reached on the amount of the loss; the remaining 20% is held until final inspection. However, some companies will not make any payments until the job is completed. On large jobs, this can mean the difference between continuing or stopping work unless the contractor can borrow against a line of credit to keep going.

Payment also can be delayed when the law requires that mortgage companies as well as property owners be parties to the insurance payment. If the mortgage company is channeling its part of the payments to you through an escrow account, delays can occur. Doing advance research to identify the parties involved and the mortgage company's payment policies can prevent such frustrations.

Finally, you should establish at the beginning of the job the property owner's deductible or other share of payment. Sometimes the property owner's policy inadequately covers the damage, or it may not cover specific types of damage. To protect yourself, you need to know if all of your contract is going to be paid by the insurance company or if the customer has to pick up some of these costs. If the latter is true, you need to make sure the customer understands what he must pay for.

In general, payment is one of the strong points of insurance restoration work. One major plus is the virtual absence of bad debts — though delays may occur, insurance companies do pay their bills. Knowing this makes the delays easier to take.

The Market Is Worth the Mess

Insurance restoration requires remodeling skills, a willingness to tangle with cleanup and salvage operations, tactful client management, and the ability to effectively market your services to the insurance world. An aptitude for technical subjects such as engineering and chemistry doesn't hurt, either. But for remodelers who want to level the peaks and valleys in their work flow and who have a genuine interest in serving people, it can be an ideal way to expand a business.

For More Information

The Association of Specialists in Cleaning and Restoration (ASCR) regularly offers seminars and courses for people interested in insurance restoration work, including a week-long "Certified Restorer Program" given once a year. For more information, contact the ASCR at 10830 Annapolis Junction Road, Suite 312, Annapolis Junction, MD 20701; 301/604-4411. ■

Walter Lumpp is the founder and president of Forest Hill Associates, Inc. and Forest Hill Enterprises, Inc., an organization of insurance reconstruction specialists, based in Charlottesville, Va.

Index